# DIGITAL COMMUNICATIONS

# Other Reference Books Available from Sams

*Computer Dictionary, (4th Edition)*
Charles J. Sippl

*Data Communications, Networks, and Systems*
Thomas C. Bartee, Editor-in-Chief

*Electronics: Circuits and Systems*
Swaminathan Madhu

*Gallium Arsenide Technology*
David K. Ferry, Editor-in-Chief

*Handbook of Electronics Tables and Formulas (6th Edition)*
Howard W. Sams Engineering Staff

*Introduction to Digital Communications Switching*
John P. Ronayne

*Image Tubes*
Illes P. Csorba

*Mobile Communications Design Fundamentals*
William C.Y. Lee

*Optical Fiber Transmission*
E.E. "Bert" Basch, Editor

*Practical Microwaves*
Thomas S. Laverghetta

*Principles of Solid-State Power Conversion*
Ralph E. Tarter

*Reference Data for Engineers: Radio, Electronics, Computer, and Communications (7th Edition)* Edward C. Jordon, Editor-in-Chief

*Semiconductor Device Technology*
Malcolm E. Goodge

*Sound System Engineering (2nd Edition)*
Donald and Carolyn Davis

To order these and other Sams books, call toll-free 800-428-SAMS.

# DIGITAL
# COMMUNICATIONS

THOMAS C. BARTEE
Editor-in-Chief

**Howard W. Sams & Co.**
A Division of Macmillan, Inc.
4300 West 62nd Street, Indianapolis, IN 46268 USA

International Standard Book Number: 0-672-22472-0
Library of Congress Catalog Card Number: 86-61451

Acquiring Editor: *Greg Michael*
Editor: *Diana Francoeur*
Interior Designer: *T. R. Emrick*
Illustrator: *Ralph E. Lund*
Cover Artist: *Stephanie Ray*
Compositor: *Shepard Poorman Communications Corp., Indianapolis*

*Printed in the United States of America*

# Contents

# Introduction

The recent explosive growth of the communications industry has been accompanied by many significant advances in communications technology. This book is about these advances and the new technologies they have engendered.

Nine leading experts have written on their specialities. These authorities bring to their subjects a valuable perspective and overview shaped by many years of experience in communications. Their contributions reflect the technical expertise that would be expected of those in the forefront of their fields. Technical accuracy and currency are assured.

*Digital Communications* is organized into nine chapters, each one treating an important technical area. For the reader's convenience, each chapter provides references. The chapters may be read in order, but they may also be read separately, since each chapter stands on its own. The chapters may be used as:

- A professional reference and source of material in the subjects covered.

- A resource providing an additional perspective for those involved in the technology discussed.

- An introduction to the subject discussed in the chapter.

Chapter 1, "Fiber-Optic Transmission Technology and System Evolution," is written by Ira Jacobs, Director of the Transmission Technology Laboratory at AT&T Bell Laboratories. This chapter describes the important characteristics of fiber optics in communications systems and discusses applications in various areas of communications, including advantages and disadvantages for each application. Historical information concerning fiber optics development is also presented.

In Chapter 2, "Satellite Communications," Irwin L. Lebow, Vice President at SRA, presents an applications-oriented view of satellite communications. The chapter begins with a comparison of the satellite medium in the context of other competing media. Then, the operating characteristics of satellite communication systems are described, and system operation and satellite applications are explained. This is followed by considerations concerning strategies for optimizing the use of satellite links in various applications.

Chapter 3, "Integrated Services Digital Networks (ISDN)," is by Eric Scace, Senior Scientist for ISDN at GTE Telenet Communications. ISDN is carefully defined in the chapter, and the various options are described. ISDN services and organizations are covered, as are typical configurations. Standards and the need for standardization are then discussed.

The fourth chapter, "A Business View of Computer-based Messaging," is by Vinton G. Cerf, former Vice President of Engineering at MCI and currently Vice President for National Research Initiatives. The history of electronic messaging is presented, along with insights into the importance of each developmental stage. An overview of current commercial systems and their characteristics is then given. Economic considerations for business use are outlined, as is the important role of electronic messaging in future office automation.

Chapter 5, "Electronic Mail Systems," by Debra Deutsch, Senior Scientist at BBN Laboratories, describes in considerable detail the functions, standards, and architectures associated with electronic mail systems. A model of electronic mail systems is presented and used to describe how the different components of a system interact. Different architectures are explained, and advantages and disadvantages are covered. A set of standards for electronic mail systems is then presented.

The sixth chapter, "Cellular Networks," by Philip T. Porter, District Research Manager at Bell Communications Research, begins with a brief history of cellular communications. The technical principles of these systems are then spelled out. At each point, the results are explained, using carefully derived formulas. The strategies for obtaining mobility are given, as well as the problems resulting from this feature. Finally, international developments are noted, along with current research and development.

Chapter 7, "Challenges in Communications for Command and Control Systems," is written by John J. Lane, Deputy Assistant Attorney General and former Director of Information Systems in the Office of the Assistant Secretary of Defense for Command, Control, Communications, and Intelligence. This chapter describes the foundations of command and control systems, using examples from present systems to illustrate each point.

An overview of the current status of these systems is given, including security considerations and the great need for interoperability.

In Chapter 8, "Digital Coding of Speech," N. S. Jayant, Head of the Signal Processing Research Department at Bell Laboratories, explains coding methods for the digital representation of speech. Before speech can be converted to digital form, transmitted through a communications system, and finally converted back into sound, it is necessary to have coding algorithms for the conversions. The choice of these algorithms affects such system properties as voice quality (at the receiving end) and complexity of equipment. In this chapter the primary coding techniques are discussed, and each is evaluated with regard to its system properties.

The ninth chapter, "Video Teleconferencing," by Richard Harkness, Director, Market Planning and Strategy, for Compression Labs, explains the technology of video teleconferencing and indicates its economic benefits, as well as other benefits. The chapter surveys current commercial systems, giving their pertinent characteristics. The technical considerations of the various subsystems are explained, as is the importance of compression and other factors. Standards and interoperability are also covered.

I am permanently indebted to the late Dr. Alexander Nedzel and to Dr. Walter Wells, Professor Peter Elias, and Professor Robert Fano of the Lincoln Lab and MIT faculty for their early encouragement and support of my work in communications.

Without Jim Hill's active involvement this book would not have been possible. His many helpful comments throughout the planning and preparation of the manuscript have been much appreciated.

THOMAS C. BARTEE

# Contributors

**Ira Jacobs** is Director of the Transmission Technology Laboratory at AT&T Bell Laboratories in Holmdel, New Jersey, where he is responsible for digital transmission and signal processing technology. Before that he was responsible for design and development of digital multiplexers and fiber-optic transmission systems.

Dr. Jacobs joined Bell Laboratories in 1955. From 1955 until 1967, he worked in electromagnetic and communication theory and was responsible for a number of military satellite and space communications projects. Since 1967, Dr. Jacobs has been involved in transmission systems engineering and development and has directed much of AT&T's development of fiber-optic transmission systems. In 1981 he was elected a Fellow of the Institute of Electrical and Electronics Engineers for contributions and leadership in the development and application of fiber-optic systems.

Dr. Jacobs received a B.S. degree from City College of New York in 1950, and M.S. and Ph.D. degrees in physics from Purdue University in 1952 and 1955, respectively. *Dr. Jacobs is the contributor for Chapter 1.*

**Irwin L. Lebow** is Vice President at Systems Research and Applications (SRA) Corporation, where his principal responsibilities are in telecommunications. SRA is a systems management and engineering firm and offers telecommunications network services for both government and commercial clients. It is currently under contract with the U.S. Treasury Department to develop candidate network architectures to meet the Treasury's future voice and data needs.

Dr. Lebow has over thirty years of experience in information technology related areas, including government and commercial applications, as well as military and nonmilitary. In more than twenty years at MIT's Lincoln Laboratory, he worked on solid-state circuit applications to digital systems, digital computer design, and communications systems. He was a designer of the OG-24 computer, one of the first all solid-state computers, and an initiator of Lincoln's satellite communications technology program. In 1970 he was a member of the Laboratory's steering committee.

In 1975 Dr. Lebow joined the Defense Communications Agency (DCA) as Chief Scientist and Associate Director, Technology. As the Agency's highest ranking civilian, he had broad technical responsibilities in the Defense Communications Systems and the World Wide Military Command and Control System. In particular, he was responsible for the Research Development Test and Evaluation program at the Agency, which included major modernization efforts in digital transmission, secure and nonsecure voice, packet-switched data, satellite communications, and command and control systems. In 1981 Dr. Lebow joined American Satellite Company as Vice President, Engineering. His responsibilities included the engineering of ASC's shared and private networks, systems engineering, acquisition of ASC satellites, and research and development on advanced techniques.

Dr. Lebow is coauthor of the book *Theory and Design of Digital Machines* (McGraw-Hill, 1962) and is the author of many technical papers in the communications and computer science fields. He is a Fellow of the IEEE and of the American Physical Society and is a member of the Scientific Advisory Group at the DCA and of the Radio Engineering Advisory Committee of the Voice of America. He is the recipient of the Defense Department's Meritorious Civilian Service Medal. He holds a Ph.D. degree in physics from MIT. *Dr. Lebow is the contributor for Chapter 2.*

**Eric L. Scace** is Senior Scientist for ISDN at GTE Telenet Communications Corporation. Over the past ten years, Mr. Scace has participated in CCITT SG II, VII, XI, and XVIII activities on packet-switching public data networks and on ISDNs. He presently chairs ANSI T1D1.2, a North American standards group on ISDN switching and signalling protocols.

Mr. Scace has also worked with the National Bureau of Standards and with General Electric on data communications matters. During this time, he participated in the International Civil Aviation Organization's Automated Data Interchange System Panel, and in ISO and ANSI activities on data communications. He developed for the U.S. National Weather Service a fully computerized network for the collection and distribution of meteorological data, maps, forecasts, and warnings.

Mr. Scace holds a B.S. degree in atmospheric sciences from Cornell University. *He is the contributor for Chapter 3.*

**Vinton Gray Cerf** is Vice President of the Corporation for National Research Initiatives (NRI). From 1983 to 1986 he was Vice President of Engineering at MCI Digital Information Services Company. He was responsible for the development and implementation of a major, nationwide data communications system employing the latest in packet communication technology for the support of large-scale commercial applications and internal operational traffic for MCI.

Prior to this, he was Principal Scientist at the Defense Advanced Research Project (DARPA) Information Processing Techniques Office (IPTO). His duties included technical oversight of all research programs and advanced planning and coordination of the IPTO research program. He was responsible for the technical management of several research programs involving digital communications and processing technology—in particular, the interconnection of packet-switched networks, mobile packet radio, local networks and packet satellite networks, network and operating system security, adaptive network techniques, and advanced multiprocessor architectures. From 1972 to 1975, Dr. Cerf was a faculty member of Stanford University in the Electrical Engineering and Computer Science Departments, where he taught courses in systems programming and machine organization.

Dr. Cerf is a member of Sigma XI, IEEE, and IFIP. From 1979 to 1980 he was ACM National Lecturer. He received a B.S. degree in mathematics from Stanford University, and M.S. and Ph.D. degrees in computer science from the University of California at Los Angeles. *Dr. Cerf is the contributor for Chapter 4.*

**Debra Deutsch** is Senior Scientist at BBN Laboratories, Inc., concentrating on computer network technologies, with specialization in system architectures, upper layer protocols, and interoperability issues. Her major activities have included the development and recommendation of electronic mail system standards to the U.S. National Bureau of Standards (NBS), the interfacing of an existing message-switching installation to a packet-switching network, research on multimedia message systems, and consultation on architectural and security aspects of message systems. In addition, she was involved in the design and development of the BBN Hermes message system.

As part of her work for NBS, Ms. Deutsch was the principal author of Federal Information Processing Standard 98, "Specification for Message Format for Computer-based Message Systems." This standard served as a major input to the development of the CCITT X.400 series of recommendations. Ms. Deutsch was an active member of Special Rapporteur's Group that produced the X.400 series. She also participated in ISO/TC 97/

SC18/WG4 during the development of their Message-oriented Text Interchange System standard, which is analogous to the X.400 series. Based on these experiences, Ms. Deutsch has consulted and written on the details and use of the CCITT X.400 series as it can be applied to new and existing systems. *Ms. Deutsch is the contributor for Chapter 5.*

**Philip T. Porter** is District Research Manager, Bell Communications Research Inc., where he has group responsibility for the study of intrasystem frequency reuse and efficient modulation and coding techniques to be used in future systems that provide untethered communications.

Prior to 1984, Mr. Porter was employed by AT&T Bell Laboratories, Holmdel, New Jersey, where he participated in early development planning for electronic station sets and Picturephone visual telephone service. In the mobile radio field, he has been involved in the development of Bellboy® paging, IMTS, and the Metroliner system. He participated in studies leading to the North American 850-MHz cellular system. From 1971 to 1977, he supervised the systems planning of network and mobile control logic. Later he was responsible for long-range mobile radio studies and for international liaison.

Mr. Porter is a member of Phi Beta Kappa and of various IEEE societies; he is also a U.S. CCIR delegate for Study Group 8A. He received B.A. and M.A. degrees in physics from Vanderbilt University in 1952 and 1953, respectively. *Mr. Porter is the contributor for Chapter 6.*

**John J. Lane** is the Deputy Assistant Attorney General, Office of Information Technology, and Special Advisor for Technology to the Deputy Attorney General, Department of Justice. He provides technical and program guidance, exercises program and budget approval responsibility for the Department's Information systems, and operates the majority of the Department's ADP and telecommunications facilities.

He is the former Director of Information Systems for the Department of Defense, in the Office of the Assistant Secretary of Defense ($C^3$). In that position, he exercised policy and program oversight of strategic information systems, data communications networks, Service and Defense Agency Communication programs, and major portions of the Defense Communications System.

Mr. Lane is the author of the book *Command and Control and Communication Structures in Southeast Asia,* in which he describes and analyzes the evolution of $C^3$ systems during the Vietnam conflict. Mr. Lane's professional recognitions include an Air Force Project Manager of the Year Award, the Air Force Communications-Electronic Professionalism Award, and the AFCEA Honor Award for $C^3I$ Research. He holds graduate degrees in law, business administration, and computer science. *Mr. Lane is the contributor for Chapter 7.*

**N. S. Jayant** is Head of the Signal Processing Research Department at AT&T Bell Laboratories, Murray Hill, New Jersey. He was a Research Associate at Stanford University from 1967 to 1968 before joining the technical staff of AT&T Bell Laboratories in 1968. Dr. Jayant has worked in the field of digital coding and transmission of waveforms, with special reference to speech signals and problems relating to speech privacy and packetized voice. He holds several patents on these subjects.

Dr. Jayant is the author of technical papers on digital coding and waveform transmission. He is the editor of *Waveform Quantization and Coding* (IEEE Press, 1976), the coauthor of *Digital Coding of Waveforms—Principles and Applications to Speech and Video* (Prentice-Hall, 1984), and the author of a book chapter, "Digital Coding of Speech," in *Data Communications, Networks, and Systems*, vol. 2 (Macmillan, 1986). He is a Fellow of the IEEE and was the first editor-in-chief of the *IEEE-ASSP Magazine*. He has organized and taught several courses on coding and digital communications.

He received a B.S. degree in physics and mathematics from Mysore University (India) in 1962, and B.E. and Ph.D. degrees in electrical communication engineering from the Indian Institute of Science, Bangalore, in 1965 and 1970, respectively. *Dr. Jayant is the contributor for Chapter 8.*

**Richard C. Harkness** is Director, Market Planning and Strategy, for Compression Laboratories, Inc. (CLI). He is responsible for establishing plans and strategies for product evolution, market positioning, and business opportunities. Before joining CLI, Dr. Harkness was a Senior Strategic Planner with Satellite Business Systems, where he defined product strategies in relation to market requirements and technology advances. As an expert on teleconferencing, he has been involved in both the technical and marketing aspects. Prior to 1978, he was principal investigator at Stanford Research Institute of an NSF-funded Technology Assessment of Telecommunications/Travel Trade-offs and was also involved with Standford's office automation program. Before that, he was involved in strategic and diversification planning for Boeing Aerospace Company.

Dr. Harkness has published widely on the subject of teleconferencing. He holds a B.S.E.E. degree from Duke University and interdisciplinary M.S.E. and Ph.D. degrees from the University of Washington. *Dr. Harkness is the contributor for Chapter 9.*

# 1

# FIBER-OPTIC TRANSMISSION TECHNOLOGY AND SYSTEM EVOLUTION

IRA JACOBS

In a fiber-optic communications system, information is transmitted by modulated beams of light that are guided in hairlike glass fibers. Since the first prediction in 1966 that fibers might be useful for telecommunications (Kao and Hockham), fiber has rapidly become the cable medium of choice for transmission applications. These applications range from data buses for the linking of digital switching equipment within a building, to intercontinental undersea transmission systems.

Fiber-optic technology is evolving rapidly. In 1970, the first low-loss fiber suitable for telecommunications was announced (Kapron et al.). An input signal passed through a 1-kilometer segment of this fiber was attenuated, or reduced, by 99% of its original amplitude. Today, fibers are being made that attenuate an input by less than 4.5% per kilometer (Kitayama et al. 1985).

The first semiconductor lasers having a structure suitable for use in fiber-based telecommunications systems were also announced in 1970 (Panish et al.). The mean time before failure (MTBF) was then 2 hours. In less than ten years, semiconductor laser technology had evolved to the point where lasers with a predicted room-temperature MTBF in excess of 1 million hours (more than 100 years) were being made, though not routinely (Hartman et al. 1977). There has tended to be about a five-year interval between first achievement in a research environment and similar results in standard manufacture.

In 1979 and 1980, initial fiber systems in the United States transmitted at an information rate of 45 million bits per second, with a repeater[1] spacing in the range of 5 to 10 kilometers (Cook and Szentisi 1983). Long-

---

1. A *repeater* is a device that repeats its input but in amplified form. The input may also be "cleaned" to some extent. Repeaters are discussed in section 1.4.2.

haul systems presently being installed typically have transmission rates in the range of 400 to 600 million bits per second, with repeater spacings of about 40 kilometers (Jacobs 1986b). System experiments have been performed achieving a transmission rate of 4 billion bits per second through 117 kilometers of fiber (Korotky et al. 1985).

In addition to the increases in transmission capacity and in the distance that signals can be sent without repeaters—which has particular importance for long distance applications—there have been exciting advances in the integration of optical and electronic functions (integrated optoelectronics) and in the capability for optical signal processing (integrated optics). These latter advances provide the necessary functionality and reduced terminal costs that are so important for short distance applications.

The evolution (past, present, and future possibilities) of this dynamic and highly important technology is the subject of this chapter.

# 1.1 Advantages and Limitations of Optical-Fiber Communications

The principal advantage of communication at optical frequencies is the very large bandwidth and, consequently, the very large communications capacities achievable. For example, a wavelength of $7.5 \times 10^{-7}$ meters[2] corresponds to a frequency of $4 \times 10^{14}$ hertz.[3] (Hertz, a unit of frequency equal to one cycle per second, is abbreviated "Hz.") Thus, a bandwidth equal to 10% of the center frequency is $4 \times 10^{13}$ Hz, or 40 million megahertz (MHz). This bandwidth could, in principle, accommodate ten million 4-MHz video channels, or ten billion 4-kHz voice channels.

Another advantage of communication at optical wavelengths is small size. Extremely narrow optical beams may be generated by miniscule light sources (refer to section 1.3.3). The hairlike optical fibers are much smaller in diameter and have much lower attenuation than typical metallic conductors. Also, energy is confined entirely to the fiber—there is no radiation, noise pickup or crosstalk—characteristics that may occur in metallic systems.

The only fundamental disadvantage of optical communications is the large photon energy at optical frequencies. This results in more received

---

2. Units typically used for optical wavelengths are micrometers (1 $\mu$m = $10^{-6}$ m) or nanometers (1 nm = $10^{-9}$ m). Somewhat less common now are angstroms (1 A = $10^{-10}$ m). Thus, a wavelength of $7.5 \times 10^{-7}$ meters is also represented as 0.75 $\mu$m, 750 nm, and 7500 A.
3. The product of wavelength ($\lambda$) and frequency (f) is the speed of light: c = f$\lambda$ = $3 \times 10^8$ meters per second.

power being required per bit of information than, for example, in low-noise microwave systems. Also, practical input sources for fiber-optic systems are limited in power to typically a few milliwatts.[4] Thus, fiber-optic systems are limited in power, but bandwidth is typically plentiful. In such cases wideband modulation techniques, which trade bandwidth for power efficiency, are generally employed. Pulse code modulation (PCM) is an efficient way of making this trade with analog information. Indeed, fibers are stimulating the growing trend to the digital transmission of all types of information. (See Chapter 8 for a discussion of digital speech coding.)

# 1.2  History

Initial activity in optical communications (e.g., smoke signals) was no doubt the direct result of vision being one of the primary human senses. The general availability of light sources and the transparency of the earth's atmosphere have long stimulated interest in free-space optical communication. Indeed, in 1880, Alexander Graham Bell received a patent for the photophone, a device that used reflected sunlight to transmit sound to a receiver.[5] The reflected light was intensity modulated by the vibration of a reflecting diaphragm located at the end of a tube into which the person spoke (Fig. 1.1). The modulated light was picked up by a receiving mirror placed 200 meters away (Fig. 1.2). Although Bell was granted a patent for his device, the public was skeptical of its practicality; and Bell's report (1880a) on his photophone at a meeting of the American Association for the Advancement of Science evoked a sarcastic article in the August 30, 1880 issue of the *New York Times*:

> What the telephone accomplishes with the help of a wire the photophone accomplishes with the aid of a sunbeam.... The ordinary man, however, may find a little difficulty in comprehending how sunbeams are to be used. Does Prof. BELL intend to connect Boston and Cambridge, for example, with a line of sunbeams hung on telegraph posts, and, if so, of what diameter are the sunbeams to be, and how is he to obtain them of the required size? What will become of his sunbeams after the sun goes down? Will they retain their power to communicate sound, or will it be necessary to insulate them, and protect them against the weather by a thick coating of gutta-percha? The public has a great

---

4. Owing to the small size of the light source, power densities are very large. Indeed, thermal effects are critical in semiconductor light sources.
5. It appears that Bell's interest in this work was stimulated after he learned that selenium could be used as a photodetector (Bruce 1973). Advances in materials science have played, and continue to play, a key role in the evolution of fiber-optic systems.

deal of confidence in Scientific Persons, but until it actually sees a man going through the streets with a coil of No. 12 sunbeams on his shoulder, and suspending it from pole to pole, there will be a general feeling that there is something about Prof. BELL's photophone which places a tremendous strain on human credulity.

Fig. 1.1.
Alexander
Graham Bell's
photophone
transmitter
*(Courtesy of
AT&T Bell
Laboratories).*

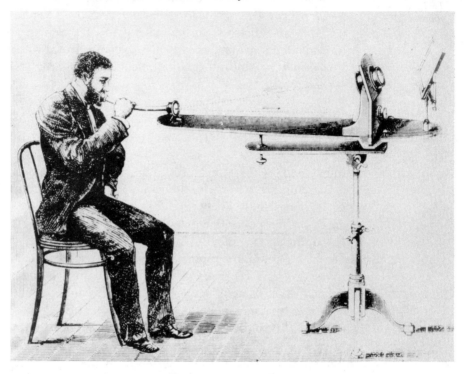

As the *Times* noted, the sun is not always available and the earth's atmosphere is not always transparent. It took nearly 100 years for these two essential ingredients of a practical optical communications system—a suitable, reliable light source and a predictable transparent transmission medium—to become available.

It was not until 1958 with the invention of the laser by Schawlow and Townes at AT&T Bell Laboratories that the prospect of a practical light source began to emerge.[6] In 1960 the first laser utilizing a ruby crystal was achieved by T. H. Maiman of Hughes Aircraft.[7] Later that same year

---

6. The laser was initially called an optical maser, with *maser* being an acronym for *m*icrowave *a*mplification by *s*timulated *e*mission of *r*adiation. The term was soon changed to *laser*, the *l* standing for light, and *laser* became an accepted term meaning "a device which produces an intense, coherent, directional beam of light by stimulating...transitions to lower energy levels" (Jay 1984).

7. The references in this historical overview are of necessity abbreviated and are perhaps somewhat subjective. The intent is to indicate the general time scale of the evolution—not to assign definitive credit. Company identifications are intended to connote widespread activity but, again, are incomplete, with perhaps greater emphasis given to U.S. efforts.

Fig. 1.2.
Alexander
Graham Bell's
photophone
receiver
*(Courtesy of*
*AT&T Bell*
*Laboratories).*

the gaseous helium-neon laser was achieved by Bennett, Herriott, and Javan (1960) of Bell Laboratories. Unlike the ruby laser, which operated only in a low duty-factor pulsed mode, the gas laser emitted light continuously, was reliable, and had high spectral purity. The gaseous laser continues to be used for many scientific and industrial applications, but it is much too large for a practical telecommunications system.

A semiconductor laser was needed. The first such lasers were reported in 1962 by Hall et al. of General Electric and by Nathan et al. of IBM. However, these were experimental laboratory devices that would operate only when pulsed with a very low duty factor or if cooled to very low temperatures. It was not until 1970 that Hayashi, Panish, and Foy at Bell Laboratories devised a heterojunction structure that led to the first achievement of continuous operation of a semiconductor laser at room temperature. Although much development work still lay ahead, this was an important breakthrough, indicating that semiconductor injection laser diodes, literally no larger than a grain of salt, might serve as practical optical communications sources.

The second ingredient unavailable to Alexander Graham Bell was a good transmission medium. The earth's atmosphere is generally quite transparent to optical signals, but there are times when fog, mist, rain, or snow can greatly reduce visibility. Although free-space optical links are used for some short telecommunication applications, both weather and obstruction considerations limit widespread use.

Limited by free-space optical transmission and spurred on by the invention of the laser, researchers turned to guided wave transmission of optical signals. Various structures, including hollow pipes with gas lenses and dielectric waveguides, were considered (Marcatili and Schmeltzer 1964; Miller 1964).

The fact that thin filaments of glass can guide beams of light is an old phenomenon,[8] and the concept has long been used in applications requiring the bending of light around corners, as in medical instrumentation. (In such applications, an entire bundle of fibers is illuminated, rather than a single fiber.) However, in the 1960s, even the best-quality optical glass would attenuate light signals below detectable levels in distances of at most a few tens of meters, rather than the several kilometers needed for practical fiber-optic communications. (Lowest attenuations, then, were in the order of 1 decibel per meter, corresponding to 1000 decibels per kilometer.[9]) In 1966, Kao and Hockman of the Standard Telecommunications Laboratory in Great Britain indicated in a landmark paper that if high silica glass were used, if impurities were sufficiently controlled, and if a lower index of refraction cladding surrounded the core, then attenuations of 20 dB/km or lower might be achieved.[10] Extensive research in achieving low-loss fibers was undertaken in Britain (largely coordinated by the British Post Office), in Japan (chiefly at Nippon Sheet Glass and Nippon Electric Company) and in the United States (primarily at Corning Glass Works and Bell Laboratories). The first success was attained in 1970 by Kapron, Keck, and Maurer at Corning, who announced the achievement of losses of 20 dB/km at a wavelength of 632.9 nm in single-mode fiber.

Thus, 1970 saw the announcement of two key technological achievements having particular significance for fiber-optic communications: the announcement by Bell Laboratories of the first semiconductor laser to op-

---

8. Conceptually, this phenomenom results from total internal reflection when light is incident from a material having a higher index of refraction to one having a lower index of refraction (e.g., from glass to air) at an angle of incidence greater than a critical angle.

9. The term *decibel*, abbreviated *dB*, is a unit used to express the relation between two amounts of power, $P_1$ and $P_2$. The number of decibels is equal to $10 \log_{10}(P_1/P_2)$. Thus, if $P_1 = 10$ watts and $P_2 = 20$ watts, the number of decibels is $10 \log_{10}(10/20)$, or $-3.01$ dB; if $P_1 = 20$ watts and $P_2 = 10$ watts, the result is $+3.01$ dB. To convert decibels to power ratios, simply divide by 10 and raise 10 to the resulting number. For example, 20 dB corresponds to a power ratio $10^{(20/10)}$, or $10^2 = 100$; and $-10$ dB corresponds to $10^{(-10/10)}$, or 1/10.

10. The quantity 20 dB/km was thought to be a benchmark, since about a 40-dB loss might be accommodated between transmitter and receiver, and 2 km is a common repeater spacing for metallic systems.

erate continuously at room temperature and the announcement by Corning of the first low-loss fiber. As Fig. 1.3 shows, the period prior to 1970 was mainly a time of research in fiber-optic communications (Millman 1984). The period following 1970 saw attention centered on technology development (i.e., bringing the technology from a laboratory curiosity to practical components) and on systems engineering (i.e., determining applications where this technology might be most economically applied). (See Gloge 1976; Jacobs 1976; and Miller, Li, and Marcatili 1973).

Fig. 1.3. Key phases and milestones in the evolution of fiber-optic transmission systems.

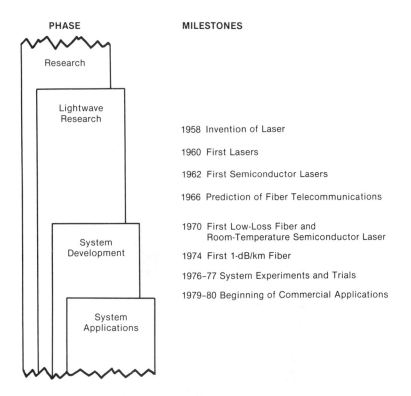

**PHASE**

Research

Lightwave Research

System Development

System Applications

**MILESTONES**

1958 Invention of Laser

1960 First Lasers

1962 First Semiconductor Lasers

1966 Prediction of Fiber Telecommunications

1970 First Low-Loss Fiber and
     Room-Temperature Semiconductor Laser

1974 First 1-dB/km Fiber

1976–77 System Experiments and Trials

1979–80 Beginning of Commercial Applications

By the mid-1970s, a number of system experiments and trials (Jacobs et al. 1978) would lead to later (1979–1980) commercial applications in North America (Cook and Szentisi 1983), Japan (Ishio 1983), and Europe (Moncalvo and Tosco 1983). Thus, it was less than ten years from the first indication in 1970 that this technology might be feasible, to the beginning of practical transmission applications. This is a remarkably short interval for the introduction of a radically new technology.

It should not be inferred from Fig. 1.3 that fiber-optic work has passed directly from research to systems engineering, to development, to application. On the contrary, it has involved the continued interaction and close coupling of all of these disciplines. Indeed, as first generation systems were being introduced, the research and development leading to subsequent generations was taking place. The continuing dynamic nature of fiber-optic transmission technology and application is a principal theme of this chapter.

# 1.3  Technology

This section will discuss the technology involved in fiber-optic communications systems, including fiber parameters, cabling, light sources, and detectors.

## 1.3.1  Fiber

An *optical fiber* is a thin filament of glass[11] with a central core having a slightly higher index of refraction than the surrounding cladding. (Light travels more slowly in the core than in the cladding.) Light is guided by total internal reflection at the core-cladding boundary (see Fig. 1.4). More precisely, the fiber is a dielectric waveguide in which there are a discrete number of propagating modes. If the core diameter and the index difference are sufficiently small, only a single mode will propagate. From a transmission system standpoint, the two principal fiber parameters are attenuation and bandwidth.

The basic attenuation mechanism in fibers is scattering caused by the random structure of the glass itself. This scattering loss decreases as a function of wavelength to the reciprocal fourth power, $1/\lambda^4$, (Rayleigh scattering). Various molecular absorption bands affect loss at wavelengths above about 1 $\mu$m. The first generation of fiber-optic systems operated at a wavelength around 0.9 $\mu$m (in the near infrared), since good sources (GaAlAs lasers and LEDs) and detectors (silicon photodetectors) were available at that wavelength. Reductions in the impurity levels, coupled with changes in the dopant materials used to vary the index of refraction, have allowed realization of much lower attenuations at wavelengths above 1 $\mu$m, with loss minima of about 0.35 dB/km and 0.2 dB/km now being achieved at wavelengths of 1.3 and 1.55 $\mu$m, respectively (see Fig. 1.5). Research results achieved in 1982, and currently being achieved in manufacture, are essentially at the theoretical minima achievable with silica-based optical fibers, with the exception of loss peak

---

The discussion in section 1.3 is based on "Fiber Optic Transmission Systems," Chapter 8, by I. Jacobs, from *Electronic Communications Handbook*, ed. A. F. Inglis (New York: McGraw-Hill, forthcoming); and is an expansion and update of material presented by the author in Jacobs 1984 and 1985.

11. Plastic fibers are used for short distance imaging applications but are typically not useful for telecommunications.

**Fig. 1.4. Types of optical fiber.**

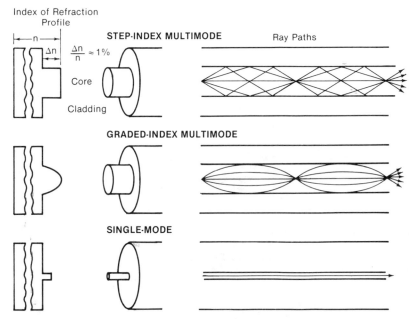

In graded-index multimode fibers, the ray paths are curved and propagation delays are nearly equalized. In single-mode fibers, the core diameter and the index of refraction difference, $\Delta n$, are smaller, and there is only one propagation path.

at 1.4 $\mu$m, due to OH-ion resonance (Li 1985). The increase in attenuation at wavelengths above 1.6 $\mu$m is associated with molecular absorptions in silica. Research is underway on new materials that in theory allow operation at longer wavelengths and the achievement of even lower attenuations.

It may be noted from Fig. 1.5 that somewhat lower attenuations are achieved with single-mode than with multimode fibers. This is a result of the lower level of dopants in the core of the single-mode fiber. Single-mode fibers are no more difficult to manufacture—and hence not intrinsically more expensive—than low-attenuation multimode fiber. However, owing to the smaller-core diameter (typically about 8 $\mu$m for single-mode fiber compared with 50 $\mu$m for multimode fiber), it is more difficult to couple light into single-mode fibers, and it is more difficult to splice and connect such fibers. However, considerable progress has been made in these areas (see section 1.3.2), and technological advances have led to increased application for single-mode fibers. The key advantage of single-mode fibers is that they allow transmission of much higher bandwidth signals than do multimode fibers.

Pulse spreading (dispersion) limits the maximum modulation bandwidth (or maximum pulse rate) that may be used with fibers. There are

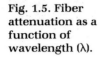

**Fig. 1.5. Fiber attenuation as a function of wavelength (λ).**

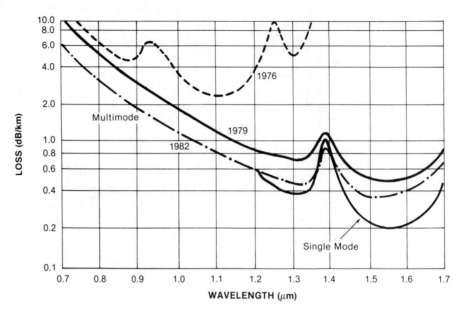

*The lowest curve is for single-mode fiber (single mode for wavelengths greater than 1.1 μm); the other curves are for multimode fiber.*

two principal forms of dispersion: *modal dispersion* and *chromatic dispersion*. In multimode fiber, the different modes have different propagation delays, and this results in modal dispersion. For example, as seen in Fig. 1.4, light rays bouncing back and forth at a steep angle travel a longer path than do rays reflected at a more grazing angle. Dispersion in multimode fiber is reduced by grading the index of refraction of the core in a nearly parabolic fashion. Now the light rays travel curved paths (Fig. 1.4). Although the outer paths are physically longer than the inner paths, the propagation delays are nearly equalized because light travels more rapidly in the regions of lower index of refraction. Graded-index multimode fibers may be made with bandwidths of the order of 1 GHz-km, whereas stepped-index multimode-fiber bandwidths (resulting from modal dispersion) are of the order of 10 MHz-km.

In addition to modal dispersion, there is chromatic dispersion, the result of different wavelengths of light travelling with different propagation speeds. Chromatic dispersion arises not only because the index of refraction varies with wavelength (material dispersion), but because propagation delay is dependent on the ratio of core dimensions to wavelength (waveguide dispersion). For silica glass, material dispersion goes through zero at a wavelength of 1.3 μm (Fig. 1.6). Since there is also a local minimum in fiber attenuation at that wavelength, 1.3 μm is a wavelength of particular interest for fiber-optic communication applications.

**Fig. 1.6.** Chromatic dispersion in single-mode fiber (pico-seconds of dispersion per kilometer of length and per nanometer of source spectral width) for various index of refraction profiles.

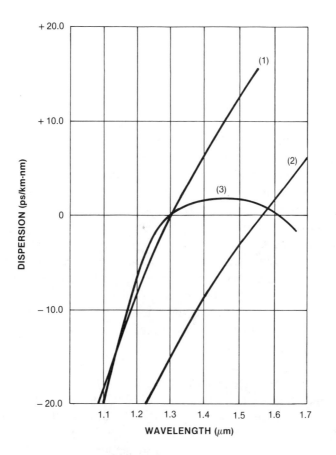

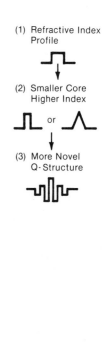

*(1) conventional single-mode fiber, (2) dispersion-shifted fiber, and (3) wideband fiber (presently in research stage).*

Waveguide dispersion is typically small when compared with material dispersion. However, there are more complex single-mode fiber structures in which the waveguide dispersion is larger and which may be used to shift the chromatic dispersion minimum to 1.55 μm (the wavelength of minimum loss) or to obtain very low dispersion over a broad wavelength range. It should be noted, though, that whereas modal dispersion in multimode fibers is dependent solely on the fiber, chromatic dispersion in single-mode fibers is also dependent on the spectral characteristics of the light source.[12] Thus, fibers may be utilized at wavelengths other than the wavelength of minimum chromatic dispersion if

---

12. Chromatic dispersion occurs in multimode fibers as well. However, in multimode fibers, modal rather than chromatic dispersion is typically dominant. An exception is the use of LEDs at 0.9 μm on multimode fiber, where chromatic dispersion tends to dominate owing to the large spectral width of LED light sources.

the light sources are sufficiently monochromatic. This is considered further in sections 1.3.3 and 1.4.3.

## 1.3.2  Cables, Splices, and Connectors

For practical application, fibers must be packaged in protective cables, and a means must be available for both permanent and demountable joining of the fibers. This section will discuss principal technology issues for cables, splices, and connectors.

Glass fibers are strong. Typically, fibers are manufactured with tensile strengths in excess of 50,000 pounds per square inch (50 kpsi), and fibers have been made with strengths in excess of 500 kpsi, greater than the tensile strength of steel.[13] What makes glass weak and brittle are flaws and cracks, which grow and propagate. Indeed, any handling of an unprotected fiber will markedly reduce the strength of that fiber. Consequently, as part of the fiber manufacturing process, fibers are coated, usually with a soft polymer, to provide a buffer and to protect against further handling. It should be noted, however, that even with a strength of 500 kpsi, a fiber having an outer diameter of 125 $\mu$m (0.005 inch) has a breaking force of only 10 pounds. Thus, fiber cables must contain additional strength elements to further buffer and protect the fibers.

There are a large variety of cable structures, ranging from simple jacketed single-fiber cables to multifiber cables. The multifiber cables, in turn, range from loose tube designs, with individually buffered fibers, to ribbon-structured cables in which multiple fibers are held between two strips of plastic. Ribbon-structured cables permit a higher packing density (larger number of fibers in a given cable diameter; e.g., a cable containing 144 fibers, 12 ribbons of 12 fibers each, is 12 mm in diameter) and also facilitate multifiber splicing. Cables can be preconnectorized with an array connector, allowing the joining of 12 fibers in a single operation (see Figs. 1.7 and 1.8).

Splicing, or connecting, fibers is conceptually simple. The two fibers to be joined need only be butted up against one another. The trick is to achieve and maintain the requisite alignment accuracy with practical, economical techniques. Fig. 1.9 illustrates the effect of lateral misalignment on splice or connector loss for typical multimode (50-$\mu$m core diameter) and single-mode (8-$\mu$m core diameter) fibers. The smaller-core diameter of the single-mode fiber results in a much greater sensitivity to misalignment. (Although lensing techniques can reduce the sensitivity to lateral and lon-

---

13. One of the few deviations from the use of metric units in this chapter is strength. The metric unit for strength, *pascal* (1 giga pascal [GPa] = 145 kpsi), is not yet widely used in the United States.

**Fig. 1.7. A 144-fiber cable containing 12 ribbons of 12 fibers each** *(Courtesy of AT&T Bell Laboratories).*

**Fig. 1.8. A 12-fiber connector** *(Courtesy of AT&T Bell Laboratories).*

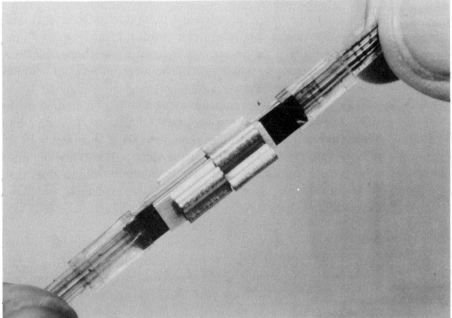

*Fibers are placed in etched chips of silicon, which are aligned with a mating silicon chip.*

gitudinal misalignment, they accentuate the sensitivity to angular misalignment.)

The loss of a connector or splice is a function not only of the alignment accuracy but also of the properties of the fibers. Nonidentical

Fig. 1.9.
Connector loss
as a function of
lateral offset
between fiber
cores.

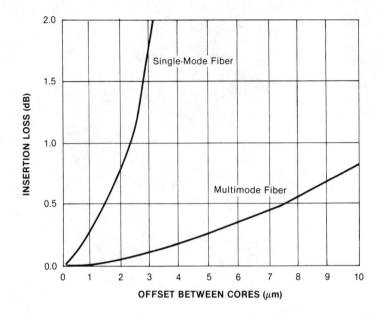

fibers (owing to geometrical or optical differences) will result in nonzero splice loss even when perfectly aligned. For multimode fiber, the loss generally is not reciprocal. For example, there will be a very small loss in going from a smaller- to a larger-core fiber, but a much larger loss in the opposite direction.

Splices are typically made by either fusing or bonding the fibers together, utilizing appropriate alignment fixtures. Since the fiber core may not be perfectly concentric with the cladding, the lowest losses in single-mode fiber splices (less than 0.05 dB) are obtained by techniques in which optical measurements (namely, minimizing scattered light at the splice) are used rather than sole reliance on geometrical alignment (Miller and DeVeau 1985). Conventional fusing or bonding techniques generally result in a splice loss of about 0.2 dB.

Demountable connectors utilize cylindrical ferrules or conic plugs that mate in biconic receptacles to achieve the appropriate centering. There is considerable variance in splice and connector losses associated with the following variables: dimensional tolerances of the connector piece parts, dimensional tolerances and index-of-refraction profile tolerances of the fibers, fiber end preparation, and cleanliness (a speck of dust could render opaque the connection between two fibers). Connector losses somewhat below 1 dB are now being achieved with both multimode and single-mode fiber, though higher precision (and generally more expensive) parts are required for the single-mode case.

Considerable progress has been made in reducing the time required to splice fibers and the difficulty of splicing them, and in reducing the cost and size of connectors. The installation time for long distance fiber-optic cables is much less than that for metallic cables of comparable capacity, owing to the smaller size and lower weight of the fiber cable, the longer distance between splices, and the higher capacity of transmission. The major thrust of the work currently being done on fiber cable, connector, and splicing technology is to make this statement true for short distance cabling (e.g., within a building) as well.

### 1.3.3  Sources (Lasers and LEDs)

Semiconductor light-emitting diodes (LEDs) or semiconductor injection laser diodes (ILDs)[14] are the primary light sources for fiber-optic communications systems. Their key advantages are that they are small in size, are well matched to the dimensions of the fiber, and may be directly modulated. In both sources, an electric current passed across a semiconductor junction excites charged carriers in the semiconductor crystal. When these carriers relax to the ground state, light is emitted at a wavelength dependent on the energy band structure of the semiconductor.

The laser is a more complex structure than the LED. Within the laser semiconductor chip is the equivalent of an optical cavity in which the emitted light is guided and channeled between two partially reflecting mirrors formed by the cleaved end faces of the chip. This light is reflected back into the cavity, where it stimulates the emission of additional light. The result is a buildup of an intense, highly directive, and narrow spectral width beam of light that emerges from the two end faces of the laser (see Fig. 1.10). Light from the front face of the laser is efficiently coupled (about 50%) into the fiber. Light from the rear face is detected within the laser package, and this detected signal is used in a feedback loop to keep the output power of the laser fixed.

Lasers are nonlinear devices. The output is low until the input current exceeds a threshold, beyond which the output increases rapidly (Fig. 1.11). Lasers are typically biased just below threshold ("off" state) and are turned "on" by a small signal current. Although the bias and signal currents are small (typically about 50 mA), the current densities are very large (greater than 1000 A/cm$^2$). Much of the electrical energy (a few tenths of 1 watt) provided to the laser goes into heat rather than light. (The total optical output is a few milliwatts, of which about 1 milliwatt [0 dBm] is coupled into the fiber.) It is critical that this heat be removed

---

14. Although the designation "ILD" is sometimes used in the literature, it does not have the universality of the designation "LED." Consequently, we will simply use the term *laser*.

Fig. 1.10. Laser
package with
laser chip
bonded to
copper heat
sink. Fiber exits
via plastic conic
plug connector
*(Courtesy of
AT&T Bell
Laboratories).*

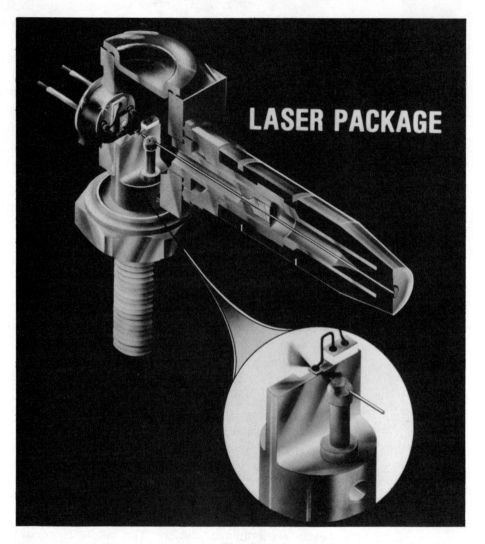

LASER PACKAGE

from the active area of the laser, since lasers will not operate reliably and
efficiently at high temperature. Thus, much of the laser package is a heat
sink to conduct heat away from the active area. Indeed, many lasers contain thermoelectric coolers to provide active cooling and enhance reliability.

Significant improvements have been made in the reliability of semiconductor lasers since 1970 (see Fig. 1.12). Although lasers with a mean
time before failure (MTBF) in excess of 1 million hours (more than 100
years)[15] were reported in 1977 (Hartman), considerably more effort was

15. Lifetime is typically estimated from accelerated life tests at high temperature or from extrapola-
tion of threshold current increase with time.

needed to achieve laser structures that were suitable for high-speed modulation and could be manufactured with satisfactory results.

**Fig. 1.11. Laser output power as a function of drive current, shown for a representative sample.**

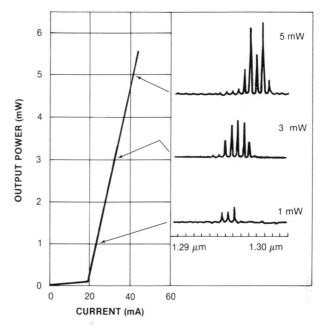

*Spectra indicate that lasers are generally multifrequency and that the spectral distribution is a function of power.*

**Fig. 1.12. Improvements in the mean lifetime of lasers.**

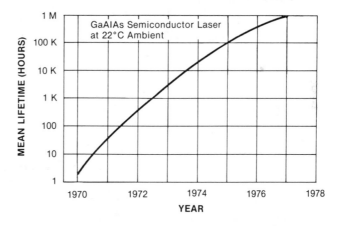

Semiconductor lasers remain as complex and rather expensive devices. LEDs are simpler and cheaper. They are less temperature sensitive and much more reliable. However, LEDs do have a disadvantage. Whereas a laser emits light in a highly focused beam that is efficiently coupled into a fiber, radiation from an LED is over a much broader angle, resulting in significantly less (typically 10 to 20 dB less) power coupled into a multimode fiber[16] than from a laser. Coupling loss from an LED into a single-mode fiber is even greater. Consequently, LEDs typically have not been used with single-mode fiber, though this situation is changing (see section 1.5.5).

In addition to radiating over a larger range of angles than lasers, LEDs radiate over a wider spectral range. This results in considerable chromatic dispersion if operation is other than at the minimum dispersion wavelength of the fiber.

Lasers, although spectrally purer than LEDs, typically are not single-wavelength devices. Fig. 1.13 illustrates a typical spectrum of an InGaAsP (the material system most commonly used for lasers in the 1.2- to 1.6-μm spectral range) laser as a function of temperature. Temperature control may be required not only for reliability but for stabilization of the center wavelength and the modal pattern.

Fig. 1.13. Typical spectrum of InGaAsP laser illustrating the effect of temperature on the radiated wavelengths.

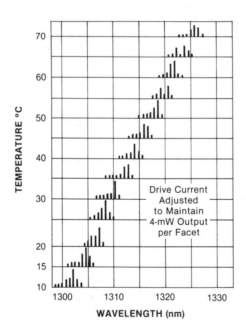

As shown in Fig. 1.13, a semiconductor laser emits a number of discrete spectral lines (also termed longitudinal modes), since the resonant cavity in the laser, though small, is large compared with the wavelength of the light. To obtain a single-wavelength output (single longitudinal

---

16. Improved coupling is achieved with multimode fibers having a larger-core diameter and a higher numerical aperture (numerical aperture is a measure of the light-gathering property of the fiber). Such fibers are finding extensive use for short distance applications.

mode [SLM] laser), either an external cavity or special resonant structures within the laser must be employed (Suematsu, Arai, and Kishino 1983). Examples of the latter are distributed feedback (DFB) or cleaved-couple-cavity (C³) (Tsang 1985) structures (see Fig. 1.14). SLM lasers are of particular interest at a wavelength of 1.5 $\mu$m, where conventional single-mode fibers do not have minimum dispersion and where narrower spectral width sources are required to avoid being limited by chromatic dispersion.

**Fig. 1.14.
A cleaved-couple-cavity laser chip in the eye of an ordinary sewing needle** (*Courtesy of AT&T Bell Laboratories*).

*The two sections of the laser have different spacings between their resonant wavelengths. Only the single wavelength, common to the two sections, is radiated.*

It should be noted, however, that DFB or C³ SLM lasers at a center wavelength of 1.5 $\mu$m (1500 nm) typically have a spectral width of about 0.15 nm.[17] This is one part in $10^4$, which, since the center frequency is 2 $\times$ $10^{14}$ Hz, translates to a frequency width of 20 GHz. This width is much larger than typical modulation bandwidths and precludes the use of coherent (heterodyne or homodyne) detection. Much current research is directed towards achieving practical laser structures with even narrower spectral widths, and these narrower spectral widths may afford opportunities for more sensitive receivers and new system architectures.

---

17. Chromatic dispersion at 1.5 $\mu$m is about 0.02 nanosecond per kilometer of path and per nanometer of spectral width (see Fig. 1.6). Thus, for a path length of 100 km, a spectral width of 0.15 nm will result in a dispersion of 0.3 ns—not negligible, for example, at a pulse rate of 1 Gbps where the pulse width is 1 ns.

### 1.3.4 Detectors

Receivers for fiber-optic communications systems employ semiconductor photodetectors in which incoming photons generate carriers (electrons and holes) by a process that is essentially the inverse of that occurring in the light source. (Indeed, there has been research on using the same device as both a source and a detector—for example, in a half-duplex data link. However, different structures are typically used for sources and detectors to achieve optimal performance.) For on-off binary modulation, an ideal receiver counts incoming photons and, depending on whether a threshold is exceeded, decides whether or not a pulse is present. There is an inherent statistical fluctuation in the number of photons within a pulse, and this results in a minimum average number of photons needed to achieve a given error probability (20 photons for $10^{-9}$ error probability). For a 100-Mbps system (with binary on-off modulation), this would correspond to a minimum of $10^9$ photons per second. If each photon generated 1 photoelectron,[18] the resultant electrical current would be $1.6 \times 10^{-10}$ amps. Considerable electronic amplification is required before standard pulse detection circuitry can be employed. In an avalanche photodetector (APD), some of this amplification is internal to the detector (photoelectrons, accelerated by a high electric field, generate additional electrons). In PIN[19] photodetectors, all of the amplification is by external electronic amplifiers, although there is much research on integrating the detection and amplification functions (integrated optoelectronics).

The amplification process introduces additional noise, and practical receivers generally require about 1000 photons per bit of information, which is 100 times (20 dB) greater than the theoretical minimum. At 100 Mbps, 1000 photons per bit corresponds to $10^{11}$ photons per second. At $\lambda = 1.3$ $\mu$m, the photon energy is $1.5 \times 10^{-19}$ joules, so that $10^{11}$ photons per second corresponds to $1.5 \times 10^{-8}$ watts ($-48$ dBm). (The photon energy is equal to Planck's constant, $h = 6.63 \times 10^{-34}$ joule-second, times the frequency of the light.) This is considerably more power than is required in microwave systems and is a consequence of the fact that the photon energy is so high at optical frequencies. (An equivalent noise temperature may be calculated by dividing the photon energy by Boltzmann's constant. At $\lambda = 1.3$ $\mu$m, this equivalent noise temperature is 11,000°K.)

---

18. For simplicity, the description is in terms of electrons, but the charged carriers can be electrons or holes, or a combination of the two.
19. *PIN* stands for *positive-intrinsic-negative*. A PIN diode has a structure particularly well-suited for photodetection.

Since theoretical receiver sensitivity may be expressed as the number of photons per bit (independent of bit rate), the minimum receiver power increases linearly with the bit rate. (There is a 3-dB reduction in receiver sensitivity for each doubling of the bit rate. Practical factors may result in a somewhat faster decrease in sensitivity.) An equivalent way of viewing this is that fiber-optic systems have a flat noise spectral density so that the noise power increases linearly with the modulation bandwidth. As noted previously, this noise spectral density (or the equivalent noise temperature) is much higher for fiber-optic than for microwave systems.

It should also be noted that photodetectors are square law devices; that is, the output photocurrent is proportional to the intensity of the incoming light (square of the incoming field strength). If a strong local optical oscillator signal, coherent in phase with the incoming optical signal, were to be added prior to the photodetector, then the output electrical signal would be proportional to the product of the local oscillator and the incoming signal field strengths. This would result in an increase in the photocurrent, relative to the direct-detection case, by the ratio of the local oscillator to the incoming field strengths. Unlike the avalanche process, this does not in principle introduce additional noise, and coherent detection offers the promise of more closely approaching the theoretical sensitivity limits (Henry 1985; Cohen 1986).

Fig. 1.15 shows receiver sensitivity at a bit error rate of $10^{-11}$ and a wavelength of 1.3 $\mu$m. The upper band indicates the sensitivities achieved with practical direct-detection receivers. The lower band indicates theoretical lower limits. Current coherent experiments are achieving sensitivities between these two bands.

The potential advantage of coherent detection is not only improved re-

**Fig. 1.15.** Receiver sensitivity needed to achieve a $10^{-11}$ bit error rate at a wavelength of 1.3 $\mu$m (Cohen 1986).

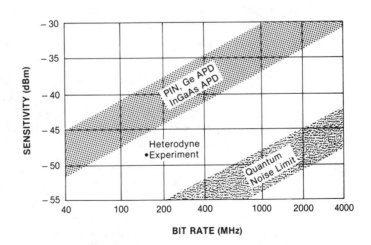

ceiver sensitivity but, by the use of heterodyne techniques, the attainment of a wavelength selectivity beyond that achievable by optical filtering techniques. This could allow, for example, the wavelength division multiplexing of an enormous number of channels within the optical band.

# 1.4   System Design

We now turn to some of the factors influencing system design. These factors include the use of digital versus analog transmission, bit rate and repeater spacing capability, and multiplexing techniques.

## 1.4.1   Digital versus Analog

Fiber-optic communications systems are typically power limited; that is, semiconductor sources couple relatively low power into the fiber, and, as noted previously, the equivalent receiver noise is very high. On the other hand, large bandwidths are generally readily available. This may be contrasted with radio systems in which bandwidth is allocated by regulatory agencies, or with metallic systems in which bandwidth is, in effect, limited by attenuation increasing with frequency.

For communication channels in which power is at a premium but bandwidth is readily available, there are well-known advantages to using wideband modulation techniques, which trade signal-to-noise ratio for bandwidth.[20] Pulse code modulation (PCM) is a particularly efficient way of making this trade, and thus fiber-optic systems are compatible and, indeed, are a further spur to the growing trend towards digital transmission of voice and other information.

Analog modulation can be used with fiber. Direct intensity modulation may be used, though there are linearity issues, particularly with laser sources. Both signal-to-noise and linearity requirements are lessened by the use of fm techniques. Since practical techniques do not presently exist for frequency-modulating an optical source, it is necessary to use subcarrier techniques. For example, an analog baseband video signal might be used to frequency-modulate an intermediate frequency (if) signal centered at about 100 MHz. The modulated if signal is then used to intensity-modulate the laser.

Analog fiber-optic systems are generally used only for short unrepeatered video links. Most fiber-optic applications use digital transmission with simple on-off modulation.

---

20. Another example of such a communication channel is communication from a deep space probe. Chapter 2 describes satellite systems, and Chapter 8 explains PCM in detail.

## 1.4.2  Bit Rate and Repeater Spacing Capability

One measure of the capability of a digital transmission system is the product of bit rate and the maximum transmission distance before pulse regeneration (a repeater) is required. The higher the bit rate, the smaller the number of fibers required to achieve a given total capacity. The larger the repeater spacing, the fewer the repeater sites required in a long-haul transmission system. Thus, for a given total capacity and distance, the number of repeaters is inversely proportional to the bit-rate repeater-spacing product.

The bit-rate repeater-spacing product has been doubling yearly. This is illustrated in Fig. 1.16, which shows approximate composites of both best research results and actual deployed systems. For example, in 1985, practical systems with a bit rate of about 500 Mbps and a repeater spacing of about 40 km were deployed (product equal to 20,000 Mbps-km. The "hero" research result (for serial transmission) was 4 Gbps and 117 km, corresponding to a product of 500,000 Mbps-km. It is interesting to observe that there is about a five-year time lag between the research achievement and the corresponding achievement in practice. This time interval is required not only to progress from research to production technology, but to achieve even more advanced technology, since deployed systems require safety margins not contained in the research experiment.

Fig. 1.16. Increases in the product of bit rate and unrepeatered transmission distance achieved in laboratory experiments (upper line) and in existing and announced future commercial systems (lower line).

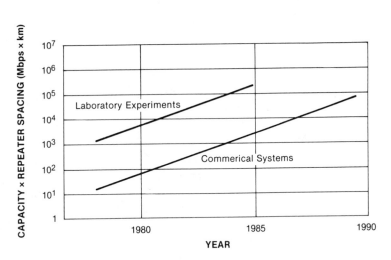

The bit rate and repeater spacing of practical systems is, of course, dependent on the technology employed. Fig. 1.17 illustrates representative practical system capability for various technologies. At low bit rates, repeater spacing is limited by receiver noise (loss limit). In this regime, repeater spacing decreases only slowly with increasing bit rate. At high bit rates, pulse dispersion may limit repeater spacing. In this regime, there is a rapid decrease in the repeater spacing as the bit rate increases.

**Fig. 1.17.**
**Approximate**
**(and somewhat**
**conservative)**
**limitations on**
**fiber-optic**
**system bit rates**
**and repeater**
**spacings**
**achievable in**
**practical**
**systems with**
**present**
**technology.**

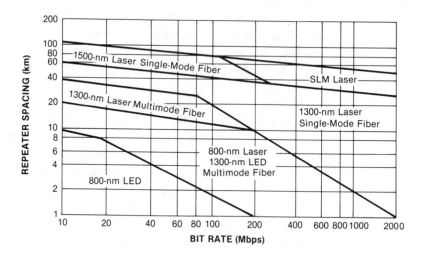

The technology with the least capability, but presently the simplest and the cheapest, is light-emitting diodes at short wavelength (0.85 $\mu$m) on multimode fiber. The bit-rate repeater-spacing product is limited by chromatic dispersion to about 200 Mbps-km, adequate for many moderate capacity intrabuilding or campus applications.

Approximately the same capability on multimode fibers may be achieved with lasers at short wavelength (0.85 $\mu$m) as with LEDs at long wavelength (1.3 $\mu$m). The laser couples more power into the fiber, but the fiber has a lower attenuation at 1.3 $\mu$m. In both cases, modal dispersion limits capability to about 2000 Mbps-km. (Chromatic dispersion is not limiting in these cases, owing to the narrow spectral width of the laser and the minimum in chromatic dispersion at 1.3 $\mu$m.) Even longer repeater spacings (about 40 km) may be achieved with lasers at 1.3 $\mu$m, but the same dispersion limit applies for multimode fiber. Even longer repeater spacings (about 60 km) may be achieved with lasers at 1.3 $\mu$m on single-mode fiber, owing to the lower attenuation achievable with single-mode fiber. The dispersion limit in this case is determined by the closeness of the match between the laser center wavelength and the zero dispersion wavelength of the fiber and by the spectral width of the laser. Technology is available today to extend this limit well beyond the range of the bit rates and distances depicted in Fig. 1.17.

Operation at 1.55 $\mu$m, where even lower fiber losses are achievable, offers the opportunity of even longer repeater spacings (about 100 km).

However, with conventional lasers and conventional single-mode fiber, chromatic dispersion limits the maximum bit-rate distance product to about 10,000 Mbps-km. To achieve the simultaneous benefits of very high bit rates and very long repeater spacings requires either single-frequency (single longitudinal mode) lasers or fibers with the dispersion minimum shifted to 1.55 $\mu$m.

It should be noted that the technology boundaries depicted in Fig. 1.17 do not represent technology limits, but rather depict representative capabilities achievable in practical systems.

One technology choice not shown in Fig. 1.17 is the use of LEDs with single-mode fiber. This was not initially considered a practical choice, owing to the very large coupling loss in going from an LED to a single-mode fiber. With the growing trend to single-mode fibers, there is increasing interest in LED structures (edge-emitting LEDs) that couple more power into a fiber than conventional LEDs. Such technology may prove useful for short distance applications requiring very high capacities but not the ultimate in repeater spacing. Experiments have been performed with LEDs achieving a bit rate of 180 Mbps and an unrepeatered distance of 35 km (Saul et al. 1985). The corresponding best result to date with lasers is 4 Gbps and a distance of 117 km (Korotky et al. 1985).

## 1.4.3 Multiplexing

Since the total capacity of the fiber far exceeds the requirements of any single communications source, many channels are typically combined, or *multiplexed*, before being transmitted on the fiber.

*Time-division multiplexing* (TDM), in which a number of low-speed digital streams are synchronized and then interleaved to create a single high-speed digital stream, has been the principal form of multiplexing used in fiber-optic systems. The multiplexing is implemented electronically with the conversion from electronics to optics (the optical transmitter) following the final stage of multiplexing, and with the conversion from optics to electronics (the optical receiver) preceding the demultiplexing. The optical interfaces (transmitters and receivers) are often incorporated into the muldem (multiplexer/demultiplexer) to avoid the expense of separate muldem and fiber-optic system terminals and to minimize the amount and complexity of the high-speed electronic circuits. Indeed, the speed capability of electronic signal-processing circuits tends to be a more limiting factor in the capacity of fiber systems than either the dispersion properties of the fiber or the speed capabilities of the light sources or detectors.

The use of *wavelength division multiplexing* (WDM), in which different wavelengths of light are used as separate carriers, is a means of achieving total capacities beyond that practically achieved solely with TDM. (This is analogous to radio systems that are channelized into manageable frequency bands.) Optical filters may be used to combine (multiplex) and to separate (demultiplex) the various channels. Relative to TDM, WDM substitutes a multiplicity of lower speed channels for a single higher-speed channel. Since this necessitates a multiplicity of sources and detectors (as well as the optical filter), this trade is generally not economical until the speed capability of a single electronic channel is exceeded. Thus, the trend has been to exploit the speed capability of TDM before utilizing the added capacity afforded by WDM. (An exception is multimode fibers in which modal dispersion may provide more stringent limits on transmission bit rate than do electronic circuit limitations; see Fig. 1.17.) However, there are reasons to expect that WDM will play a larger role in the future—both to extend the capacity of the fiber beyond that achievable with serial processing and to utilize wavelength-selective optical processing techniques to achieve functions such as channel dropping filters.

# 1.5   System Applications

Fiber optics has found important applications in telecommunications. This section will discuss general considerations for transmission media and then look at the principal system applications of metropolitan trunking, long-haul trunking, undersea cable, television, and local area networks.

## 1.5.1   General Considerations

Widespread application of any new technology is dependent either on the technology satisfying a new need or on its meeting an existing need more economically than alternative technologies. To date, there have been no applications of fiber-optic communications that could not be satisfied technologically by other transmission media. Rather, fiber is being applied in a growing number of applications because it is the economic choice for these applications. A quantitative discussion of the economics of fiber-optic systems is outside the scope of this chapter. However, some general remarks may be made.

Transmission media may be divided into two general categories: cable and free-space propagation. Fiber falls into the first category, as do wire pairs and coaxial cable. (Metallic waveguides have some special transmission applications but generally are not used for telecommunication purposes.) The second category contains a wide range of radio systems including microwave line-of-site and satellite links. Radio systems offer unique advantages for broadcast and mobile applications and also offer advantages for communications to remote regions. Cable systems are usually less interference prone, are more secure, and do not require the frequency administration and coordination of radio systems. No one technology is the right answer for all applications, and it is reasonable to expect that radio and cable systems will continue to coexist.

A direct comparison is more readily made between fiber and other cable media. Fig. 1.18 compares attenuation as a function of modulation frequency for graded-index fiber, a 22-gauge copper pair, and 0.95-cm diameter coaxial cable. It is the very low attenuation of the fiber and the independence of this attenuation with frequency that make possible the much higher communication capacities and much longer unrepeatered distances achievable with fiber than with copper cable.

**Fig. 1.18. The effective loss of 1-km lengths of various transmission media.**

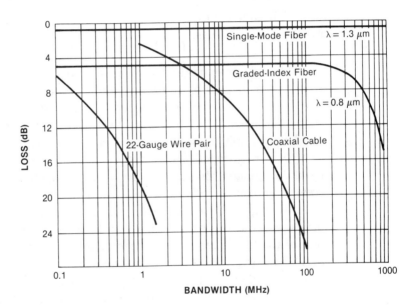

Although the cost of fiber continues to decrease, the cost is still much higher than that of a copper pair (more than ten times), though it is commensurate with the cost of coaxial cable. It is difficult for fiber to compete economically if there is a simple one-for-one replacement of copper pairs by fiber. However, if a fiber replaces many copper pairs or a coaxial cable, then the economics are more favorable.

## 1.5.2 Metropolitan Trunking

In the telephone network, initial fiber systems have proved to be most suitable for interoffice trunking (circuits between switching offices) in metropolitan areas. In 1980 AT&T deployed its first commercial metropolitan fiber-optic system in Atlanta (Jacobs and Stauffer 1980), and over the past five years there have been extensive applications in metropolitan areas throughout the United States. Fiber-optic systems have proved so attractive for the following reasons:

1. The large capacity of the fiber matched the large circuit cross sections between metropolitan switching offices.

2. The large repeater spacing allowed office-to-office transmission without the need for intermediate repeaters in manholes (metallic systems have repeaters at 2-km spacing).

3. The small-diameter cable made efficient use of space in crowded metropolitan ducts.

4. Digital transmission (T1) systems were already extensively used in metropolitan areas.

Although telephone networks are transmitting a growing variety (voice, data, image) of communication traffic, voice channels remain the dominant component. Consequently, such networks must provide an economic means for the transmission of voice circuits. Metropolitan trunks were initially (and in many cases still are) provided for by a separate pair of copper wires for each trunk (two pairs for "four-wire trunks"). Longer distances were accommodated by using thicker-gauge copper and/or amplifiers (voice frequency repeaters) at periodic intervals. Analog multiplexing (frequency-division multiplexing) systems were developed and used for longer trunks in metropolitan areas. The invention of the transistor made digital transmission of voice practical, and the first PCM transmission system for telephony (T1) was introduced by AT&T in 1962. In T1, twenty-four voice circuits are encoded (64 kbps each), time-division multiplexed, and transmitted at 1.544 Mbps over two pairs of copper wires. Compared with analog multiplexing, digital multiplexing resulted in lower-cost terminals and better transmission performance than that achievable with analog transmission on wire pairs. As the cost of installing wire pairs increased and the cost of digital terminals decreased, T1 proved in at shorter and shorter distances relative to simple voice-frequency trunks.

With time, the congestion of voice-frequency trunks was replaced with the congestion of large numbers of T1 systems. This congestion is measured not only in the number of wire pairs but in the number of repeaters in manholes, and in the effort required to install, operate, and maintain very large numbers of T1 systems. By the mid-1970s, higher capacity systems were needed. Although some improvements have been made in the capacity of wire-pair digital systems, and there has been some application of digital coaxial systems, this need has been met largely with fiber-optic systems.

Initial systems transmitted at 45 Mbps. This is one of the standard rates (DS 3) in the North American digital hierarchy (see Table 1.1), and it accommodates the multiplexing of twenty-eight 1.544-Mbps T1 streams, corresponding to 672 digital voice channels. Subsequent systems, from a wide range of suppliers, have become available, with capacities being an integral multiple of the DS 3 standard. Trunking systems with capacities of 45, 90, 135, 180, 405, and 540 Mbps are in service, and systems with capacities of 810, 1080, and 1620 Mbps have been announced.[21]

TABLE 1.1.
DIGITAL
HIERARCHIES

| | | North American | European (CEPT) | Japanese |
|---|---|---|---|---|
| Level 1 | Bit Rate (Mbps) | 1.544 | 2.048 | 1.544 |
| | Voice Circuits | 24 | 30 | 24 |
| Level 2 | Bit Rate (Mbps) | 6.312 | 8.448 | 6.312 |
| | Voice Circuits | 96 | 120 | 96 |
| Level 3 | Bit Rate (Mbps) | 44.736 | 34.368 | 32.064 |
| | Voice Circuits | 672 | 480 | 480 |
| Level 4 | Bit Rate (Mbps) | 274.176 | 139.264 | 97.728 |
| | Voice Circuits | 4032 | 1920 | 1440 |
| Level 5 | Bit Rate (Mpbs) | Not defined | 565.148 | 397.20 |
| | Voice Circuits | | 7680 | 5760 |

Initial systems operated at short wavelength (in the range of 0.82 to 0.87 $\mu$m) on multimode fiber, but there was quick evolution to long wavelength (1.3 $\mu$m) and then to single-mode fiber. Systems at 405 Mbps and above have been used exclusively on single-mode fiber, since modal dispersion in multimode fiber would greatly restrict repeater spacing at such bit rates. The trend has been to utilize single-mode fiber and systems with capacities of 405 Mbps or higher in new metropolitan applications. Such systems typically have a highly modular architecture, and there is little cost premium in utilizing such systems at less than full capacity, relative to installing lower capacity systems. This affords the opportunity of meeting subsequent growth needs by adding circuit packs in the terminals but not replacing any of the line equipment. Fig. 1.19 shows such a terminal. The 33-cm-high by 66-cm-wide shelf provides all

---

21. The above capacities are integral multiples of the 45-Mbps DS 3. The actual line rates are typically somewhat higher, owing to added bits for synchronization, performance monitoring, telemetry, maintenance communication, etc. Although there are standard interfaces and standard capacities, the actual line bit rates and formats tend to be unique to each supplier's product.

of the circuitry necessary for interfacing nine DS 3 signals to and from a 417-Mbps optical line. Each DS 3 has separate circuit packs so that the shelf need not be fully equipped if fewer than nine DS 3s are required.

**Fig. 1.19. The Terminating Muldem Assembly of the AT&T FT-Series G system** *(Courtesy of AT&T Bell Laboratories).*

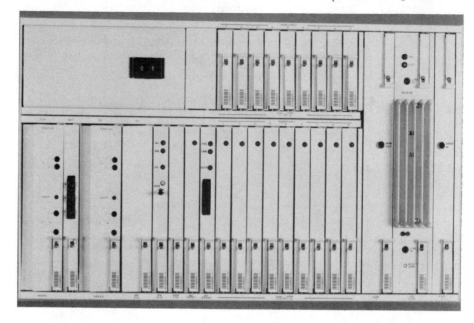

Fiber systems in metropolitan areas tend to be used for high-capacity backbone routes, with existing T1 metallic lines being used to provide the access to the fiber hubs. In such networks it is often economical to route circuits over the fiber backbone even if this results in a longer physical path than that of a direct route. Thus, the fiber "highway" takes much of the traffic off the "local roads." The key, however, for metropolitan networks, where the hubs are relatively close together, is to reduce the cost of access to, and egress from, the fiber highway. This is the present thrust of much of the multiplex, digital cross-connect, and fiber-optic system terminal design.

### 1.5.3  Long-Haul Trunking

For long-haul trunking, terminals are much farther apart, and the line haul costs (installed cable and repeaters) are the dominant cost element. Also, unlike short-haul trunking where wire pairs were the dominant transmission medium, long-haul was dominated by microwave radio and, to a lesser degree, by coaxial cable.[22]

---

22. For example, in 1980 there were approximately 48,000 route miles of radio, and 21,000 route miles of coaxial cable in the AT&T Communications (then known as Long Lines) network. In terms of equipped circuit capacity, about 75% was on microwave radio and 20% on coaxial cable.

There have been continued increases in the capacity of analog microwave radio and coaxial cable systems (Tables 1.2 and 1.3). In the case of microwave radio, this has been done by utilizing existing towers and antennas, with advances in higher-powered solid-state microwave sources, more sensitive receivers, and more bandwidth-efficient modulation techniques (single-sideband modulation). It has been difficult for any cable system to compete economically with augmentations in the capacity of the analog microwave network.

**TABLE 1.2. TECHNOLOGICAL PROGRESS IN COAXIAL CABLE SYSTEMS**

| System | Year | Circuits per Coaxial | Repeater Spacing (Miles) | Typical Coaxials per Sheath | Sheath Capacity (Circuits) |
|--------|------|----------------------|--------------------------|-----------------------------|----------------------------|
| L-1 | 1941 | 600 | 8 | 4 | 600 |
|  |  |  |  | 8 | 1,800 |
| L-3 | 1953 | 1,860 | 4 | 8 | 5,580 |
|  |  |  |  | 12 | 9,300 |
| L-4 | 1967 | 3,600 | 2 | 20 | 32,400 |
| L-5 | 1974 | 10,800 | 1 | 22 | 108,000 |
| L-5E | 1978 | 13,200 | 1 | 22 | 132,000 |

**TABLE 1.3. PROGRESS IN 4-GHZ FM MICROWAVE RADIO SYSTEM (TD) CAPACITY**

| Year | Voice Circuits per 20-MHz Channel | Route Capacity (Circuits) |
|------|-----------------------------------|---------------------------|
| 1950 | 480 | 2,400 |
| 1953 | 600 | 3,000 |
| 1960 | 600 | 6,000 |
| 1967 | 900 | 9,000 |
| 1968 | 1,200 | 12,000 |
| 1973 | 1,500 | 16,500 |
| 1980 | 1,800 | 19,800 |

There were several factors in the early 1980s that began to make fiber economically viable for long-haul transmission. Foremost among these was the introduction of digital switching—and the significant terminal cost savings that resulted when digital transmission systems were interfaced with the digital switch.

In January 1980, AT&T-Long Lines filed an application with the FCC for construction of a fiber-optic system between Washington and Boston for the Northeast Corridor (see Fig. 1.20). The proposed system was about 20% less expensive than the alternative, which was to augment existing analog facilities (AT&T 1980). The savings were a consequence of the large concentration of digital 4ESS™ switching offices along this route. Since the circuit lengths were shorter than those typical of the long-haul network, the terminal cost savings more than compensated for an increase in line-haul costs.

Although there were the usual "Petitions to Deny" from other carriers, and requests from the FCC for more information, the application appeared to be progressing normally through the regulatory process. A key event then occurred in June 1980 when Corning Glass Works filed with the FCC a document entitled "Comment in Support of Northeast Corridor Lightguide Fiber Installation." The concluding paragraph of this sixteen-page document stated:

> Corning supports the installation of a lightguide fiber communications link in the Northeast Corridor; Corning believes that the Commission in passing on AT&T's petition should seriously consider the long-term structure and vigor of the lightguide fiber industry and should set forth guidelines that, both in concept and in practical detail, will preserve the competitive nature of the industry.

Much of the ensuing discussion centered on procurement issues, a consequence of which was the FCC's approval (November 25, 1980) of the first phase of the Northeast Corridor (New York to Washington required for January 1, 1983 service) based on AT&T's agreement that the second phase (New York to Boston required one year later) would be let out for competitive procurement.

In April of 1981 the Purchased Products Division of AT&T issued a "Request for Proposal" for the second phase of the Northeast Corridor from New York to Boston, as well as an extension from Washington to Richmond (Fig. 1.20). Western Electric (now AT&T Network Systems), the manufacturing entity of AT&T, was one of several U.S. manufacturers responding to this RFP. There were also several responses from Europe and Japan.

After considerable input from both the executive and the legislative branches of government, on October 3, 1981 AT&T-Long Lines selected Western Electric—who had submitted the lowest bid of the domestic suppliers—to provide the New York to Boston and Washington to Moseley links. The FCC construction application was filed in November; Fujitsu (who was the lowest bidder overall) filed a "Petition to Deny." The FCC solicited comments from the Defense, State, and Commerce Departments, following which, in April 1982, it approved the AT&T application.[23]

---

23. During this same time, Pacific Telephone issued an RFP for a Pacific Corridor system between Los Angeles and San Francisco. Western Electric was the low bidder and was awarded the contract for this application also.

**Fig. 1.20. Map of the AT&T Northeast Corridor System showing terminal locations spaced along the length of the 776-mile system.**

Service was provided in the Northeast Corridor on schedule with a 90-Mbps system on multimode fiber at a wavelength of 0.825 $\mu$m.[24] The system was installed along an existing coaxial system right-of-way and utilized above-ground repeater huts from that system, spaced approximately 7 kilometers apart. The cable was designed to support transmission at longer wavelengths to allow a capacity upgrade with wavelength-division multiplexing. The original plan was to add two additional 90-Mbps channels at the same repeater spacing—one using lasers at 0.875 $\mu$m and the other using LEDs at 1.3 $\mu$m. The better-than-specifications fiber loss and bandwidth, coupled with laser and electronics advances, allowed, in practice, a much more attractive upgrade plan. A single 180-Mbps channel is being added with lasers at 1.3 $\mu$m, skipping every other repeater station (see Fig. 1.21). Thus, the augmentation to a capacity of 270 Mbps (4032 digital voice circuits) per fiber is being achieved with the addition of one-fourth the number of regenerators originally planned.

---

24. Actually, the initial filing with the FCC called for 45-Mbps transmission for Phase 1, with subsequent conversion to 90 Mbps. However, development of the 90-Mbps system was advanced to meet the Phase 1 as well as Phase 2 application.

Fig. 1.21.
Wavelength-
division-
multiplex
(WDM) upgrade
plan for the
Northeast
Corridor.

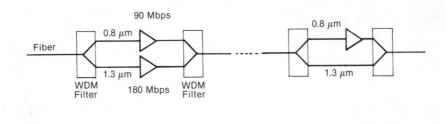

WDM filters are added at every repeater station, and 180-Mbps regenerators, operating at 1.3 μm, are added at alternate repeater stations.

In its original 1980 order approving the Northeast Corridor application, the FCC recognized the evolving nature of fiber-optic technology:

> Continuing improvements in fiber technology, like most other new technologies, should produce innovations and increased capacity in fiber systems and reduce future per circuit fiber system costs relative to existing technology. We believe, however, that the experience to be gained from this project outweighs whatever theoretical benefits might arise from delay in its implementation. This experience is necessary to foster the technological developments that will lead to "learning curve" decreases in cost. Without experience in the early stages of development, we may never see some of the advances that fiber technology promises to bring to telecommunication users. Thus, we find that Applicants should be authorized to begin implementation of this new technology in the long haul network at this time.

Although the Northeast Corridor multimode fiber-optic system was considerably more economical than metallic or radio alternatives, it was clear that even greater economies could be achieved with single-mode fiber, owing to longer repeater spacings and higher capacity transmission which would result in fewer fibers and fewer repeaters. Such economies were essential for the widespread economic prove-in of long-haul fiber systems.

In July of 1982, AT&T issued a Request for Proposal (RFP) for five additional routes (1900 route kilometers) for service beginning January 1, 1986. The RFP specified that "critical optical components" had to be manufactured in the United States, but this did not deter Japanese and European responses. However, Western Electric submitted the lowest bid and was awarded the contract.

Procurement of fiber-optic systems continues to be highly competitive. AT&T Communications has purchased systems from manufacturers other than AT&T Network Systems (Western Electric), and similarly AT&T Network Systems has sold fiber-optic systems to long-haul carriers other than AT&T Communications.

The United States has been the principal market for long-haul fiber-optic systems, and indeed many Japanese and European suppliers have focused on this market. In addition to a large number of suppliers, there are also a large number of competing carriers installing, or planning to install, fiber-optic systems. Right-of-way has become a valuable asset, and right-of-way owners (railroads, pipelines, etc.) are entering the telecommunications business. Although many commentators predict a glut of long-haul communications capacity, a more optimistic view is that this added capacity (and lower long-haul rates) will stimulate additional uses of communications.

## 1.5.4 Undersea Cable

Overseas telecommunications now consists primarily of a mix of undersea coaxial cable and satellite circuits. Both of these are relatively new technologies. The first transatlantic undersea cable for telephony (TAT-1) went into service in 1956, and the first international satellite (Intelsat-1) went into service in 1965. These technologies have almost completely supplanted the limited capacity and poor quality high-frequency radio circuits of previous years.

There has been, and continues to be, considerable debate on the economies of satellite versus cable for overseas circuits (Guterl and Zorpette 1985; Rutkowski 1985). Suffice it to say here that performance, economy, and diversity factors argue for a mix of satellite and cable facilities. What is incontrovertible is that fiber has supplanted coaxial cable as the medium of choice for future undersea cable systems.

Each new generation of undersea coaxial cable system has achieved higher-capacity transmission at the expense of a larger-diameter cable and shorter repeater spacings (see Table 1.4). System studies in the late 1970s indicated that TAT-8, which would be required for service in 1988, would need a new generation of technology. These same studies indicated that a fiber-optic system was potentially considerably less costly than a coaxial system, and AT&T began exploratory development on a high-capacity undersea system termed "SL" (Anderson et al. 1980; Runge and Trischitta 1984).

There are three principal differences between an undersea and a terrestrial fiber-optic system. The major one is component reliability. Ultrareliable components must be used in an undersea system to achieve continuity of service. However, continuity of service is equally important for terrestrial and undersea systems. In the case of terrestrial systems, service continuity is achieved by the use of automatic protection switching in which, in addition to a number of service lines between terminal locations, there is a protection line. Any fault in a service line results in

**TABLE 1.4.
GENERATIONS
OF UNDERSEA
CABLE SYSTEMS**

| System | First Service | Capacity (3-kHz VCs) | Number of Cables | Coax. Dia (cm) | Repeater Spacing (km) |
|--------|---------------|----------------------|------------------|----------------|-----------------------|
| SA | 1950 | 32 | 2 | 1.6 | 70 |
| SB | 1956 | 48 | 2 | 1.6 | 70 |
| SD | 1963 | 138 | 1 | 2.5 | 37 |
| SF | 1968 | 845 | 1 | 3.8 | 19 |
| SG | 1976 | 4200 | 1 | 4.3 | 9.5 |

an automatic switch to the protection line. Test equipment locates the fault, and a craft person is dispatched to make a repair. The repair must be done quickly, since a second fault during the time that the protection line is in use would result in an outage.

For an undersea system, the time and cost for a repair is huge compared with that for terrestrial systems. Thus, economics dictate that ultrareliable components be used in an undersea system (or, if necessary, that there be spares with automatic switching for individual critical components). The requirement for TAT-8 is that there be no more than three failures requiring repair in the twenty-five-year service life of the system.

A second difference between an undersea and a terrestrial system is that, in the undersea case, repeater powering is carried over the cable. Thus, the undersea cable contains metallic conductors for the powering current fed from the system terminals. In terrestrial lightwave systems, repeaters are generally in accessible locations and are powered by commercial power with battery backup.

A third difference is that, unlike terrestrial systems, undersea systems cannot be upgraded in capacity by adding WDM or by replacing repeaters.

Similar to the terrestrial case, TAT-8 was also subject to competitive procurement. The AT&T SL system was chosen over systems proposed by British and French firms. AT&T will provide the terminal in the United States, as well as the repeatered cable from the United States to a branching point off the coast of Europe (Fig. 1.22). Standard Telephones and Cables, plc., will provide the link to England, and Submarcom (a consortium of CIT Alcatel and Les Cables de Lyon) will provide the link to France, using the AT&T SL design. The SL system is also planned for the Hawaii 4/Transpac 3 cable slated for service in 1988. This system, joining California, Hawaii, Guam, and Japan, will be provided by AT&T and KDD (the Japanese international telecommunications entity).

The SL system employs single-mode fibers, with laser transmission at 1.3 $\mu$m and a repeater spacing of about 50 km. The transmission bit rate

**Fig. 1.22. Route of the TAT-8 transatlantic cable planned for service in 1988.**

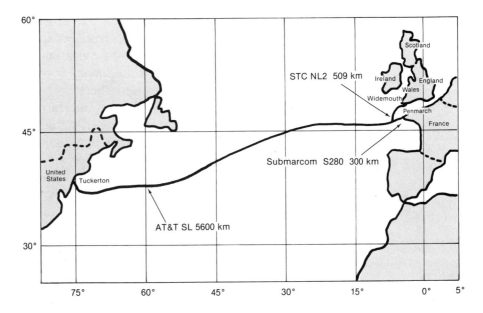

is 296 Mbps, corresponding to about four thousand 64-Kbps digital voice circuits per pair of fibers.[25] Bit compression and statistical multiplexing techniques may be used to expand the voice circuit capacity by a factor of five.

In addition to repeatered undersea cable systems, there are many potential applications for unrepeatered links (e.g., between islands). When all of the electronics are at the terminal locations, the burden of ultrareliability is removed, and it is possible to use more advanced technology to achieve even longer unrepeatered distances. For example, in 1985, AT&T provided the Department of Defense with a low-bit-rate (3-Mbps) unrepeatered fiber link from a Pacific island to an offshore platform about 150 km away. This system utilized lasers at a wavelength of 1.55 $\mu$m, the wavelength of minimum attenuation in the fiber.

## 1.5.5 Loop Feeder

Although trunking applications of fiber are important, loop applications are viewed as a potentially much larger market. The loop portion of the telephone plant is the connection between the subscriber and the serving switching office. There are more than 100 million subscriber loops in the United States.

Historically, loops have been provided by a dedicated pair of copper wires from the subscriber to the serving central office. Large cross-section

---

25. The capacity is 3840 voice circuits utilizing the European multiplexing hierarchy, or 4032 voice circuits utilizing the North American hierarchy.

"feeder cables" are used close to the central office, and smaller cross-section "distribution cables" close to the subscriber. In recent years, digital multiplexing techniques have been used to check the large amount of copper needed in the feeder portion of the loop. For example, in the AT&T SLC® 96 carrier system, ninety-six subscribers are served using ten pairs of copper wires from the main office to an interface point. Final distribution to the subscriber is on dedicated copper pairs. In 1983, a version of SLC 96 became available that serves the ninety-six subscribers on four fibers (Bohn et al. 1984). (Two fibers are used for service, one in each direction, and two are used for protection. Transmission is digital at 6 Mbps.)

The principal advantages of fiber in this application are the absence of intermediate repeaters (the metallic system requires repeaters at spacings of 2 km) and the capability for meeting growth. Similarly to the trunking applications, additional stages of multiplexing may be used to expand the capacity; and, indeed, loop multiplexers[26] with capacities of 672 (45-Mbps) and 1344 (90-Mbps) subscriber voice channels are available.

In 1985, subscriber loop carrier systems served about 40% of the loop growth in the United States, and about half of these applications employed fiber transmission. Where copper transmission is used with loop carrier, it is invariably on copper that is already in place. There are very few copper cables being installed for loop feeder applications.

Initial loop feeder applications utilized multimode fiber with LED transmission at a wavelength of 1.3 $\mu$m. LEDs are particularly advantageous for loop application, owing to their lower cost and their ability to operate reliably over a wider temperature range than lasers. The longer transmission distances achievable with lasers are generally not required for the loop application. In fact, by utilizing multimode fiber optimized for LED transmission, transmission distances of 20 km can be achieved.

Although multimode fiber with LED transmission meets the present loop bit rate and distance needs, the trend to single-mode technology is extending to loop feeder as well. There are several arguments for the use of single-mode fiber in the loop plant:

1. Single-mode technology is becoming pervasive.
2. The distinction between loop and trunk applications is blurring.
3. Access beyond the nearest central office may be desired.
4. Options for greatly increased capacity (e.g., multiple video channels) are desirable and should be kept open.

---

26. Although the basic transmission functions are the same in trunk and loop applications, some of the maintenance and environmental requirements are different.

Although present loop feeder applications run the fiber to an interface point near the subscriber, the stage is set for extension of fiber to the subscriber's premises when communications requirements warrant. Residential voice and data needs may well be satisfied with copper pair distribution, and it may take wideband services (such as television) for fiber to be extended to the residential premises. For business subscribers having a large number of telephone channels, the carrier terminal may already be on the business premises. In such cases, the fiber cable may provide a mix of multiplexed telephone channels, wideband private-line data channels, and video conferencing channels serving the business subscriber.

## 1.5.6 Television

The one telecommunication service that is inherently wideband is television. A standard TV channel has a bandwidth of 4 MHz and, when encoded digitally with PCM, requires about 90 Mbps. There are a variety of techniques for TV bandwidth reduction, but these are typically very complex and/or result in performance degradations.[27] Fiber offers the potential for high-quality full-bandwidth television transmission.

One of the earliest uses of fiber for television transmission was at the 1980 Winter Olympics in Lake Placid, New York. What started as an experimental system, paralleling the primary video metallic cables, was pressed into service when an additional TV link was required and the telephone poles could not accommodate another metallic cable. The fiber was to be used initially for an auxiliary channel, but measurements indicated that the fiber system had better performance than the metallic system.[28] Consequently, the fiber link became the primary TV channel from the Olympic ice arena to the American Broadcasting Company's Television Control Center two miles away. The historic broadcast of the United States ice hockey team's win over the U.S.S.R. was carried on its first leg by fiber.

A much more extensive application of fiber came in the 1984 Summer Olympics in Los Angeles, where fiber was used to provide television transmission to the ABC Television Control Center from fourteen locations spread over a wide geographic area. Transmission was at 90 Mbps, utilizing an already in-place 90-Mbps metropolitan backbone trunking

---

27. Sophisticated video encoding techniques take advantage of the redundancy within a frame and/or between frames. Such techniques may result in loss of resolution when frames have considerable detail or when there is rapid motion.

28. A characteristic of metallic cables, not shared by fiber, is that attenuation is a function of frequency. Equalization in the terminal equipment compensates for this. However, owing to the large temperature variations in this aerial application, there was a small misequalization in the metallic transmission. The metallic transmission was good, but the fiber was better.

system that had been installed primarily for voice trunking. By installing fiber extensions to the Olympic arenas and by adding the digital video terminals, this network was available to meet the extensive video transmission needs of the Olympics. This was by far the most economic means of meeting these needs, since only a small fraction of the facilities had to be provided solely for the television application. Performance was excellent.

Both the Lake Placid and the Los Angeles Olympics applications were essentially point-to-point applications. Several CATV providers have used fiber for similar point-to-point "supertrunk" applications, such as connecting a satellite terminal with the head-end of the CATV system. Fiber, however, has not found extensive application for CATV distribution to the subscriber.

CATV distribution systems are optimized for "broadcast" transmission; that is, transmission of the same information (often more than 100 television channels) from one source (at the system head-end) to a multiplicity of receivers at different locations. A coaxial-cable, tapped-bus architecture is ideal for this application. All of the channels are carried, frequency-division multiplexed, on a single coaxial cable. Taps off this common bus provide access to all the channels at each household. A set-top converter (essentially a tuner) selects the channel to be watched. Premium channels are scrambled, descramblers being provided at households subscribing to these channels. Amplifiers are required at periodic intervals along the cable to compensate for the reduction in signal power resulting from the taps. Indeed, tap distribution loss, not attenuation in the cable, is the dominant loss mechanism.

Fiber is not well suited for a tapped-bus architecture. Taps for a fiber-optic cable are more difficult to make and are more expensive than taps in a coaxial cable. This is a technology problem that could possibly be overcome. More to the point, a system with passive taps requires the ability to accommodate large signal loss between amplifiers and/or the availability of simple low-cost amplifiers, neither of which is available with fibers. Because of the relatively low power that may be transmitted on the fiber, and the relatively high power that is required at the receiver (owing to the high energy of the photon at optical frequencies, as described in section 1.3.4), fiber systems cannot accommodate as much signal loss as coaxial cable can. For point-to-point links this is not a problem because the attenuation of the fiber is much lower than the coax. However, it is a major problem with a tapped-bus architecture where tap loss dominates.

Fiber is much better suited for a star architecture (Fig. 1.23) in which there is a central switch with a distinct fiber (or perhaps several fibers) dedicated from the switch to the subscriber. The "channel selection" is now a data set, which sends a request to the switch to connect the desired channel to the subscriber's line. This affords obvious advantages in providing services such as pay-by-view. It also simplifies the provision of premium services and protects against theft of service; further, it should result in simpler and less expensive set-top units. However, for a dense distribution of subscribers, the total cost of a fiber star network is invariably greater than that of a tapped-bus coaxial network.

Fig. 1.23. Differences in the configuration and service capability of star and tapped-bus (T) configurations.

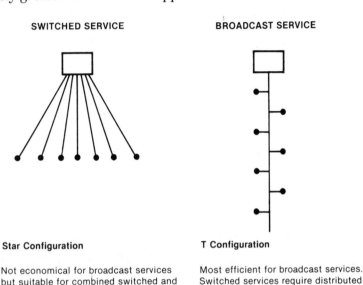

**SWITCHED SERVICE**

**BROADCAST SERVICE**

**Star Configuration**

Not economical for broadcast services but suitable for combined switched and broadcast services.

**T Configuration**

Most efficient for broadcast services. Switched services require distributed processing.

If one wishes simply to provide only standard telephone service to a household, it is hard to compete with a wire-pair distribution network (star configuration with a switch at the serving central office). Similarly, if one wishes to provide only one-way broadcast television distribution, it is hard to compete with a tapped-bus coaxial network. The question is, if one wants to provide both of these services, plus some additional services (for example, switched video services), might not a single-fiber star network be more economical and provide more add-on service capability than separate metallic telephone and CATV networks? The question is, of course, complicated by the existence of telephone copper wire distribution to almost all households, and by the fact that CATV coaxial cable already passes by about 50% of the households in the United States. New housing communities offer perhaps the best opportunity to test the attractiveness of wideband fiber distribution networks. There have been a number of such applications throughout the world—HI-OVIS in Japan, Biarritz in France, Elie Manitoba in Canada, and Milton-Keynes in England, to name a few (Chang 1983; Sakurai 1983). However, it is only with the recent removal of regulatory restrictions on joint provision of telephone and CATV service that significant trials are beginning in the United States.

### 1.5.7 Local Area Networks

A *local area network* (LAN) is usually defined as a communication system designed to interconnect heterogeneous computers, terminals, and office machines in a geographically bounded area for the purposes of resource and data sharing (Anderson et al. 1980). LANs share some of the characteristics of both telephone and CATV networks. Like telephone networks, they must provide two-way interconnection (transmission) between a large number of terminals, with the capability of establishing connections (switching). Like CATV networks, they frequently employ tapped-bus, coaxial-cable architectures with distributed rather than centralized switching.[29]

For a discussion of LANs, the reader is referred to Chapters 5, 6, and 7 of *Data Communications, Networks, and Systems* (Bartee 1985). These three chapters cover PBX local networks, baseband local networks, and broadband local networks, respectively. Each chapter has a discussion of factors affecting the choice of "the wiring medium." There is an important distinction, however, between the use of fiber in LANs and the fiber transmission applications discussed in earlier sections of this chapter. Implicit in the telephone trunking and loop applications to date is the separation of the transmission and switching functions. Fiber is used for transmission. However, the very essence of LANs is the intermixing of the transmission and switching functions. The question for LANs, then, is not how fiber technology might be used to provide the transmission function, but rather how fiber technology might be used to affect the basic architecture and structure of LANs. Unfortunately, it is easier to ask this question than to answer it.

Several factors appear clear. The interference immunity of fiber offers performance advantages and allows operation in electrically noisy environments. The very large bandwidth capability of fiber is potentially advantageous for serving very large numbers of terminals, allowing much higher transmission rates between terminals and allowing communication protocols that are wasteful of bandwidth. The small diameter and low weight of fiber cables are clearly well suited for the office envi-

---

29. In a CATV network the "switching" is done by a tuner at the television set. Somewhat analogously, in a LAN utilizing a coaxial cable bus, the receiver selects from the common data stream the signal intended for it by utilizing address characteristics contained within the signal (e.g., a fixed time slot, a fixed frequency band, or a coded address pattern).

ronment, but many factors require improvement, such as minimum bend radius of cables, cost and size of connectors, and ease of splicing. Short distances between terminals and the very large number of terminals argue for simple, highly integrated, low-cost transmitters and receivers, preferably integrated directly into the computer terminals. This requires definition of standard interfaces. Short distances, low-cost connectors, and simple transmitters and receivers all argue for multimode technology, and indeed there is a push in some quarters for the use of even larger-core diameter and higher numerical aperture (NA) fibers.[30] On the other hand, there is the general push to single-mode technology and to future use of integrated optics and other optical processing techniques—which are limited largely to single-mode technology. Simple and inexpensive optical taps, splitters, couplers, etc., are desired. These would be especially needed in tapped-bus architectures. However, as indicated in the previous section, tapped-bus architectures are not particularly well suited for fiber technology.

Fiber technology is already being used selectively to provide specialized functions in some LANs; for example, as a high-speed link between two well-separated nodes. As in telephone networks, fiber is particularly well suited for such point-to-point applications. The challenge remains how best to utilize this technology for the more complex task of local distribution (Koyama 1985).

# 1.6 Summary and Look at the Future

As indicated in the previous sections, fiber-optic transmission applications began with metropolitan trunking. These were point-to-point applications of moderate distance and capacity (Fig. 1.24). Technological advances, primarily higher-capacity transmission and longer repeater spacings, extended the application to long-haul transmission, both terrestrial and undersea. For long-haul transmission, the technology thrust continues towards longer repeater spacing and higher-capacity transmission.

The attenuation of present single-mode fiber is essentially at the theoretical limits of that achievable with silica-based glass. Although improvements continue to be made in higher-powered lasers and in more sensitive receivers, such techniques are unlikely to extend repeater spacings of practical systems much beyond the 200 km of present experi-

---

30. The initial standard multimode fiber has a 50-$\mu$m diameter core, 125-$\mu$m outside diameter, and an NA = 0.23. AT&T also manufactures a 62.5-$\mu$m diameter core, 125-$\mu$m od, NA = 0.29 fiber for application with LEDs. Corning manufactures an 85-$\mu$m diameter core, 125-$\mu$m od, NA = 0.26 fiber. Others have argued for even larger diameter fiber for LAN use.

Fig. 1.24.
Schematic
representation
of the capacity-
length domain
for initial fiber-
optic system
applications,
and the
technology and
market forces
resulting in the
expansion of
this domain.

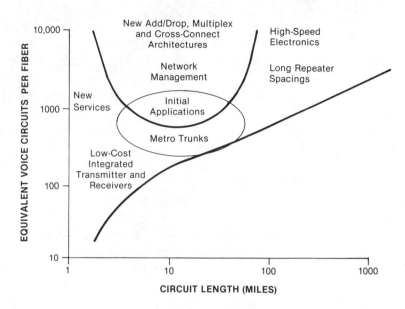

ments. An order of magnitude increase in repeater spacing, which would allow achieving repeaterless intercontinental undersea spans, would require an order of magnitude decrease in fiber attenuation. Research is underway on new materials for making fibers, allowing transmission at even longer wavelengths, which in theory could achieve attenuations of 0.001 dB/km or less. The requirements on impurity levels and dimensional controls make the achievements of such attenuations questionable. On the other hand, the achievement of practical systems, with attenuations of less than 0.5 dB/km, was questionable in some minds not that many years ago.

Technologically, the continued increase in capacity is more readily achieved than increases in repeater spacing. Experiments have been performed with serial bit rates as high as 8 Gbps, but this requires extremely sophisticated electronic circuitry. Although advances continue to be made in the speed capability of electronic integrated circuits (particularly gallium arsenide circuits), one may question whether there will be sufficient application for circuits operating in this speed range to justify their development. Indeed, the future thrust may be for high-speed multiplexing and other signal-processing operations to be performed optically rather than electronically.

The question "How low a bit rate can be economically transmitted on a fiber?" may be more important than "How high a bit rate can be transmitted technologically?" There are many more applications calling for low to moderate capacity than there are for extremely high capacities. As optical technology costs continue to be driven down, economic prove-in occurs at lower capacities and for shorter distances, greatly increasing the range of applications. If fibers can be economically deployed for moderate capacities, then their presence can stimulate additional uses of communications.

Although fiber-optics transmission has experienced the greatest growth for long-haul transmission, this market (in the United States) may well plateau over the next several years. Potentially, the largest growth market is local and metropolitan area networks. The direction for such applications is not for fiber to be simply a transmission medium, but for optical technology to influence the basic architecture of the information network. Integrated optics offers the potential for performing many signal-processing functions directly in optical form. Coherent techniques offer the potential for multiplexing extremely large numbers of optical channels and, by using heterodyne wavelength filtering techniques, for adding and dropping channels. The available bandwidth is so broad that it may be possible to assign a fixed wavelength to each terminal location. Optical interconnection, optical signal processing, optical switching, optical multiplexing, all of these—not just optical transmission—are likely to be part of our future vocabulary.

# 1.7  References

### *Cited References*

Anderson, C.D., et al. 1980. An undersea communication system using fiberguide cables. *Proceedings of the IEEE* 68, no. 10 (October): 1299–1303.

AT&T. 1980. File no. W-P-C3071. Submitted to the Federal Communications Commission, January 23.

Bartee, T. C., ed. 1985. *Data communications, networks, and systems.* Indianapolis: Howard W. Sams.

Bell, A. G. 1880a. Paper delivered at meeting of American Association for the Advancement of Science, August 27, at Cambridge, Massachusetts.

———. 1880b. U.S. Patent 235,199: Apparatus for signaling and communicating, called "Photophone." Filed August 28, 1880. Issued December 7, 1880.

Bennett, W. R., Jr., et al. 1960. U.S. Patent 3,149,290: Gas optical maser. Filed December 28, 1960. Issued September 15, 1964.

Bohn, P. P., et al. 1984. The fiber SLC carrier system. *AT&T Bell Laboratories Technical Journal* 63, no. 10, pt. 2 (December): 2389–2416.

Bruce, R. V. 1973. *Bell: Alexander Graham Bell and the conquest of solitude.* Boston: Little, Brown.

Chang, K. Y. 1983. Broadband integrated fiber-optic distribution field trial; system design and performance. *IEEE Journal on Selected Areas in Communications* SAC-1 (April): 439–444.

Cohen, L. G. 1986. Trends in U.S. broadband fiber optic transmission systems. *IEEE Journal on Broadband Communication,* forthcoming.

Cook, J. S., and O. I. Szentisi. 1983. North American field trials and early applications in telephony. *IEEE Journal on Selected Areas in Communications* SAC-1 (April): 393–97.

Corning Glass Works. 1980. Comment in support of Northeast Corridor Lightguide Fiber Installation. Filed with Federal Communications Commission, June 2.

Corwin, W. L. 1984. A communications network for the Summer Olympics. *IEEE Spectrum* (July): 38–44.

Federal Communications Commission. 1980. File no. W-P-C-3071. Memorandum, opinion, order, and authorization. Adopted November 25, 1980. Released January 27, 1981.

Gloge, D., ed. 1976. *Optical fiber technology.* The Institute of Electrical and Electronics Engineers, Inc. New York: IEEE Press.

Guterl, F., and G. Zorpette. 1985. Fiber optics: Poised to displace satellites. *IEEE Spectrum* (August): 30–37.

Hall, R. N., et al. 1962. Coherent light emission from GaAs junctions. *Physical Review Letters* 9 (November): 366.

Hartman, R. L., et al. 1977. Continuously operated AlGaAs DH lasers with 70° lifetimes as long as two years. *Applied Physics Letters* 31:756.

Hayashi, I., M. B. Panish, and P. W. Foy. 1970. Junction lasers which operate continuously at room temperature. *Applied Physics Letters* 17:109.

Henry, P. S. 1985. Lightwave primer. *IEEE Journal of Quantum Electronics* QE-21 (December): 1862–79.

Ishio, H. 1983. Japanese field trials and early applications of fiber optics in telephones. *IEEE Journal on Selected Areas in Communications* SAC-1 (April): 404–12.

Jacobs, I. 1976. Fiberguide transmission system capability and needs. *International Communications Conference,* June.

_____. 1984. Basics of lightwave transmission. *Proceedings of the National Communications Forum* 38:131–35.

_____. 1985. Applications of fiber optic communications. *SIGNAL* 39, no. 10 (June): 51–55.

_____. 1986a. Fiber optic transmission systems. Chapter 8 in *Electronic communications handbook,* edited by A. F. Inglis. McGraw-Hill, forthcoming.

_____. 1986b. Design considerations for long-haul lightwave systems. *IEEE Journal on Selected Areas in Communications,* forthcoming.

Jacobs, I., and J. R. Stauffer. 1980. FT-3—A metropolitan trunk lightwave system. *Proceedings of the IEEE* 68, no. 10 (October): 1286–90.

Jacobs, I., et al. 1978. Atlanta fiber system experiment. *Bell System Technical Journal* 57, pt. 1 (July-August): 1717–1895.

Jay, Frank, ed. 1984. *IEEE standard dictionary of electrical and electronics terms.* New York: The Institute of Electrical and Electronics Engineers, Inc.

Kao, C. K., and G. A. Hockham. 1966. Dielectric-fiber surface waveguides for optical frequencies. *Proceedings of the IEEE* 113 (July): 1151–58.

Kapron, F. P., et al. 1970. Radiation losses in glass optical waveguides. *Applied Physics Letters* 17 (November 15): 423.

Kasson, J. M., and R. S. Kagan. 1985. PBX local area networks. Chapter 5 in *Data communications, networks, and systems*, edited by T. C. Bartee. Indianapolis: Howard W. Sams.

Kitayama, K., et al. 1985. Design and performance of ultra-low-loss single-mode fiber cable in 1.5 $\mu$m wavelength region. *IEEE Journal of Lightwave Technology* LT-3 (June): 579–85.

Korotky, S. K., et al. 1985. Four Gb/s transmission experiment over 117 km of optical fiber using a Ti: Li NbO$_3$ external modulator. Optical Fiber Conference '85 Post Deadline Paper, February.

Koyama, M., ed. 1985. *Journal of Lightwave Technology*. Special issue on fiber-optic local area networks. LT-3, no. 3 (June).

Li, T., ed. 1985. *Optical fiber communications*. Vol. 1, *Fiber fabrication*. New York: Academic Press.

Maiman, T. H. 1960. Stimulated optical radiation in ruby. *Nature* 187 (August): 493.

Marcatili, E. A. J., and R. A. Schmeltzer. 1964. Hollow metallic and dielectric waveguides for long distance optical transmission and lasers. *Bell Systems Technical Journal* 43 (July): 1783–1809.

Miller, C. M., and G. F. DeVeau. 1985. Simple high-performance mechanical splice for single-mode fibers. *Optical Fiber Conference Proceedings*, 6.1–6.3.

Miller, S. E. 1965. Light propagation in generalized lens-like media. *Bell Systems Technical Journal* 44 (November): 2017–64.

Miller, S. E., T. Li, and E. A. J. Marcatili. 1973. Research toward optical-fiber transmission systems. *Proceedings of the IEEE* 61 (December): 1703–51.

Millman, S., ed. 1984. Lightwave communications. Chapter 7 in *A history of engineering and science in the Bell System, Communications Sciences 1925–1980*. AT&T Bell Laboratories.

Moncalvo, A., and F. Tosco. 1983. European field trials and early applications of fiber optics in telephony. *IEEE Journal on Selected Areas in Communications* SAC-1 (April): 398–403.

Nathan, M. I., et al. 1962. Stimulated emission of radiation from GaAs p-n junctions. *Applied Physics Letters* 1 (November): 62–64.

Panish, M. B., et al. 1970. Double-heterostructure injection lasers with room temperature thresholds as low as 2300 A/cm$^2$. *Applied Physics Letters* 16 (April): 326.

Potter, D. 1985. Baseband local area networks. Chapter 6 in *Data communications, networks, and systems*, edited by T. C. Bartee. Indianapolis: Howard W. Sams.

Runge, P. K., and P. R. Trischitta. 1984. The SL undersea lightwave system. *IEEE Journal of Lightwave Technology* LT-2 (December): 744–53.

Rutkowski, A. M. 1985. Satellite competition with optical fiber. Paper presented at Satellite Summit, April.

Sakurai, K. 1983. Broadband fiber system activity in Japan. *IEEE Journal on Selected Areas in Communications* SAC-1 (April): 428–35.

Saul, R. H., et al. 1985. One-hundred eighty Mb/s, 35 km transmission over single mode fiber using l.3 $\mu$m edge-emitting LEDs. *Electronic Letters* 21:773.

Schawlow, A. L., and C. H. Townes. 1958. U. S. Patent 2,929,922: Masers and maser communication system. Filed July 30, 1958. Issued March 22, 1960.

Suematsu, Y., S. Arai, and K. Kishino. 1983. Dynamic single-mode semiconductor lasers with a distributed reflector. *IEEE Journal of Lightwave Technology* LT-1 (March): 161–76.

Summers, J. K. 1985. Broadband local area networks. Chapter 7 in *Data communications, networks, and systems,* edited by T. C. Bartee. Indianapolis: Howard W. Sams.

Tsang, W. T. 1985. The cleaved-couple-cavity ($C^3$) laser. Chapter 5 in *Semiconductors and semimetals,* 22 part B: 257–373.

### General References

Kao, C. K. 1982. *Optical fiber systems: Technology design and applications.* New York: McGraw-Hill.

Midwinter, J. E. 1979. *Optical fibers for transmission.* New York: Wiley.

Miller, S. E., and A. G. Chynoweth, eds. 1979. *Optical fiber telecommunications.* New York: Academic Press.

Mims, F. M., III. 1982. The first century of lightwave communications. *International Fiber Optics* 3, no. 4 (February): 10–26.

Personick, S. D. 1981. *Optical fiber transmission systems.* New York: Plenum Press.

——. 1985. *Fiber optics technology and applications.* New York: Plenum Press.

Schwartz, M. I. 1984. Optical fiber transmission—From conception to prominence in 20 years. *IEEE Communications Magazine* 22, no. 5 (May): 38–48.

# 2

# SATELLITE COMMUNICATIONS

Irwin L. Lebow

When a new technology makes its appearance on the scene, there is often a tendency among its partisans to paint it as a miracle that will solve all the heretofore unsolved problems in its field. This is what happened when communications satellites first appeared in the early sixties. In the years since, communications satellites have indeed made a major impact on long distance communications in many ways, but of course not in all ways. With the emergence of a still newer technology, that of fiber optics, there are those who would pit the satellite against the optical fiber and ascribe all advantages to the latter.

Partisans of fiber optics are no more correct in their predictions than were the early partisans of satellites. Each technology has a place in the market, and the purpose of this chapter is to treat satellite communications as a medium having both strengths and weaknesses and, consequently, having more applicability in some areas than in others.

This chapter is therefore applications-oriented, and only that technical material having some bearing on the applications areas is presented in depth. We set the stage in the first three sections: sections 2.1 and 2.2 introduce the satellite medium in the context of other competing media, while section 2.3 focuses on the advantages and disadvantages of the satellite medium relative to the competition.

Sections 2.4 through 2.7 are technical. Section 2.4 is a qualitative overview of the elements of satellite communications. Sections 2.5 and 2.6 are quantitative, presenting the mathematics of satellite links and computing the capacities of various links. Section 2.7 introduces methods of sharing a satellite among different users (i.e., the techniques for multiple access).

The fundamentals having been addressed, sections 2.8 and 2.9 treat applications. Section 2.8 provides some overall perspective, and section

2.9 concludes the chapter with a more quantitative treatment of private networks, emphasizing the emerging area of micro earth station networks.

# 2.1   Transmission Media

There are many media that can be used to propagate communications signals. Perhaps the most fundamental way to categorize these media is to divide them into two classes: *guided* and *unguided*. By guided media, we mean media such as copper wires and optical fibers in which physical paths define the communications propagation paths. By unguided, we mean, in essence, radio media in which signals are radiated from antennas into space and are received by remote antennas.

## 2.1.1   Guided Media

Propagation distances for guided media, both electrical and optical, are limited by losses in the guiding medium to the tens of miles, depending on the medium and the bandwidths involved. To span longer distances, amplifiers or repeaters are inserted at appropriate intervals to compensate for these losses. In this way the guided media can achieve very long distance communications over land and under water.

The essential difference between optical communications signals propagating through glass fibers and electrical signals propagating through cables is bandwidth. An optical fiber can support many times the capacity of a copper wire. A major portion of the cost of guided communications is the cost of installing the cable: purchase of rights, digging trenches, installing poles, etc. These costs are similar whether one copper wire or twenty fibers are installed. Thus, the economics of guided communications improve markedly as the installed capacity on a given path is increased. This is the essential reason for the great economies in modern high-density fiber-optic installations, which are now becoming so prevalent.

## 2.1.2   Unguided Media

All radio signals travel in straight lines in a vacuum. The earth's atmosphere is, of course, not a vacuum. In particular, it contains bands of ionized particles some 50 to 200 miles above the earth's surface, and these can have important effects on propagation, depending on frequency.

At frequencies above 30 MHz or so, the ionosphere (as the bands are called) is essentially transparent. At low frequencies, say below 30–50 kHz, the ionosphere and the earth display many of the characteristics of a waveguide and can support long distance communications in a quasi-guided mode. At somewhat higher frequencies, between 3 and 30 MHz (the high-frequency or shortwave band), the ionosphere behaves more like an imperfect mirror, permitting propagation over long distances depending on the height of the ionosphere and the angle at which the signal transmitted from earth strikes the ionosphere.

At the higher frequencies where the ionosphere has no effect, terrestrial propagation is limited to "line-of-sight" paths around 20 miles or so in length, depending on the height of the antennas and the nature of the intervening terrain. As in the case of guided transmission, long distances can be achieved by cascading such line-of-sight links over land, but not, of course, across the ocean.

## 2.2   Satellite Communications

A *satellite communications path* is simply a two-hop cascade in which the relay point is located far above the earth on a spacecraft in earth orbit. The distance on earth that can be spanned by such a relay depends on the height of the repeater. For the most common orbit at 22,300 miles, terrestrial distances as great as 10,000 miles can be spanned. As the previous section pointed out, until the advent of the communications satellite, the only long distance unguided media were: (1) cascaded radio relays, (2) very low frequency radio (below 30 kHz) and (3) high-frequency or shortwave radio (3–30 MHz). The latter two are inherently low-capacity media suitable only for specialized applications, and the first is limited to overland spans. The satellite therefore fills a huge void in that it is capable of transmitting high capacities over long distances, either over land or water. Moreover, because of its unique geometry, it is inherently a broadcast medium with a natural ability to transmit simultaneously from one point to an arbitrary number of other points within its coverage area.

## 2.3   The Role of Satellite Communications

Within the guided and unguided categories of media, several play significant roles in the vastly expanding long-distance communications world of today.

*Long-haul domestic communications* are dominated by three media: terrestrial microwave relay, fiber-optic cable, and satellite relay. Some coaxial copper cable is still in use, but it is rapidly being replaced by the more economical fiber-optic cable and microwave relay.

*Long-haul overseas communications* are limited to two media: under-sea cable and satellite relay. These are the only two ways of spanning an ocean. (As noted before, shortwave radio is excluded because of its lim-ited capacity.) Undersea copper cables have relatively limited capacities; hence, from the viewpoint of capacity, the satellite stands alone. It is because of the large capacity of satellite relay versus the limited capacity of undersea cable that transoceanic television transmission became a reality only with the advent of satellites. When the undersea optical cable makes its first appearance, it will provide for the first time an alternative way of transmitting wide-bandwidth signals across the ocean.

## 2.3.1   Advantages of Satellite Communications

In the previous section we cited the long-distance domestic and overseas communications alternatives. In each case, the satellite competed against one or more terrestrial relay media. The fundamental advantages of sat-ellite relay derive from three distinctions between satellite relays and ter-restrial relays, whether the relays are over land or water, guided or unguided, electrical or optical.

1. Satellite relays are inherently wide-area broadcast; that is, point to multipoint. In contrast, terrestrial relays are inherently point to point. They have point-to-multipoint capabilities only to the extent that all of the destination points lie along the relay routes.

2. Satellite circuits can be installed rapidly. Assuming, of course, that the satellite is in position, earth stations can be installed and commu-nications established in days, or even hours. Subsequently, a station may be removed relatively quickly from one location and reinstalled else-where. In contrast, terrestrial circuits of any kind require time-consum-ing installations.

3. Satellite communications are applicable to mobile communica-tions. Although terrestrial networks may also interconnect mobile vehi-cles having line-of-sight access to the networks (e.g., by cellular radio), the satellite relay has a unique degree of flexibility in interconnecting mobile vehicles. Indeed, the satellite has become an alternative to short-wave radio in this specialized area and has significant reliability ad-vantages.

## 2.3.2   Disadvantages of Satellite Communications

There is one inherent disadvantage to satellite communications that dom-inates all others. When the satellite is in geosynchronous orbit (see sec-tion 2.4), the communications path between the terrestrial transmitter and the receiver is approximately 75,000 kilometers (km) long. Since the speed of electromagnetic propagation is $3 \times 10^5$ km/s, there is a delay of $\frac{1}{4}$ second between the transmission and the reception of a signal. In the case of voice communications, $\frac{1}{2}$ second elapses between the end of a talk spurt by one of the participants and the point at which the person begins to hear the response. Some people find this delay annoying.

Another phenomenon exacerbated by delay is echo. The usual connection between a local telephone and the long-haul telephone network is two-wire; that is, the same pair of wires carries sending and receiving signals. All long-haul trunks are four-wire, with separate pairs for the two transmission directions. Imperfect impedance matching at the two-wire/four-wire junction can result in a portion of a speaker's transmission being reflected back to the speaker, and when an audible reflection occurs with a $\frac{1}{2}$-second delay, it can be extremely annoying. In the early days of satellite communications, devices called "echo suppressors" were used to mitigate this problem. They were only partially successful, however. Later, more-sophisticated echo cancellers replaced the older echo suppressors, and these devices generally eliminate echoes very satisfactorily. But the old reputation has been hard to overcome.

Data transmission over satellite is not troubled by echo effects, but other delay effects have to be considered. Interactive computer communication requires a response within a few seconds. The response delay is due to both communications and computing delays. When a satellite link is used, the time permitted for computing delays is shortened by $\frac{1}{2}$ second.

When long file transfers are performed, the accuracy of the transmission must be very high, so high that the normal technique is to encode the data in order to detect transmission errors with very high probability and then to initiate a retransmission of the data block containing the errors. When a satellite link is used, retransmissions are delayed by the path delay. The satellite path is thus idle for some periods, and overall efficiency is reduced. Path delay can be overcome by *pipelining* the data flow; that is, by continuing to transmit continuous data streams until the receipt of a retransmission request, at which point the data stream is broken to insert a retransmission of an old block of data. While pipelining can be achieved relatively easily, it contributes to the delay disadvantage of satellite data transmissions.

### 2.3.3 Economic Trade-offs

In applications suitable for either satellite or terrestrial media, the ultimate choice often rests heavily on the relative costs of the media. Terrestrial costs are proportional to distance, while satellite costs are independent of distance. We can therefore compare the media with a crossover distance; that is, the distance at which the costs are equal.

Generally speaking, the crossover distance is between 1000 and 1500 miles depending on the capacity of the link: the higher the capacity, the greater the crossover distance. At extremely high capacities (multi-gigabits per second), the multifiber cable, when used between pairs of points, appears to be in an economic class by itself as compared to either terrestrial microwave relay or satellite relay.

# 2.4   Elements of Satellite Communications

This section begins the discussion of satellite communications fundamentals. It contains many diverse topics that fall into two categories: (1) those relating to communications using an orbiting spacecraft as a relay vehicle and (2) those relating to the spacecraft itself.

## 2.4.1   Satellite Orbits

The most commonly used orbit for the communications satellite is the so-called *geosynchronous orbit,* an equatorial orbit that is located at the distance at which the orbital velocity of the satellite equals the velocity of a point on the earth's equator. The satellite is therefore stationary relative to the earth.

It is easy to compute the geosynchronous altitude from elementary physics. In order for a mass $m$ to move in a circular orbit of radius $R$, it must be acted on by a centripetal force of magnitude $MR\,\omega^2$, where $\omega$ is the angular velocity of the circular motion expressed in radians per second (see Fig. 2.1). The source of this centripetal force is, of course, the gravitational force $GmM/R^2$, where $M$ is the mass of the earth and $G$ is the gravitational constant. Thus:

$$mR\omega^2 = \frac{GmM}{R^2} \qquad\qquad (2.1)$$

At the surface of the earth, the gravitational force on a body of mass $m$ is $mg$, where $g$ is the acceleration due to gravity. Thus:

$$\frac{GmM}{R_E^{\,2}} = mg \qquad\qquad (2.2)$$

where $R_E$ is the earth's radius.

Fig. 2.1.
Dynamics of a
circular orbit
about the earth.

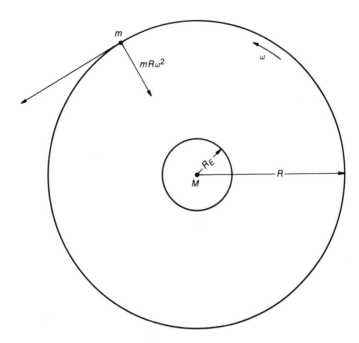

Substituting equation (2.2) into equation (2.1) and solving for $R$, we obtain:

$$R = \sqrt[3]{\frac{g\,R_E^2}{\omega^2}} = \sqrt[3]{\frac{g\,R_E^2}{4\pi^2 f^2}} \tag{2.3}$$

where $f = \omega/2\pi$ is the rotation rate in revolutions per second.

If we set $f = 1$ revolution/day, and we use the values $R_E = 6370$ km and $g = 9.9$ m/s, we obtain $R = 42{,}208$ km for the orbital radius and $R - R_E = 35{,}838$ km for the orbital height above the equator, which is very close to the precise value of 35,860 km, or 22,300 mi.

As we noted earlier, the geosynchronous orbit is the most common orbit because the satellite is stationary with respect to a point on earth. Since the satellite is stationary, the earth station that is using the satellite as a relay point is relieved of the necessity of constantly having to move its antenna to point its beam at the satellite. The geosynchronous altitude is high enough that approximately one-third of the earth is visible to the satellite; put another way, as few as three satellites spaced uniformly around the geosynchronous arc can provide full global coverage.

Of course, the polar regions are not illuminated by a geosynchronous satellite. Looking at Fig. 2.2, we observe that the angle subtended by the earth is 17.3° and that the highest latitude at which the satellite is visible is 81.3°. Practically speaking, an earth station requires a minimum elevation angle of at least 5° above the horizon, which reduces the highest latitude to about 76° at the longitude of the satellite.

**Fig. 2.2. Maximum illumination of the earth by a geosynchronous satellite.**

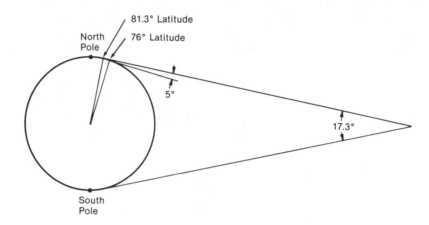

The maximum coverage along the equator provided by a geosynchronous satellite is 18,080 km. Applying the 5° minimum elevation angle criterion, this is reduced to about 16,900 km, corresponding to about 42% of the equatorial distance.

Fig. 2.3 shows the coverage patterns on earth for a geosynchronous satellite on the Greenwich Meridian, assuming a minimum elevation angle of 5°.

Because of the deficiency of the geosynchronous orbit in providing polar coverage, the Soviet Union has deployed its Molniya series of satellites. The *Molniya orbit* is elliptical, inclined 65° with respect to the equator, with a 40,000-km apogee in the northern hemisphere and a 500-km perigee in the southern hemisphere. A satellite in this orbit has a 12-hour period, almost all of which is spent at or near apogee. This orbit provides excellent east-west coverage at the high latitudes, precisely what is needed to cover a very large northern land mass. Three such satellites are used to provide 24-hour coverage.

## 2.4.2   Frequency Bands

Table 2.1 lists the frequency bands that have been allocated for the use of satellite communications. Of the six bands, three are commercial and three are military.

### Commercial Bands

Of the three commercial bands, the one lowest in frequency, *C band*, has been used the most widely. From the viewpoint of propagation, it is ideal

Fig. 2.3.
Synchronous-
satellite cov-
erage pattern,
assuming a 5°
elevation angle.

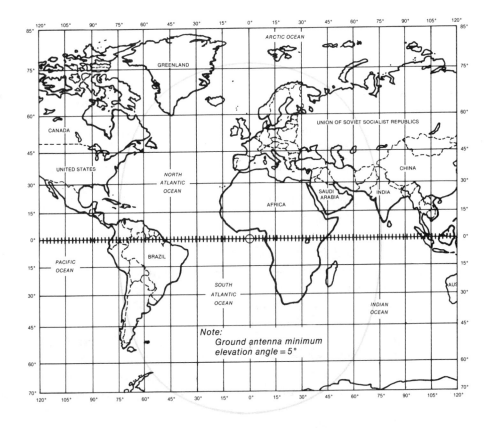

| TABLE 2.1. FREQUENCY BANDS FOR SATELLITE COMMUNICATIONS | Band | Downlink Bands (MHz) | Uplink Bands (MHz) |
|---|---|---|---|
| | Uhf—Military | 250–270 (approx) | 292–312 (approx) |
| | C band—Commercial | 3700–4200 | 5925–6425 |
| | X band—Military | 7250–7750 | 7900–8400 |
| | Ku band—Commercial | 11700–12200 | 14000–14500 |
| | Ka band—Commercial | 17700–21200 | 27500–30000 |
| | Ka band—Military | 20200–21200 | 43500–45500 |

because rain attenuation is minimal in this region. It has one major disadvantage, however: it is shared with terrestrial microwave. Since C band terrestrial microwave and satellite links can interfere with one another, obtaining frequency authorizations for satellite earth stations in populated areas is a severe problem, so severe that it is close to impossible to clear a wide bandwidth in an urban area for satellite use. For this reason, commercial earth stations serving major cities are typically located thirty miles outside of the city and are connected to their urban users by microwave relay.

*Ku band* is now becoming more and more common. Since there is no domestic microwave in this band, earth stations can be located almost anywhere, even on rooftops for private network service. Its main disadvantage is rain attenuation, particularly in the southeastern regions of the country. Because of this, attenuation margins of 5–10 dB have to be provided, depending on the application.

*Ka band* will be the next to be exploited, but probably not until the domestic orbital arc is so crowded at the Ku and C bands that a new frequency band is required simply to permit total capacity to grow. Ka band suffers still greater rain attenuation than Ku band. This can be partly compensated for by the use of narrower satellite beams to concentrate the satellite's transmitted power.

## Military Bands

The workhorse band for the military is *X band*, with properties very similar to C band. The *uhf band* is also widely used to serve mobile platforms. The reason for this is the fact that the effectiveness of a small, very broadbeam earth station antenna is proportional to the square of the wavelength. Aircraft, ships, and small land vehicles using broadbeam antennas, which don't require pointing, can take advantage of this property of radio transmission to obtain low-capacity services. This topic will be discussed further in section 2.8.4. The widespread use of the uhf band is, however, limited by the fact that it is also used for line-of-sight ground to air services.

There has been some commercial maritime application using the same principles in the 1.5-GHz band, and there promises to be domestic commercial application around 1.0 GHz for mobile platforms (see section 2.8.4).

## Frequency Reuse

Because of the limited bandwidth available for use by a satellite system, most satellites transmit and receive on two orthogonal polarizations, either vertical and horizontal or right-circular and left-circular. We show the vertical, horizontal case in Fig. 2.4, with plane waves propagating out of the paper, with electric field vectors E in one case horizontal and in the other vertical. Polarizing elements in the antennas and their waveguide-feeds produce this effect. In the ideal case of perfect orthogonality, the waves can occupy the same frequency band and not interfere with each other. In reality, some 30 dB of isolation can be achieved, making it practical to use the same frequency spectrum in the two polarizations—albeit with some mutual interference.

In the latest-generation satellites using narrow beams, a third degree

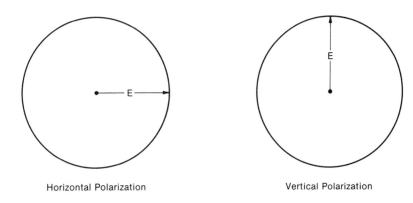

Fig. 2.4.
Orthogonal
polarizations.

of orthogonality is introduced. This third orthogonal axis is possible because beams whose footprints on earth are nonoverlapping can use the same spectrum, each with two polarizations.

### 2.4.3 Transponders

The most widely used satellite transponder is a simple frequency-translating repeater shown in Fig. 2.5. A signal captured by the satellite's receive antenna is first amplified in a so-called low-noise amplifier (LNA), then translated in frequency, amplified in a high-power amplifier (HPA), and finally retransmitted to earth through the transmit-antenna. Such a simple system is called in the vernacular a "bent-pipe" transponder.

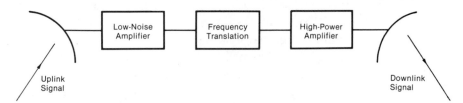

Fig. 2.5.
Frequency-
translating
repeater.

A typical second-generation C-band domestic satellite transponder will contain twenty-four such transponders, each having 36-MHz bandwidth separated by 4 MHz and using two polarizations. The transponder will usually contain two receivers (one as a backup) and will have in excess of twenty-four power amplifiers. Typically, four amplifiers might be supplied for three transponders, arranged so that a backup amplifier can be switched in to replace a failed amplifier when a ground command is given.

A satellite is expected to remain in service for at least seven years. Redundant components have to be supplied to support useful service over this period.

## 2.4.4   Earth Stations

A typical earth station will contain the elements shown in Fig. 2.6. Since a satellite may serve large numbers of earth stations, each station generally will not require the full satellite bandwidth capability. However, for flexibility purposes it is desirable that an earth station have tuning capability over the entire bandwidth, even though it will not be using all of the bandwidth at any one time.

Fig. 2.6.
Elements of an
earth station.

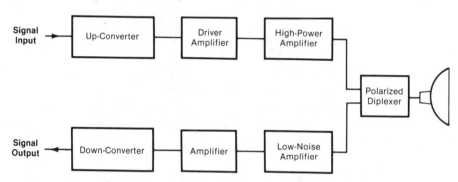

The fundamental operations are similar to those in space. The signal to be transmitted is converted to the uplink frequency before being amplified and directed to the appropriate polarization port of the antenna feed. The signal received from the satellite is amplified in an LNA before being down-converted from the downlink frequency. Critical components will often be installed redundantly, with automatic switchover in the event of failure so that uninterrupted operation is maintained.

Because of the greater flexibility on the ground than in space, the components that *look* the same on a block diagram of a transponder and an earth station will have different performance levels. An LNA in space is typically a tunnel diode amplifier with a noise temperature around $1000°K$. On earth, an LNA is typically a field-effect transistor amplifier or a parametric amplifier, which can achieve noise temperatures as low as the $50°K$ range. Similarly, satellite power amplifiers generate powers in the tens of watts using travelling-wave-tube amplifiers (TWTAs). Earth station antennas using either TWTAs or klystrons can generate power levels in the kilowatt range.

The same is true with antennas. Satellite antennas will have beams broad enough to cover wide areas—from "full earth coverage" of $17.5°$ beams down to U.S. coverage with roughly $6°$ beams and pencil beams of $2°$ to $3°$. Earth station antennas, which can be much larger, generate beams well under $1°$ in width. The relationship between beam width $\theta$, frequency $f$, and antenna diameter $D$ ($\theta \sim \lambda/D$) relates the antenna sizes needed to generate beams of various widths at the various frequencies or wavelengths. With some exceptions, satellite antennas are under 10 feet in diameter, while earth station antennas can be as large as 60 feet, although 10 to 30 feet is more common.

The newest applications (which are discussed in section 2.9) alter this traditional relationship between earth station and satellite parameters. In an effort to broaden applications for earth stations, the earth stations are becoming smaller and cheaper and are now being used with lower performance parameters than those for satellites: 3- to 6-foot dishes, 1- to 2-watt power amplifiers. In fact, the satellites that would service these disadvantaged earth stations are very high performance, with larger antennas and higher-power amplifiers.

## 2.4.5  Spacecraft Service Functions

As spacecraft designed to house transponders, satellites must provide a host of service functions. Among these functions are power, stabilization, propulsion, station keeping, attitude control, and telemetry, tracking, and command.

### Power Subsystems

Satellites are powered by the sun. Arrays of photovoltaic solar cells convert solar radiation to electricity, which is then converted to usable voltage levels in a power supply. Today's solar cells have a conversion efficiency of 12% to 15%. Glass covers partially protect the cells from space particle radiation of various kinds. This protection limits the degradation of power output to about 35% over seven years. Typical satellite solar arrays will generate 1 to 2 kW of power.

A geosynchronous satellite will be in shadow for a period of time each day when the sun is eclipsed by the earth around the equinoxes. In order to permit continuous operation, nickel-cadmium batteries are typically provided on board the satellite. The batteries have sufficient power to fill the void during the eclipse period. When the satellite is not eclipsed, a small fraction of the solar array power keeps the batteries charged.

### Stabilization

The space platform defined by the communications satellite must maintain a precise orientation in space in order for its antennas to be directed toward their intended areas on earth and for its solar arrays to be pointed at the sun. The space platform is oriented by defining an axis in space that is established by a rotational motion about the axis. The rotational movement is accomplished by using one of two different types of spacecraft: (1) In a *spin-stabilized satellite*, the main body of the satellite spins about its axis. Such a satellite is shown in Fig. 2.7. The satellite is cylindrical, spinning about the axis of the cylinder. Solar cells cover the cylindrical surface, and antennas are mounted on a *de-spun* platform. The platform maintains an earth-facing orientation that is derived from earth-seeking sensors. (2) In a *three-axis-stabilized* or *body-stabilized satellite*, a spinning momentum wheel within the satellite establishes the same frame of reference as the spinning cylinder of the spin-stabilized spacecraft. In this case, however, the entire spacecraft is effectively the de-spun platform, which maintains its orientation in space relative to the momentum wheel axis. Fig. 2.8 shows this type of satellite. Note that the solar cells are mounted on panels external to the spacecraft body, allowing them to be oriented toward the sun.

**Fig. 2.7. Spin-stabilized satellite.**

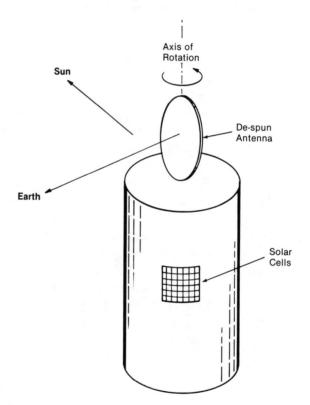

## Propulsion

A satellite contains a propulsion system with a supply of fuel to move it to its assigned position in orbit, to maintain it in that position (station keeping), and to maintain the direction of spin axis attitude control) in the face of forces that perturb it. The propulsion system usually consists of a supply of hydrazine ($N_2H_4$) gas with properly placed jets to develop

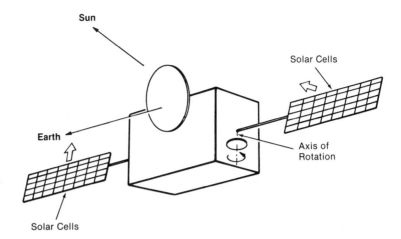

**Fig. 2.8. Three-axis-stabilized satellite.**

impulses of the proper magnitude and direction. Typically a satellite will need 200–300 pounds of fuel on board to carry out all of its functions over its lifetime.

## Station Keeping and Attitude Control

In station keeping and attitude control, the position of the earth and/or the sun or other celestial object is sensed, and then impulses at the appropriate locations and of the appropriate magnitude are applied to maintain the satellite's position and orientation.

## Telemetry, Tracking, and Command (TT&C)

A satellite is a very complex system that must survive over several years to perform its functions. It has many switches on board, enabling its communications parameters to be changed to conform to an altered situation or to be adjusted in the event of component failure. Sometimes, a change in situation requires that the satellite be moved from one position in the orbit to another. Controlling all of these highly intricate maneuvers is the *telemetry, tracking, and command system*, which has both an earth and a space subsystem.

The telemetry system provides a data stream to the ground that reports on each of the satellite subsystems, independent of any other communications links. For example, if a transponder has failed, the telemetry data contains information that permits diagnosis of the cause of the failure. The command system permits a specially configured earth station to modify the satellite configuration (e.g., to move the satellite). In this it is assisted by the telemetry system, which reports on the effectiveness of the command that was inserted.

The satellite contains both a transmitter for telemetry only and a receiver for command only. The TT&C earth station contains the necessary equipment to insert commands, receive and interpret the telemetry, and compute orbital parameters. It will track the position of the satellite either by using a special beacon signal on the spacecraft or by inserting tracking tones through the main communications system.

### 2.4.6 Satellite Weight Budget and Cost

A typical second-generation satellite with twenty-four transponders weighs between 1100 and 1200 pounds. About 50% of this weight is consumed by the communications package, including the antennas—for about 350 pounds—and by the power system (solar cells, batteries, etc.)—for another approximately 250 pounds. The physical structure and the housekeeping subsystems constitute the other half of the weight load. On top of this the satellite must carry about 200 to 300 pounds of fuel. Such a satellite will cost between $35 million and $40 million (1984). The total cost to place two such satellites in orbit, including a ground spare and all of the necessary TT&C facilities, is about $230 million to $250 million, or approximately $5 million per transponder.

### 2.4.7 Launch Vehicles

In order for a satellite to achieve its desired position on the orbital arc, it must be launched by a primary vehicle, either a reusable vehicle, such as the space shuttle, or an expendable vehicle, such as Ariane. The launch process is complex and intricate. An expendable vehicle will have two or three stages to boost the satellite payload to the altitude of the geosynchronous arc, but in a highly elliptical orbit. Once in position, the expendable vehicle releases the satellite, which contains a small rocket called the "apogee kick motor." The effect of this rocket is to circularize the highly elliptical orbit. The space shuttle carries the satellite to a low earth orbit of about 300 miles. The satellite is then released from the shuttle bay, whereupon it fires a so-called perigee motor, which is equivalent to the final stage of the expendable vehicle. The apogee kick motor is then fired as it was with the expendable vehicle. (Note that the weight figures quoted in the previous section include the weight of the apogee kick motor. The perigee motor is not included, since it is really a component of the launch system.)

# 2.5   Radio Transmission Links

This section continues the discussion of satellite communications fundamentals by introducing the methodology for computing satellite link performance.

## 2.5.1 General Link Equations

Fundamental to an understanding of satellite communications links is an understanding of the behavior of an elementary radio link, such as that shown in Fig. 2.9. This elementary link is representative of an arbitrary radio link. It is therefore applicable to both the uplink and the downlink of the satellite circuit.

**Fig. 2.9. Radio link geometry.**

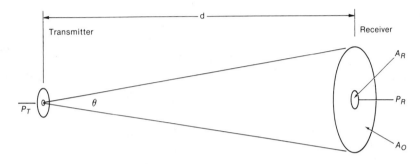

The transmitter in Fig. 2.9 is characterized by the output power $P_T$ of its power amplifier and by the directivity of its antenna, which is represented in the figure by a finite beam of width $\theta$. The receiver is characterized by the effective area $A_R$ of its antenna and by the noise temperature $T$ of its LNA.

The transmitting antenna beam is shown to illuminate an area $A_O$ at the receiver, which is a distance $d$ from the transmitter. The receiving antenna intercepts a fraction $A_R/A_O$ of the transmitted power. Thus, the power $P_R$ received by the antenna is given by the simple relationship:

$$P_R = P_T \frac{A_R}{A_O} \tag{2.4}$$

The parameter used to describe the directivity of the antenna is its gain $G_T$, defined as:

$$G_T = \frac{4\pi d^2}{A_O} \tag{2.5}$$

The quantity $4\pi d^2$ is the area of a sphere at the receiver, which is centered at the transmitter. If the transmitting antenna were isotropic, it would illuminate this sphere uniformly. Thus, the gain of an antenna is defined as the ratio of the area illuminated by an isotropic antenna to that illuminated by the antenna in question.

Substituting equation (2.5) into equation (2.4) we obtain,

$$P_R = P_T \, G_T \frac{A_R}{4\pi d^2} \qquad\qquad (2.6)$$

The product $P_T G_T$ is referred to as the *effective isotropic radiated power* (EIRP) of the transmitter and is a figure of merit for a transmitter. Equation (2.6), then, has a simple intuitive interpretation: the power to the receiver is that of an isotropic antenna radiating a power $P_T G_T$.

If the receiving antenna is a parabolic reflector, its effective area $A_R$ is proportional to its geometric cross section with proportionality constant 0.54. By analogy with optics, the directivity or gain of the antenna is proportional to $A_R/\lambda^2$, where $\lambda$ is the wavelength. The exact relationship may be shown to be:

$$G_R = \frac{4\pi A_R}{\lambda^2} \qquad\qquad (2.7)$$

In Fig. 2.10 we plot the gain of a parabolic antenna against its diameter for several frequencies. Substituting the expression for $A_R$ in equation (2.7) into equation (2.6), we obtain:

$$P_R = P_T \, G_T \, G_R \left(\frac{\lambda}{4\pi d}\right)^2 \qquad\qquad (2.8)$$

an expression symmetric with respect to the two antenna gains. The factor $(\lambda/4\pi d)^2$ is called the free-space path loss.

Equation (2.8) is the link equation most often used in practical computations. It has the virtue of containing system parameters that are readily measurable. Of course the free-space path loss is not really a "loss." It is rather a modified spreading factor. The fact that it is frequency-dependent is counterintuitive. This comes about solely because the receiving antenna parameter in the equation is the gain rather than the effective area. Equation (2.6) shows us that for a given EIRP the received power depends on receiving antenna size and not on frequency. Thus, in most applications of equation (2.8), the frequency dependence of the spreading factor is cancelled out by the frequency dependence of the antenna gain. In Fig. 2.11, we plot the free-space path loss in decibels against frequency.

The significant receiver parameter is not received signal power but its ratio is noise power. If the dominant source of noise in the receiver is thermal noise in the preamplifier, then the noise power in the receiver is given by:

Fig. 2.10. Gain of a parabolic antenna versus antenna diameter for several frequencies.

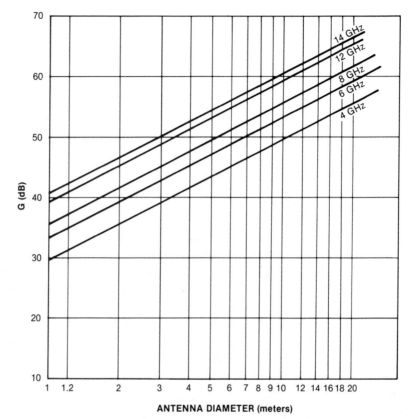

Fig. 2.11. Free-space path loss versus frequency.

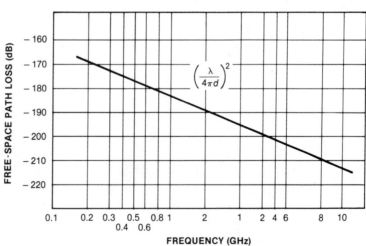

$$N = kTW$$

where, $T$ is the receiver noise temperature, $W$ is the bandwidth, $k$ is Boltzmann's constant.

In this case the signal-to-noise ratio is:

$$\frac{P_R}{N} = \frac{1}{kW}\, P_T\, G_T \left(\frac{\lambda}{4\pi d}\right)^2 \frac{G_R}{T} \tag{2.9}$$

and is proportional to the three factors: $P_T G_T$, the EIRP of the transmitter; $G_R/T$, the receiver figure of merit; and $(\lambda/4\pi d)^2$, the free-space path loss.

## 2.5.2  Satellite Links

A satellite path consists of two links in tandem: the *uplink*, from the ground transmitter to the satellite receiver, and the *downlink*, from the satellite transmitter to the ground receiver. Each link in itself behaves according to the preceding discussion. They are, however, coupled according to the block diagram of Fig. 2.12. A relatively weak signal received at the satellite is translated in frequency and then amplified before being transmitted back to earth. The noise introduced in the satellite receiver is also amplified and transmitted along with the signal, as shown in Fig. 2.12.

Fig. 2.12. Uplink-downlink coupling.

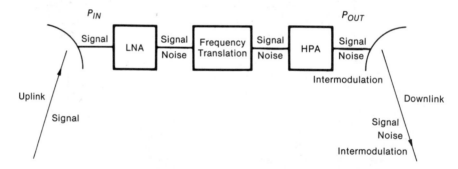

The transponder will have a gain curve relating its input and output powers. Fig. 2.13 shows a typical power amplifier input-output relationship. The earth station EIRP is carefully controlled so as to develop the desired output power from the HPA of the transponder. In general, several signals will occupy the transponder, all of which have to share this power. When several signals are present in a perfectly linear system, each signal is unaffected by the presence of the others. However, if the system is nonlinear, intermodulation products will result, and these constitute another source of noise. Thus, whenever the transponder is to support more than one signal, its operating point must be backed off from the maximum point shown in Fig. 2.13 to a point linear enough to keep intermodulation noise below tolerable limits, usually 4–6 dB below the maximum. We will refer to this again later when we treat multiple access.

**Fig. 2.13. Transponder output power versus input power.**

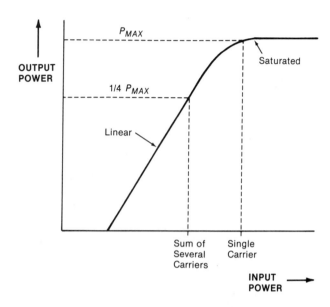

For now we will assume a single signal sharing the transponder with noise from the transponder's LNA. If the signal-to-noise ratio in the LNA is very high, then the HPA can be operated at its maximum point and still maintain the total uplink noise, including intermodulation noise below tolerable levels. Then, using equation (2.8), the signal power into the downlink receiver is given by:

$$P_R = P_{MAX} \, G_T \left(\frac{\lambda}{4\pi d}\right)^2 G_R = P^1 \, G_R \qquad (2.10)$$

And the uplink noise at the same point is:

$$P_{NU} = P^1 G_R / (S/N)_U \qquad (2.11)$$

where the uplink signal-to-noise ratio $(S/N)_U$ includes both the satellite receiver and the intermodulation noise. Then the ground receiver signal-to-noise ratio,

$$S/N = \frac{P^1 G_R}{P^1 G_R / (S/N)_U + kTW} \qquad (2.12)$$

shows explicitly two noise components: the first, the uplink noise as seen on the downlink, and the second, the downlink receiver noise.

Several interesting observations may be made from equation (2.12). If the uplink noise as received by the downlink receiver is very small compared with the downlink noise, then the first term in the denominator can be neglected. The resulting equation is simply the link equation for the downlink decoupled from the uplink. The other limiting case occurs when the uplink noise is high and/or the downlink is very strong (e.g., as a result of a large $G_R$). Then the signal-to-noise ratio is simply the uplink signal-to-noise ratio, and increasing the receiving antenna gain or decreasing its noise temperature has no effect.

Uplink noise can be caused by interfering signals, as well as by thermal noise in the satellite receiver. If the uplink EIRP is moderate to large, then the signal to thermal noise in the satellite is usually very large. The dominant noise will come from the side lobes of uplinks directed at adjacent satellites (adjacent satellite interference) or from the incomplete rejection of signals with the orthogonal polarization (cross-polarization interference). These represent the ultimate limitation on the downlink signal-to-noise ratio since, according to equation (2.12), whatever the uplink noise, increasing the receiving antenna gain will at some point cease to improve the signal-to-noise ratio.

In the important case of microterminals (which we will consider subsequently), the receiver noise from all sources can be large enough to make the uplink signal-to-noise ratio low enough that no additional downlink noise can be tolerated if a respectable throughput is to be achieved. In this case we are led to require a high enough $G_R$ that, according to equation (2.12), the downlink noise is rendered negligible.

# 2.6   Satellite Link Capacities

In this section we will apply the results of the previous section to typical satellite links. We will first compute power and bandwidth limitations and then give a simple example.

## 2.6.1   Power and Bandwidth Limitation

We will assume the use of *quaternary phase shift keying* (QPSK) modulation. Thus, for a data rate $R$ bits per second (bps), we transmit $R/2$ symbols per second, with 2 bits carried by each symbol. The maximum symbol rate is limited to something less than the transponder bandwidth $W$. Typically for a 36-MHz transponder, the maximum symbol rate is 30 megasymbols per second, corresponding to a maximum data rate $R$ of 60 megabits per second (Mbps).

The figure of merit of a modulation system is the required value of

$E_b/N_0$, where $E_b$ is the received energy per bit and $N_0 = kT$ is the noise power per hertz. The required value of $P_R/N_0$ is related to $E_b/N_0$ and $R$ by

$$P_R/N_0 = R\ E_b/N_0 \qquad\qquad (2.13)$$

or the lower the required $E_b/N_0$ for a given rate, the less $P_R/N_0$ is required.

In Fig. 2.14 we plot $R$ against $P_R/N_0$ for two values of bandwidth, 36 MHz and 72 MHz, and for $E_b/N_0 = 10$. The left portion is simply a graph of equation (2.13). When $R$ approaches its maximum value, the curve saturates. The left region is referred to as the power-limited region; the saturated region at the right is called the bandwidth-limited region.

**Fig. 2.14. Data rate versus $P_R/N_0$.**

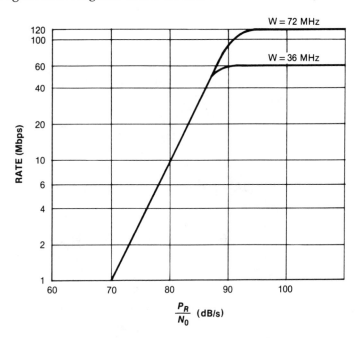

## 2.6.2   A Single Access Example

As a simple case, we will assume that there is a single uplink signal in the transponder and that this signal is large enough to permit us to achieve maximum power from the power amplifier and to neglect uplink noise, thus permitting the downlink to be decoupled from the uplink. Then equation (2.9) can be used as is.

For computational ease, the various quantities in equation (2.9) are usually expressed in decibels, which permits us to add and subtract quantities instead of multiplying and dividing. The resulting table of decibel values is often called a *power budget*.

We will first compute the satellite EIRP and then incorporate the other link parameters. We assume a 20-watt power amplifier and a transmitting antenna with a 23-dB gain corresponding to coverage of the continental United States. Thus,

| | | |
|---|---|---|
| $P_T$ = 20 watts | 13 dBW | |
| $G_T$ | 23 dB | |
| EIRP | 36 dBW | |
| Path loss at 12 GHz | − 205 dB | |
| Power to antenna | − 169 dBW | |
| $G_R$ at 12 GHz (7 meters, 5 meters) at 12 GHz | 56 dB | 53 dB |
| Margin (including rain) | 6 dB | 6 dB |
| $P_R$ | − 119 dBW | − 122 dBW |
| $kT$ (120°K) | − 208 dBW/Hz | − 208 dBW/Hz |
| $P_R/kT = P_R/N_0$ | 89 dB | 86 dB |

Locating these points on the curves in Fig. 2.14, we see that with a 5-meter antenna we are power-limited at 72 MHz and just bandwidth—limited at 36 MHz, while the 7-meter antenna brings us to bandwidth limitation at 72 MHz as well. We will expand on this link later in section 2.9 when we consider private networks.

# 2.7   Multiple-Access Techniques

If the transmitting station in the previous example has 60 Mbps to send to a single receiving station, then we have the simplest case of a satellite providing a single point-to-point link. Satellites are not often used in this way, however. More often we want to take advantage of the unique broadcast or point-to-multipoint capabilities of a satellite transponder.

The first step in this direction occurs when a single station sends a full transponder's capacity to several receiving stations. The next step in complexity occurs when a pair of stations fill a transponder with data for each other. The most general case occurs when several stations interchange data in an arbitrary way using a single transponder. In this case the stations have to share the available capacity in a cooperative way constituting what is called *multiple access*. In the following sections we describe the most important ways in which multiple access is accomplished.

## 2.7.1   Time-Division Multiplexing

The example of the previous section represents the simple case of a single, strong uplink signal. The data on this uplink may be shared among a number of downlinks by *time-division-multiplexing* (TDM) the data on the uplink, as shown in Fig. 2.15. Each receiving station receives the same downlink signal; it simply picks off the bits intended for it and ignores the others. This scheme is easy to implement. It is applicable in high-capacity networks when one station has enough traffic destined for other stations to fill up a transponder.

Fig. 2.15. Time-division multiplexing.

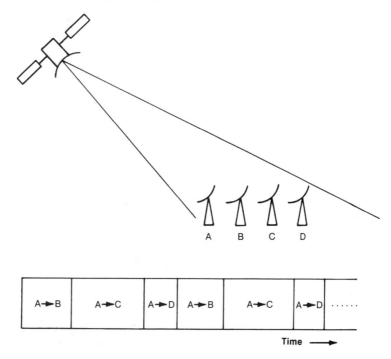

## 2.7.2 Time-Division Multiple Access

*Time-division multiple access* (TDMA) is conceptually just a step beyond time-division multiplexing, but in practice it is a great deal more complex. The concept is shown in Fig. 2.16. During any given time interval, the uplink from a single station drives the transponder, with its downlink shared among several other stations as in TDM. Here, however, the uplink is shared in time with all of the stations. A *super frame* is defined by a *master station*, with designated transmit times for each network station. Each of the transmitters synchronizes itself to a special pulse that defines the beginning of the super frame, and thus prevents overlapping or underlapping transmissions. Each transmitter also precedes its data transmission with a preamble to permit receiver synchronization. With TDMA, the transponder throughput is close to that achieved with single access. The small difference is the result of the synchronization process.

Fig. 2.16. Time-
division multi-
ple access.

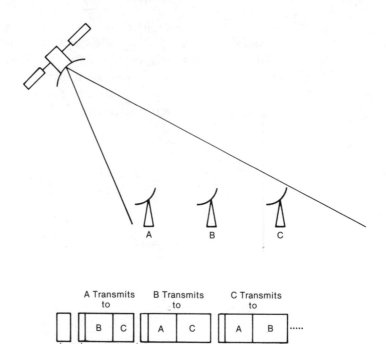

In TDMA systems each transmitter receives a fraction of a super frame long enough for the transmission of its data. The station loads will vary with time, and periodically the time allocations will be changed to accommodate changing loads.

The most sophisticated form of TDMA occurs when the transmission time allocations are based on instantaneous loads. This is called *demand assignment* and is the most efficient way to use a channel. Demand assignment (DAMA) is complex and expensive to implement, a fact that partially offsets its very high transmission efficiency.

## 2.7.3   Frequency-Division Multiple Access

The dual of TDMA is frequency-division multiple access (FDMA). In this case, shown in Fig. 2.17, the frequency passband of the transponder is allocated to the transmitting stations according to the data rate required. FDMA, unlike TDMA, necessitates simultaneous uplink signals in the transponder. To avoid generating modulation products, the transponder must be operated in its linear range. This means backing the operating point off by 4–6 dB from its maximum point. Because of this, the throughput of an FDMA transponder will generally be about 25% to 40% of that of a single uplink transponder.

**Fig. 2.17.
Frequency-
division multi-
ple access.**

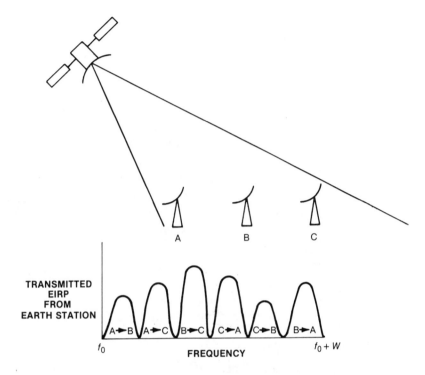

The power of each uplink signal must be closely controlled so that the intended downlink receives only as much power as needed to achieve a specified level of performance. If one downlink receives excessive power, another may not receive enough, since all are sharing the same transponder.

The coordination required for FDMA systems is generally easier and cheaper to achieve than the tight timing control in TDMA systems. This compensates for the additional throughput degradation of FDMA systems.

FDMA systems are often called single-channel-per-carrier, or SCPC, systems.

# 2.8  Applications

The communications satellite constitutes a very versatile medium, one that can be used in a wide variety of applications. These can range from applications that are similar to those in other media to ones that require the satellite's unique properties. When the first satellites were placed into orbit, the medium was looked on as a cure for all the ills of long-haul transport. Some twenty years later there are those who would deprecate the medium's advantages in all but a few specialty areas. As is usually the case, the truth tends to be somewhere in between, and in this section we will review the most important of these application areas.

## 2.8.1  "The Cable in the Sky"

It should be clear from earlier sections that a satellite transponder can be used to establish one or more point-to-point links. If these links constitute intermachine trunk groups in a large network, such as a public switched network from one of the interexchange carriers, they play the same role as trunks derived from terrestrial media, microwave or fiber.

As noted in section 2.3.4, satellite trunks can be shown to be economical for distances in excess of 1000 or 1500 miles, depending on the size of the trunk group. This is, of course, due to the fact that satellite trunk costs are distance-insensitive, whereas terrestrial trunk costs are proportional to distance. Despite these economic advantages, the satellite has made only a small impact in this area on the domestic scene. The primary reason for this is that some 85% of the traffic carried is voice traffic, and there are some who feel that the delay inherent in satellite relay is objectionable in voice conversations. The two largest domestic carriers, AT&T and MCI, use satellite trunking in their networks only minimally. This use will decrease, if anything, as time goes on, as a result of the increased economies resulting from fiber-optic long-haul trunking. But it is a mistake to state that the primary reason for the small impact is the growth in fiber optics.

Overseas communication has been quite different. Ever since the first Intelsats were placed in orbit, they have become the principal medium for transoceanic trunking because their capacities soon surpassed those available with the only other competitive medium, undersea cable. Apparently in the hierarchy of things, delay and its effects are less objectionable when the competing media have lower capacities and are more expensive. But as optical cables with their large capacities begin to make their appearance, we may well see a change in the dominant position now held by satellites for overseas trunking.

## 2.8.2  Broadcast Applications

The inherent broadcast capability of satellites has led to a large business area in the point-to-multipoint class of applications. By far the largest of these areas is television broadcasting and distribution, which constitutes from one-quarter to one-third of the domestic satellite utilization. The service is one-way, consisting of a single, relatively expensive transmitting station and a large number of relatively inexpensive receive-only stations associated with local broadcasting facilities and cable heads. These facilities contain tape storage capability so that programming material can be broadcast at any time for use by the local stations according to their individual schedules. The extension of this to very inexpensive home receivers using very powerful transponders (Direct Broadcast Satellite) had not taken place as of 1985.

Another broadcast application is remote printing. Newspapers or other periodicals with nationwide circulation find it more economical to send the copy in electrical form via satellite to widely distributed printing plants rather than to print it centrally and then physically distribute the periodicals over long distances.

Several companies distribute data bases of various kinds electronically. Both satellite and terrestrial media are used. Whenever the bandwidths required for the transmission are large, the satellite has the edge and will continue to.

## 2.8.3  Private Networks

A corporation or government organization will often find that using a private network for its intracompany voice and data communications is more economical than using tariffed services obtained from the public networks. One potential advantage of such private networks is the ability to access the facilities of the network directly, bypassing the local exchange carrier in the process.

Private corporate networks have been popular for some time. Traditionally the network is supplied by a carrier using dedicated facilities for the network. A modern version of this is the *virtual private network*, in which a carrier creates what looks like a private network to the customer, but all the facilities for the network are time-shared with other services on the public network.

Satellite-based networks have also been used to furnish both physical and virtual private networks. One of the virtues of the satellite medium in such applications is that, in principle, the earth station can be located at the customer location, making bypass of local facilities particularly easy.

A large impediment to such satellite-based networks has been the fact that most satellites have operated at C band, and frequency clearance for C-band earth stations has been difficult to obtain. The introduction of Ku band, in which the frequency clearance problem is practically nonexistent, and the implementation of new high-power satellites permit small customer-premise Ku-band stations to be used as economical networks. There are potentially many applications for such networks. A few of the more obvious are point-of-sale terminals, automatic teller machines, and direct facsimile transmission. We shall elaborate on this area in section 2.9.

### 2.8.4  Mobile Applications

Until the arrival of satellite technology, long-distance connectivity to mobile platforms was provided at various radio frequency bands by using some form of ionospheric propagation to extend beyond the horizon. The most commonly used type of propagation has been high-frequency radio. Of course, land vehicles and aircraft flying over land can communicate over long distance with a short range line-of-sight radio link extended by terrestial facilities.

The Defense Department has made widespread use of the uhf band for mobile communication. An aircraft is a particularly constraining vehicle. A narrow-beam antenna requires accurate pointing in the face of aircraft motion. It also requires housing under a radome (*radar dome*) to keep from altering the flight characteristics of the aircraft. Going down in frequency permits the use of very simple antennas with beams broad enough not to require pointing. To see why that is so, we refer to equation (2.9) and Fig. 2.10. The free-space path loss varies with $\lambda^2$ so that even a simple antenna with a hemispheric pattern ($G_R = 3$ dB) can receive enough power to support a data channel or perhaps a compressed voice channel.

The uhf band is widely used for line-of-sight applications, and frequency clearance is accordingly a significant problem in many locations. However, its advantages over high-frequency radio are very great for ships, aircraft, and small man-pack ground terminals, so that it has become a Department of Defense mainstay. A similar kind of capability has been developed for commercial use using the Marisat satellite.

Several interesting new areas involve the use of multiple satellites for combined position location and low-data-rate transmission. Once again, the lower end of the spectrum around 1 GHz will be used for this as a result of the fundamental path loss dependence on wavelength.

# 2.9  Private Networks and Micro Stations

One of the most promising new areas in which satellite systems are being exploited is private data networks that employ hundreds or even thousands of small, low-cost earth stations. The networks start out with the premise that the dominant system cost is that of the earth segment because so many stations are used. Therefore, the starting point in designing the network is a very low cost earth station. Such a station will have relatively weak uplinks and downlinks, and compensation for these weaknesses will have to be introduced elsewhere in the system.

Clearly, a more complex satellite can compensate in part for earth station weaknesses, and with the maturation of space technology certain improvements can be made here at relatively low incremental cost and risk. But there are limits to the extent that this avenue can be pursued, and other system level techniques, such as the use of a small number of powerful earth station relays, can compensate for satellite limitations. We will discuss some of these techniques in this chapter. But before doing so, we will consider an example of a more conventional network using larger earth stations. This example will point out all the differences between the two approaches.

## 2.9.1 A Standard Ku-Band Private Network

We will assume a network of five identical stations with the same parameters as those used for the example in section 2.6.2. In that example, we computed the downlink throughput, assuming a single uplink access that was powerful enough to permit us to neglect uplink noise. We will assume that the stations have 7-meter antennas and 3-kW HPAs and that the multiple-access technique is TDMA. The network geometry is the same as that shown in Fig. 2.16.

First let us examine the assumption about uplink noise. To do this we compute an uplink power budget:

| | |
|---|---|
| $P_T$ (3000 w) | 35 dBW |
| $G_T$ (7 m) | 57 dBW |
| EIRP (uplink) | 92 dBW |
| Path loss (14 GHz) | − 206 dB |
| Power to antenna | − 114 dBW |
| Antenna gain | 23 dB |
| $P_R$ | − 91 dBW |
| $kT$ (1000°K) | − 199 dBW/Hz |
| $P_R/kT = P_R/N_0$ | 108 dB/s |

This value is seen to be 19 dB, or 100 times larger than the downlink signal-to-noise ratio of section 2.6.2. Thus, our uplink assumption is valid to within a few percent, even in the presence of rain.

We can therefore use the downlink value in section 2.6.2 of 89 dB. As shown, the total satellite throughput distributed among the stations is obtained by dividing it by the $E_b/N_0$ of the demodulation system required to yield an adequately low-bit-error probability. Let us assume that this is

10 dB. This yields a total throughput of $89 - 10 = 79$ dB, or a rate of about 100 Mbps. Since we are using TDMA, about 15% of the frame will be used for synchronization, leaving a net data rate of 85 Mbps to be shared among the links in the network.

Suppose that the network is carrying only voice traffic, using 32 Kbps per voice link. Then a duplex (four-wire equivalent) voice channel uses 64 Kbps, and the satellite can support about 1300 voice channels, or an average of 65 voice channels per node pair.

As noted before, we can do several things to increase the total throughput. We can use digital speech interpolation, a technique that multiplexes several voice channels in such a way that all the silences in a speech conversation are filled. This technique can improve the total voice throughput by a factor of 1.5 to 2. We can use demand assignment of trunks to nodes to gain another factor of 20% to 30%. We might also use a more sophisticated coding and modulation system to buy another factor of 2. If all of the preceding were done, the network capacity would be increased by about a factor of 4. We should also point out that by our initial choice of TDMA over FDMA we have gained a factor close to 2, since we can use more of the satellite's output power. All of these techniques are in the same category: they represent ground station technology, of varying degrees of cost, designed to improve the throughput efficiency of a space asset.

As we pointed out in section 2.9, the philosophy of the microstations network is the opposite. There we start out with a low-cost earth station and design a system around it.

## 2.9.2   Microstation Networks

The earth station used as the "standard" in the previous section will cost in the million dollar range, the precise amount depending on the degree of sophistication in the earth station processing. A microstation must cost in the $10,000 range as an upper bound. How do we buy two orders of magnitude?

First and foremost, we must accept a drastically reduced total network throughput. If nothing else changes, reducing the earth station antenna diameter from 5 m to 1 m reduces the throughput by a factor of 25. Practically speaking, this essentially eliminates voice traffic from consideration.

Second, we must use a low-power transmitter in the earth station. As we shall show later, this results in the uplink noise becoming dominant. There are techniques for mitigating this effect which the following discussion will describe.

Third, the multiple-access technique must be simple. TDMA is out of the question because its cost is in the $100,000 range. Some way of using FDMA or TDM must be found.

To examine the possibilities, first we postulate some earth station parameters that are in the appropriate cost range. We select an antenna diameter of 1.2 meters (4 feet) and a power of 1 watt. Then we postulate satellite parameters. In all the satellites examined so far, we have assumed an antenna shaped to cover the continental United States— therefore one having a gain of about 23 dB. In this case we will postulate a larger antenna, one that would cover one-third of the continental United States. Its gain would therefore be 5 dB larger, or 28 dB. We will also assume 40 watts of output power from the HPA.

Given these fundamental parameters, we compute the uplink signal-to-noise ratio in the absence of rain:

| | |
|---|---|
| $P_T$ (1 watt) | 0 dBW |
| $G_T$ (1.2 m) | 42 dB |
| EIRP | 42 dBW |
| Path loss (14 GHz) | −206 dB |
| Power to antenna | −164 dBW |
| $G_R$ | 28 dB |
| $P_R$ | −136 dBW |
| $N_0 = kT$ (1000°K) | −199 dBW/Hz |
| $P_R/kT$ | 63 dB |
| W (72 MHz) | 78 dBHz |
| $P_R/kTW = S/N$ | −15 dB |

With these parameters the noise power in the satellite receiver is approximately 30 times the signal power. If nothing else were done, the downlink would be supporting noise transmission rather than signal.

Now let us examine the downlink, assuming that something can be done to eliminate the uplink noise. The uplink and downlink use the same antennas. They differ in that the downlink has a 40-watt transmitter as compared with the 1-watt uplink, and a 300°K receiver as opposed to a 1000°K receiver. The downlink will therefore be about 120 times more powerful, yielding a $P_R/N_0$ of 84 dB. Using a typical low-cost demodulator, this will yield a total data rate of about 70 dB, or 10 Mbps. If the application requires 56 Kbps, this link can support about 175 stations. If 2.4 Kbps are sufficient, then 4000 stations can be supported.

One way to compensate for a weak uplink is to use FDMA, since the total signal power in the transponder is the sum of the individual powers from each uplink. However, in this case the satellite transmitter power has to be backed off to about one-fourth of its peak value so that the intermodulation distortion can be reduced to an acceptably low value. In this case the downlink capacity is reduced to 45 stations at 56 Kbps and 1000 stations at 2.4 Kbps.

With 45 uplinks from the microstations, the total uplink signal-to-noise ratio is increased by 16 dB to make the total signal and noise powers about equal. With 1000 uplinks, the total uplink signal-to-noise ratio is 6 dB, just on the verge of acceptability.

A more positive way of achieving the full downlink capacity is to decouple the up- and downlinks by introducing a processor between them. In principle, this processor may be in the spacecraft or in a ground relay station with a downlink powerful enough not to degrade the already very weak uplink. The processor is complex, and using a ground station can eliminate the cost and risk of an on-board satellite processor, compensating for it with the cost of a second satellite transponder.

Both schemes are shown conceptually in Fig. 2.18. In each case the microstations transmit to the satellite using FDMA and receive a TDM satellite signal. The processor, whether on board the satellite or on the ground, receives a number of FDMA signals, demodulates them, and then remodulates them in time-division-multiplex form to constitute the downlink.

Each uplink with its $P_R/kT$ of 64 dB can support a maximum data rate well in excess of 10 Kbps using a simple modulation scheme. This can be mapped into the 10-Mbps downlink in any way desired by using the TDM multiplex format (e.g., one thousand 9.6-Kbps circuits, two thousand 4.8-Kbps circuits, or four thousand 2.4-Kbps circuits). Of course, rain or interfering signals from adjacent transponders will reduce the overall capacity.

We also observe that the use of a narrow antenna beam on the satellite necessitates using multiple beams on the satellite in order to cover the whole country and also necessitates switching among the beams to provide universal connectivity. In the effort to increase the throughput of highly disadvantaged earth stations, we have disturbed the fundamental broadcast capability of the communications satellite.

We also make the obvious observation that even if the links could support voice rates, using two satellite accesses would double the delay and render the links unacceptable to many, if not most, people.

Despite these drawbacks, the Ku-band microstation shows promise of opening a whole new class of data applications for satellite communications.

## 2.10   Concluding Comments

Communication by satellite relay has matured in the short span of twenty years. It has proved its value in a host of application areas and

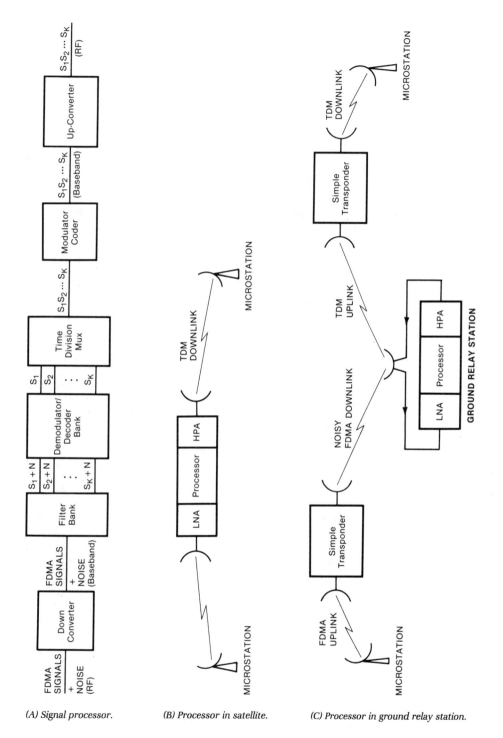

Fig. 2.18.
Satellite
processing for
microstations.

(A) Signal processor.    (B) Processor in satellite.    (C) Processor in ground relay station.

promises to continue to do so in the face of competition from still newer technologies.

From the viewpoint of technology, these twenty years have seen the spacecraft move from the realm of exotica to that of the commonplace, not only for communications but for many other applications. As might be expected, the first years have seen relatively simple spacecraft support relatively complex ground stations. As time has progressed, satellites have grown larger, carrying more and more transponders, and often in more than one frequency band (e.g., C and Ku). The trend is inexorably toward putting more complexity into the spacecraft to support increasingly inexpensive earth stations, both fixed and mobile.

But now complexity means not simply more of the same but such things as higher-power, signal processing of the kind referred to in section 2.9.2 and *antenna processing*, the generation and interconnection of multiple narrow beams. We are indeed heading from the "cable in the sky" toward the "switch in the sky." Some of this technology is being developed for military satellites and is bound to reach the commercial world in the 1990s. We may well see an associated move to Ka band during this time that will permit smaller physical structures to develop multiple narrow beams.

Thus, while such older bread and butter applications as television broadcast will continue to thrive, we can look forward to the emergence of new application areas made economically feasible by the advance of space technology.

# 2.11  References

Dicks, J. L., and M. P. Brown, Jr. 1974. Frequency division multiple access (FDMA) for satellite communications systems, *IEEE-EASCON '74* (IEEE Press, New York): 167–178.

Fthenakis, E. 1984. *Manual of satellite communications.* New York: McGraw-Hill.

Gabbard, O. G., and P. Kaul. 1974. Time-division multiple access. *IEEE-EASCON '74.* (IEEE Press, New York): 179–184.

Lebow, I. L., K. L. Jordan, Jr., and P. R. Drouilhet. 1971. Satellite communications to mobile platforms. *Proceedings of the IEEE* 59:139.

Lucky, R. W., J. Saltz, and E. J. Weldon, Jr. 1968. *Principles of data communications.* New York: McGraw-Hill.

*Reference data for satellite communications and earth stations.* 1972. Ramsey, NJ: ITT Space Communications.

Schwartz, M., W. R. Bennett, and S. Stein. 1966. *Communication systems and techniques*. New York: McGraw-Hill.

Spilker, J. J., Jr. 1977. *Digital communications by satellite*. Englewood Cliffs, NJ: Prentice-Hall.

Van Trees, H. L., ed. 1979. *Satellite communications*. New York: IEEE Press.

**3**

# INTEGRATED SERVICES DIGITAL NETWORKS (ISDN)

Eric L. Scace

Computers, digital transmission systems, digital switching systems, xerography, international jet travel, and an expanding worldwide communications market have all stimulated a new approach to telecommunications: *integrated services digital networks* (ISDNs). One part evolution, one part revolution, and one part philosophy, ISDNs represent a unique historical development.

Unlike previous advances in telecommunications, ISDNs did not spring from research and development by one individual, company, or country. Nor did ISDNs step out directly from the laboratory. Instead, ISDNs grew from the ideas of a few dozen key individuals working for different companies and governments in Europe, North America, Japan, and Australia. Meeting regularly under the framework of the International Telegraph and Telephone Consultative Committee (CCITT), these people created technical standards that now guide all ISDN implementations. In the past, technical standards for telecommunications were negotiated during or after deployment of new systems. The ISDN documents provide *anticipatory* standards—describing new network architectures, interfaces, and protocols before actual implementation and deployment has begun.

ISDN national, regional, and international standards continue to evolve, especially in those aspects heavily influenced by rapidly evolving new technology. This chapter, therefore, outlines the state of agreements as of January 1986. The chapter also explores new areas not yet fully standardized; where possible, the text identifies likely outcomes of ongoing discussions.

Public telecommunications networks in the United States operate in a environment radically different from that existing in most other countries. This chapter approaches ISDNs from the following perspectives:

- Where U.S. public and private ISDNs expect to apply international standards, this text discusses the international agreements.

- Where international agreements do not apply, or are applied differently in different regions of the world, the text focuses on the anticipated North American approach. Significant deviations from international practices are explained.

# 3.1   General Aspects of ISDN

In the last two years, the very words "integrated services digital networks" have assumed a power of their own. If a product or service cannot be classified as "ISDN," it seems to lack modern or forward-thinking solutions. This ISDN "jihad" sweeping through the industry has attracted the attention of not only the trade press, but also the business press and the general public. Articles extolling ISDN have appeared in the *Wall Street Journal* and *Newsweek*, among other surprising places. New product announcements must bear these magic four letters. This section shall separate true ISDN fact from marketing mumbo jumbo.

## 3.1.1   Quasi-ISDNs and the Real Thing

A distinction exists between "integrated services digital networks" (quasi-ISDNs) and "Integrated Services Digital Networks" (ISDNs). Many equipments and networks provide integrated services over a common digital network in some fashion. For example, a packet-switching public data network can provide multiple services, such as switched virtual circuits and message handling service (electronic mail), integrated over a common network backbone; surely this qualifies as "integrated services." Since data is transmitted, surely this must be a digital network. In this chapter, however, such a network will not be called ISDN unless it also meets the interface standards defined for ISDNs.

For a network to be an ISDN, it must:

- Provide multiple services (see the discussion in section 3.2 of service definitions).

- Offer digital interfaces to users and to other networks based on ISDN standards (see the discussion in sections 3.3 through 3.7 of ISDN interface standards).

## 3.1.2   Varieties of ISDNs

An ISDN is not monolithic. There will not be a single, worldwide ISDN. Rather, many ISDNs will be created from the networks that exist today.

Within the United States, many public ISDNs will evolve from the existing public telephony and data networks. Even more private ISDNs will evolve from existing private networking arrangements.

ISDNs are not expected to offer identical services. For example, an ISDN that evolves from a local exchange carrier's telephony network may initially be focused on circuit-switched services. Nevertheless, ISDN standards assume certain common features in order to allow the interconnection of different networks.

### 3.1.3 Reasons for Evolution

Existing networks are expected to evolve into ISDNs for two principal reasons: cost and service. The economies associated with digital operation encouraged the introduction of digital transmission and switching technology into networks during the last twenty years. Digital systems have proved easier to install, maintain, and modify. Inherent high quality of service makes it easier to engineer digital technologies. For data services, underlying digital transmission and switching offers more than twenty times the information-carrying capacity of the fastest voice-band modems used today on the public switched telephone network.

Additional digital technologies are associated with ISDN standards. The major digital changes include:

Digitization of the loop connecting the subscriber to the serving local switch (PABX for private networks, central office for public networks). Digital loops allow more information to be carried over the existing plant without an increased capital investment in new cable systems.

Digital signalling protocols for controlling calls and ancillary services.

ISDNs also expand on existing public network digital trends:

Increased reliance on digital transmission and switching.

Centralized control of the operation, administration, and maintenance of networks.

The introduction of additional digital technologies and the expansion of existing digital trends reduce network operating costs.

The service perspective also exerts pressure to evolve towards ISDN implementation. The organization and structure of ISDN interfaces and network architecture simplifies the introduction of new network services. For example, wider-bandwidth circuit-switched services can be introduced with only very minor additions to the ISDN signalling proto-cols and with an alteration in switch characteristics. In practice, the more difficult work and correspondingly larger expenses lie in the changes to the switching machine, rather than in the changes to the sig-nalling procedures to set up, modify, and tear down a wider-band-width call.

Similarly, an ISDN architecture can more easily accommodate variations of existing network services to meet the particular needs of subscribers.

### 3.1.4  Summary

- The International Telegraph and Telephone Consultative Commit-tee (CCITT) developed initial technical standards for ISDNs.

- ISDN technical standards are still evolving in some areas.

- In a formal sense, equipments and networks that claim to be "ISDN" should support the standardized subscriber and inter-network interfaces specified for ISDNs.

- Many ISDNs will exist. These ISDNs will be both public and private.

- Network operators are expected to implement ISDN technology because of its long-term reduction in costs and its increased abil-ity to offer new services.

## 3.2  ISDN Services and Supplementary Services

In the following discussion, special ISDN terminology is introduced. These terms have a unique meaning in the context of ISDNs, and for this reason they first appear in italic type in the text, as follows: *special term*. After a term has been introduced, however, it is no longer set in italic type; but the term continues to be used in its specific ISDN context.

### 3.2.1  Organization

*ISDN telecommunications services* form a superset of existing communica-tions offerings. CCITT standards establish a unifying framework for viewing these services. Although this framework does not impose partic-ular technical obligations on a network or user, it does provide a consis-tent philosophical approach to categorizing existing and new services.

This approach does NOT partition responsibility for telecommunica-tions. For example, the description of *message handling service* (i.e., electronic mail) does not presuppose which organization (user, private network, public network, information service provider, etc.) provides the various elements that make up a complete message handling service.

At the grossest level, telecommunications services divide into two broad categories: *bearer services* and *teleservices*. Bearer services describe the transportation of information between locations. Teleservices combine this transportation function with other information-processing functions. These additional information-processing functions may be located within a network (public or private), an external information service provider (e.g., data base service), or within user terminal equipment (such as a personal computer). For example, the transportation of a 64-Kbps (kilobits per second) bit stream between specified locations and without modification is a bearer service. A message handling service, however, is a teleservice; it combines bearer services (such as the 64-Kbps transportation just mentioned) with information processing (message composition, storage, retrieval, directory search, etc.).

A *supplementary service* adds a certain twist to a teleservice or bearer service, but it cannot stand on its own as an indivisible service offering. For example, *reverse charging* (collect calling) is a supplementary service. Reverse charging may be used with a bearer service, such as the 64-Kbps transportation already mentioned. Reverse charging may also be used with a teleservice, such as the message handling service, to create "collect messages."

As an objective, an ISDN supplementary service is defined and implemented in a manner independent from the associated bearer services and teleservices. This allows each supplementary service to be used in combination with many bearer services and teleservices, without requiring a special implementation of that supplementary service in each case. An example would be the methods of requesting and authorizing reverse charging, which are intended to be the same for both an electronic message and a simple voice call.

## 3.2.2  Attributes

Standard *attributes* describe in more detail bearer services, teleservices, and supplementary services. Attributes make it easier to compare different services. Bearer, tele-, and supplementary services all have their own list of attributes. At present, the bearer service attributes are the most precisely defined and shall be used to illustrate this organization.

Thirteen attributes define a bearer service, broken into three groups as shown in Fig. 3.1. Fig. 3.2 lists possible values of each attribute. The *dominant attributes* shown in Fig. 3.1 provide the basic names and categories for bearer services. Examples of ISDN bearer services are described next.

Fig. 3.1.
Categories of
attributes.

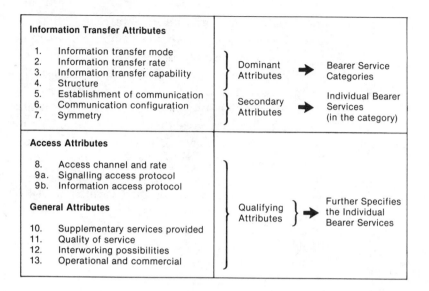

## Circuit-Mode, 64-Kbps, Unrestricted, 8-kHz Structured

The *circuit-mode, 64-Kbps, unrestricted, 8-kHz structured* bearer service provides information transfer between users without any alteration of the bit stream. Users may choose any application for this bit stream, including their own private submultiplexing system. For instance, two PABXs (private automatic branch exchanges) may employ this bearer service to transmit two 32-Kbps ADPCM voice channels; see Fig. 3.3. As another possible application, transparent access from a CCITT Recommendation X.25 DTE through an ISDN to a packet-switching public data network employs this bearer service to exchange X.25 packets; see Fig. 3.4.

Certain North American T1 network transmission systems cannot support such a bearer service, since these systems may lose synchronization when transmitting a large number of consecutive "0" bits. Several methods of approaching this problem have been devised. One method simply acknowledges this limitation and redefines the bearer service to be "restricted" rather than "unrestricted"; the restriction is that "0000 0000" cannot be transmitted. Other methods substitute another bit pattern for "0000 0000" within the transmission system. The substitution is identified at the transmission system terminating receiver by a unique violation of the bipolar electrical transmission rule used in T1 or by other special transmitted indicators. With these substitution techniques, the full 64-Kbps unrestricted bearer service can be provided.

The phrase "8-kHz structured" refers to the inherent structure of the information transferred by an ISDN in this bearer service. When one user transmits information to another user, the transmission is accompa-

**Fig. 3.2. Values for each bearer service attribute.**

| Possible Values of Attributes | | | | | | | | | | Attributes[f] |
|---|---|---|---|---|---|---|---|---|---|---|
| | | | | | | | | | | **Information Transfer Attributes** |
| Circuit | | | | | | Packet | | | | 1. Information transfer mode |
| Bit rate (kbit/s) | | | | | | Throughout | | | | |
| 64 | 384 | 1536 | 1920 | Other values for further study | | Options for further study | | | | 2. Information transfer rate |
| Unrestricted digital information | | Speech | 3.1 kHz audio | 7 kHz audio | 15 kHz audio | Video | Others for further study | | | 3. Information transfer capability |
| 8 kHz integrity | | Service data unit integrity | | | | Unstructured | | | | 4. Structure |
| Demand | | Reserved | | | | Permanent | | | | 5. Establishment of communication[e] |
| Point-to-point | | Multipoint | | | | Broadcast[a] | | | | 6. Communication configuration |
| Unidirectional | | Bidirectional symmetric | | | | Bidirectional asymmetric | | | | 7. Symmetry |
| D(16) | D(64) | B | H0 | H11 | H12 | Others for further study | | | | **Access Attributes** 8. Access channel and rate |
| I.440 | I.451 | I.462 | | Others for further study | | | | | | 9a. Signalling access protocol |
| G.711 | G.721[c] | I.460 | I.451[d] | X.25 | Others for further study | | | | | 9b. Information access protocol |
| Under study | | | | | | | | | | **General Attributes** 10. Supplementary services provided 11. Quality of service 12. Interworking possibilities 13. Operational and commercial |

*Notes:*

[a]*The characterization of the information transfer configuration attribute "broadcast" is for further study.*

[b]*The need for a "data sequence integrity" attribute is for further study.*

[c]*The use of Recommendation G.721 as an information access protocol is for further study.*

[d]*The use of Recommendation I.451 as an information access protocol is for further study.*

[e]*A provisional definition of the establishment of communication is given in Recommendation I.130. Further clarification is required.*

[f]*The attributes are intended to be independent of each other.*

**Fig. 3.3. Use of circuit-mode, 64-Kbps unrestricted, 8-kHz structured bearer service to carry two 32-Kbps ADPCM voice channels.**

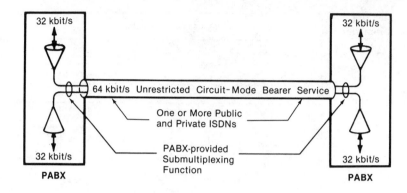

**Fig. 3.4. Use of circuit-mode, 64-Kbps unrestricted, 8-kHz structured bearer service to access X.25 packet service.**

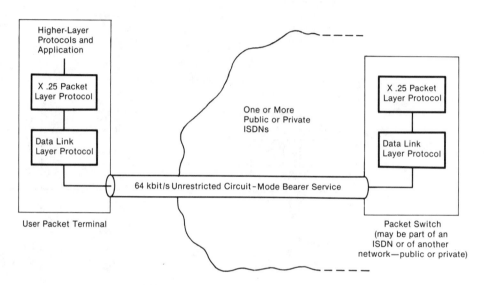

nied by 8-kHz timing information. Each cycle of timing information delimits a group of bits within the circuit; at 64-Kbps with 8-kHz timing, each time interval delimits 8 bits (an *octet*). Each octet is always delivered by the network within a corresponding time interval; that is, the octet is never split across a time interval boundary. This makes it possible for the users to exchange characters, 8-bit speech samples, etc., without an in-band, user-user synchronization scheme within the circuit.

### Circuit-Mode, 64-Kbps, 8-kHz Structured, Usable for Speech

The *circuit-mode, 64-Kbps, 8-kHz structured, usable for speech* bearer service is similar to the previous one, except that ISDNs assume the information in the circuit is human speech. An ISDN may route such a connection in any fashion appropriate for speech—even over analog lines. Furthermore, automatic conversion of the digital bit stream may be done; for example, between the $\mu$-law speech encoding algorithm (used in North America and Japan) and the A-law algorithm used on public networks in the rest of the world. (See Chapter 8 for an explanation of these algorithms.)

This brings up an important point about the access attributes. Unlike the information transfer attributes, access attributes may be different at each end of a connection. For example, a speech connection between the United States and Europe may have an access protocol of $\mu$-law for the user's speech at the U.S. end, and A-law at the European end. The access protocol attribute value is different at each end.

### Circuit-Mode, 64-Kbps, 8-kHz Structured, Usable for 3.1-kHz Audio Information Transfer

The *circuit-mode, 64-Kbps, 8-kHz structured, usable for 3.1-kHz audio information transfer* bearer service differs from the previous bearer service in that "3.1-kHz audio," rather than "speech," is supported. Calls with the 3.1-kHz audio bearer service may still be routed over analog circuits, and the network still performs conversion between digital encoding laws.

A speech call, however, may also be routed over special multiplexing systems, such as time-assigned speech interpolation (TASI), which use the bursty characteristics of human speech to squeeze more conversations on a given transmission system. For example, forty "speech" calls can be squeezed into a T1 twenty-four–channel system without a noticeable degradation in any particular call by removing the natural pauses and silences in each conversation. Significant degradation will appear, however, if some of these calls are voice-band data modems or frequency-division-multiplexed telegraph packages, for instance, rather than true human speech. The degradation occurs because modems and FDM packages transmit continuously, rather than in conversational bursts.

If analog lines are used, echo control systems may be included in the transmission facilities in order to suppress undesirable echoes. These echo control systems often must be disabled when voice-band data modems are used. The modems will transmit a special tone in this case.

By distinguishing speech and audio bearer services, the network can more easily route each call through the appropriate transmission facilities. Calls using speech bearer services may then be routed through analog circuits and TASI systems, and may employ echo control techniques. Voice-band data, frequency-division-multiplexing (FDM) packages, etc., utilize the audio bearer service and thereby avoid TASI-like systems, echo control devices, and related equipment that may impair the quality of information transmission.

### 3.2.3  Essential and Additional Bearer Services

CCITT has categorized bearer services into *essential* and *additional*. Essential bearer services are expected to be available internationally from public networks. All other bearer services are considered additional; that is, provided by some public networks at their discretion.

In this sense, the following bearer services are considered essential:

- Circuit-mode, 64-Kbps, 8-kHz structured, unrestricted.
- Circuit-mode, 64-Kbps, 8-kHz structured, speech.
- Circuit-mode, 64-Kbps, 8-kHz structured, 3.1-kHz audio.
- Packet-mode, virtual call and permanent virtual circuits.

For each of the essential circuit-mode bearer services, a standard method of obtaining these services at the user-network interface has also been selected as essential. However, such an essential access method for packet-mode bearer services has not been selected to date. Instead, a number of alternative packet-mode bearer service access techniques have been partially described in standards. Within the United States, an urgent effort is being made to finalize and select one or more essential methods for packet-mode service access.

The following bearer services are considered additional:

- Circuit-mode, alternate speech/64-Kbps nonspeech, 8-kHz structured.
- Circuit-mode, 384-Kbps, 8-kHz structured, unrestricted.
- Circuit-mode, 1536-Kbps, 8-kHz structured, unrestricted.
- Circuit-mode, 1920-Kbps, 8-kHz structured, unrestricted.

### 3.2.4  Teleservices

The description methodology for teleservices is not as well established as that for bearer services. At present, teleservice attributes focus on the type of user information and on the various higher layer protocols employed.

### 3.2.5  Summary

- ISDN standards describe services by categorizing them as bearer, tele-, and supplementary services.
- Bearer, tele-, and supplementary services are further described by their particular service attributes. The attribute values are presently most detailed for bearer services.

- Certain bearer services are classified as essential, meaning that they are expected to be available internationally from public ISDNs. All others are additional (optional).

# 3.3   ISDN Organization on the User's Premises

ISDN standards also embody an organizational philosophy for user equipment. This philosophy in turn suggests standard interface locations, and it partitions the functions performed by user equipment and private networks.

## 3.3.1   Functional Groupings

As shown in Fig. 3.5, user equipment divides into four principal functional groups:

1. Terminal equipment (TE).
2. Terminal adaptor (TA).
3. Network termination 2 (NT2).
4. Network termination 1 (NT1).

**Fig. 3.5.**
**Reference**
**configurations**
**for the ISDN**
**user-network**
**interfaces.**

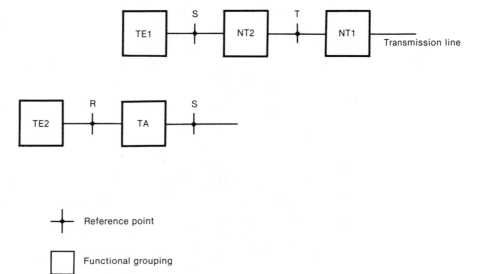

*Terminal equipment* (TE) encompasses those devices that actually generate and receive information: telephones, personal computers, data terminals, thermostats, mainframe computers, point-of-sale systems, etc. The type of interface employed by each TE further categorizes it as a *TE1* (which has a CCITT-standardized ISDN-compliant interface) or *TE2* (everything else with a non-ISDN interface).

A *terminal adaptor* (TA) converts non-ISDN interfaces to a CCITT-standardized ISDN interface. Any TE2 can, therefore, be attached to a public or private ISDN through a suitable TA. *Reference point R* refers to the division between a TE2 and its TA. An interface at reference point R may be a standardized interface, such as that described in CCITT Recommendation X.25. It may also be an analog interface, such as that to an analog telephone. Interfaces at reference point R may also be non-standardized.

*Network terminations 1 and 2* are technobabble left over from early phases of ISDN discussions. Network termination 2 (NT2) refers to on-premises switching and other intelligence employed by the user for communications. NT2s have ISDN-standard interfaces to TE1s and TAs. A PABX and a LAN (local area network) may contain NT2 functions. A computer system that processes electronic mail for local users (e.g., user agent or message transfer agent, as defined by message handling system standards) may also be an NT2. An NT2 function is separated from a TA or TE1 function by *reference point S*. Interfaces present at reference point S are always ISDN-standard.

Network termination 1 has a very restricted function: to connect the user's equipment to the digital-subscriber line transmission system. *Reference point T* designates the separation between NT1 and NT2 functions. Reference point T interfaces are also ISDN-standard. The NT1, therefore, converts the electrical signals received from reference point T into the appropriate digital subscriber line electrical systems. NT1s do not include higher-layer intelligence; in the terminology of the Open Systems Interconnection reference model, an NT1 is strictly a physical layer device, as far as user information is concerned.

The term "network termination" was originally selected because, in many countries, the equipment performing these two sets of functions is owned by the government's monopoly telecommunications network. In the early days of ISDN deliberations, only one network termination existed; but, at the insistence of the U.S. delegates, this functional group was split into NT1 and NT2. For public ISDNs, delegates envisioned the local public ISDN providing the NT1 in the United States. But in the last few years, FCC regulatory decisions decreed that the NT1 function could be customer-provided equipment. As a result, a standards-development activity within North America has tackled "reference point U," the attachment between the NT1 and the digital-subscriber transmission line system. While "reference point U" is not an internationally sanctioned term for this interface location, it is convenient and will be used in this chapter.

## 3.3.2 Relation between Functional Groupings, Reference Points, and Hardware

NT1, NT2, TE1, TE2, and TA only describe groups of functions, not actual equipment. A real piece of equipment may implement all of a functional group, a portion of a functional group, or more than one functional group. Some examples follow to illustrate this point.

A PABX may implement NT1 functions by terminating T1 carrier digital transmission systems between it and the local public network's central office. The PABX may also implement NT2 functions by providing on-premises switching for its users. Such an ISDN PABX would use standardized ISDN interfaces at reference point S to connect to individual user devices such as digital telephones. An ISDN PABX may go even further and implement TA functions in order to attach to TE2s, user devices without a standard ISDN interface (e.g., analog telephones). See Fig. 3.6.

Fig. 3.6.
Example of an ISDN PABX and its relation to standard ISDN functional groups and reference points.

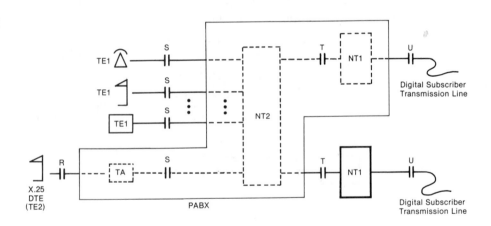

Another example uses a LAN to connect a user data terminal to a PABX. Assume that the LAN supports an ISDN standard interface. A data terminal with such an interface would connect to the LAN, and that interface could be viewed as reference point S. The LAN may attach to the PABX using a special internal bus interface. In this case, the LAN and the PABX together implement the NT2 functions; the interface between the LAN and the PABX is internal to the NT2 functional group and need not be standard. See Fig. 3.7.

**Fig. 3.7.
Example of LAN
plus PABX used
to implement
NT2 functional
group.**

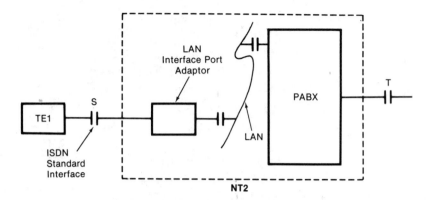

In some configurations, the allocation of equipment to functional groups, and of real interfaces to reference points, depends on the view-point adopted. For example, a multinational corporation's private ISDN may be described the same way that a public ISDN would be described, with NT1, NT2, etc., at the internal corporate users' locations and with an internal network organization of switching, transmission, and control facilities. The same international assembly of equipment, viewed from the perspective of a public network, would be interpreted as implementing NT2 functions. See Fig. 3.8.

**Fig. 3.8. Private
network as an
NT2.**

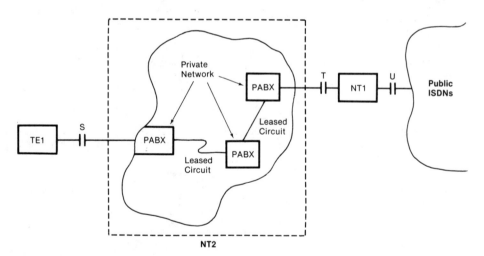

NT2 and TA functions are optional. In a simpler configuration, the user may not have any on-premises equipment implementing these func-tions. A digital telephone could be connected directly to an NT1 without any intervening equipment implementing NT2 functions; see Fig. 3.9. In this case, both reference points S and T refer to the interface between the digital phone and the NT1; see Fig. 3.10.

**Fig. 3.9. No NT2 at customer premises.**

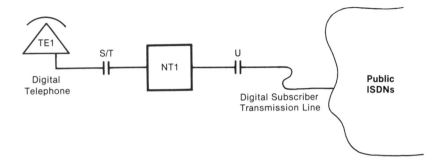

**Fig. 3.10. Both reference points S and T represented by a single interface.**

User terminal equipment may be mono-media (e.g., speech-only) or *multimedia*, such as a combined voice and data terminal. Multimedia ISDN calls can be created by combining several services. The signalling methods for this combination are still evolving; however, integrated user terminal equipment can achieve the same effect simply by establishing multiple calls, each with its own service characteristics.

### 3.3.3  Summary

- Customer-premises equipment is categorized according to function(s): terminal equipment, terminal adaptor, network termination 1, and network termination 2.

- Standard ISDN interfaces are utilized at reference points S and T.

- Within the United States, the FCC has decreed that the customer may own the NT1. A family of standard interfaces at reference point U is presently under development.

- An individual piece of equipment may implement some or all of the functions of a particular functional group.

- An individual piece of equipment may implement one or more functional groups. When one piece of equipment incorporates more than one functional group, the reference point between those groups becomes internal to that equipment. Such an internalized reference point may not have a corresponding physical interface.

- NT2 and TA functions are optional.

## 3.4   General Aspects of User-Network Interface Standards

The marketplace has yet to see the first equipment implementing ISDN standards for reference points S and T in a fully compliant manner, in part because some aspects of the call control signalling procedures are still maturing. However, many terminal, PABX, and central office equipment manufacturers are already developing equipment that is expected to eventually comply with these standards. Integrated-circuit manufacturers are also preparing devices to implement aspects of these standards.

Various ISDN pilot trials are being conducted in many countries around the world. Almost all of these trials are quasi-ISDN, in the same sense that fully compliant ISDN standard interfaces have not been used throughout the pilot systems. The ISDN interface standards may be further refined, based on the results of these trials. Fully compliant equipment will probably begin to appear after 1988.

## 3.5   User-Network Interface Channels and Structures

To avoid a proliferation of standards, a conscious effort to restrict the number of ISDN standard interfaces has occurred. At present, there are only three such user-network interfaces. The *basic-rate interface* has an information-carrying capacity of 144 Kbps, full duplex. The *primary-rate interfaces* have an information-carrying capacity of either 1536 Kbps (expected to be prevalent in the United States, Canada, and Japan) or 1920 Kbps (for other parts of the world).

*Information-carrying capacity* describes the aggregate information rate available to the user and differs from the raw, physical transmission line rate across the interface. For example, T1 transmission systems operate at a transmission line rate of 1544 Kbps; their information-carrying capacity totals 1536 Kbps (twenty-four channels each at 64 Kbps). The remaining 8-Kbps overhead provides synchronization, error detection, alarm, and other maintenance supervision functions.

The following table compares the information-carrying capacity with the standard S/T interface transmission line rates:

|  | Information-Carrying Capacity | Interface Transmission Line Rate |
|---|---|---|
| Basic Rate | 144 Kbps | 192 Kbps |
| Primary Rate | 1.536 Mbps | 1.544 Mbps |
|  |  | (North America and Japan) |
|  | 1.920 Mbps | 2.048 Mbps (rest of world) |

CCITT standards describe the channelization of basic-rate and primary-rate interface information-carrying capacity. Specific *channel types* define a speed and application for this capacity. Various channel types are combined to form an interface's *channel structure.*

## 3.5.1 Channel Types

The *B-channel* operates at 64 Kbps and carries user information during the active phase of a call. The type and format of the user information depend on the nature of the user terminal equipment and the type of call. The B-channel itself does not impose any restrictions on the type and format of information, as long as it is digital and provided at a 64-Kbps rate.

The *D-channel* operates at 16 Kbps (for basic-rate interfaces) or 64 Kbps (for primary-rate interfaces). The D-channel carries signalling information to control all calls on the interface (and possibly other associated interfaces as well). In addition, the D-channel may also carry packetized data or low-speed (100 bits per second) telemetry at times when no signalling information is waiting.

The *H-channel* family describes a set of higher-speed channels available for user information during the active phase of a call. H-channels differ from B-channels only in their size. Presently defined are: *H0-channel*—384 Kbps; *H11-channel*—1536 Kbps (sometimes shown as *H1a-channel*); and *H12-channel*—1920 Kbps (sometimes shown as *H1b-channel*). The exact definition of channels for broadband operation (above 2 Mbps) is presently under study.

## 3.5.2 Basic-Rate Channel Structures

The basic-rate interface's 144-Kbps information-carrying capacity separates into two B-channels and one D-channel. Standard notation describes this channel structure as *2B + D*. On a particular interface, however, one or both of the B-channels may be unused. In this case, the corresponding information-carrying capacity remains idle, and the notation changes to B + D, or just D, respectively. Even when some channels are not activated, the basic-rate interface continues to operate at the 192-Kbps transmission line rate.

For many user TEs, a B + D channel structure would provide sufficient capacity. The technology for producing a 2B + D interface within the subscriber's premises did not seem significantly more difficult or expensive than for a B + D interface, however. Since greater economies of scale would be realized by implementing large quantities of the same interface, ISDN standards have focused on limiting the set of interfaces. Therefore, all TEs that require up to 144-Kbps information-carrying capacity will employ the same 2B + D channel structure at the interface.

A different story exists on the digital-subscriber transmission line. Depending on the distance between the user's premises and the local central office, the type of wiring, etc., several different transmission technologies might prove economical in various circumstances. For instance, if a particular user requires only that B + D be activated, then a simpler and less costly digital-subscriber transmission line system could be possible. This difference (among others) is hidden from the user by the NT1, which converts the standard interface at reference point T with its 192-Kbps line rate into such a lower line rate for the two-wire digital-subscriber transmission line.

### 3.5.3   Primary-Rate Channel Structure

Because of existing differences in digital-transmission-rate hierarchies around the world, agreement on a single primary-rate channel structure could not be obtained. Within Japan, Canada, and the United States, standard structures build on the 1.544-Mbps transmission line rate. Most early implementations use the 23B + D channel structure. Other combinations constructed with H-channels can also be configured on a specific interface. Examples include: 4H0, H1a, and 3H0 + 5B + D.

As with the basic channel structure, a particular device may not require the full information-transfer capabilities of a primary-rate interface. Again, a smaller structure would be used in this case, and the remaining capacity at a particular interface would lie idle. An example of such a smaller structure would be 10B + D; the other thirteen 64-Kbps time slots would remain unused.

Primary-rate interfaces in other parts of the world employ the 2.048-Kbps transmission system. Typical structures include: 30B + D, 5H0 + D, and H1b + D.

In some cases more than one primary-rate interface connects between equipment; for example, a PABX may use several primary-rate interfaces between it and the local central office. *Consolidated signalling* allows all signalling traffic to be concentrated on a single 64-Kbps D-channel selected from the entire set of interfaces. For example, with three primary-rate interfaces, use (23B + D) + (24B) + (24B) for a total of seventy-one B-channels and one D-channel. Typically, however, one additional 64-Kbps time slot on a different interface sits in reserve as a backup D-channel in case of failures: (23B + D) + (23B + backup D) + (24B). The D-channel signalling procedures control changeovers to the backup.

The preceding channel definitions assume 64-Kbps clear channel capability. Many North American 1.544-Mbps transmission systems cannot transmit 64-Kbps information containing octets of zeros (0000 0000) because the receiver may lose synchronization. The CCITT standards for 1.544-Mbps primary-rate interfaces at reference points S and T employ B8ZS, a technique that substitutes a unique transmission sequence for the all-zeros octet. The unique transmission sequence contains a bipolar violation, a further complication because some existing North American 1.544-Mbps transmission systems cannot pass bipolar violations. Various clever approaches to overcoming this further limitation have been invented.

### 3.5.4 Other Channel Structures

At present, standards have not been developed for channel structures faster than the primary rate. Evaluation of *broadband channels* continues for rates in the order of tens of megabits per second and in the order of hundreds of megabits per second. These wideband channels will support applications such as digital television. CCITT discussions continue to pursue a uniform, worldwide set of wideband channel sizes and related interface structures. in order to avoid at these higher speeds the present unfortunate historical and geographical division existing at the primary rate.

Other discussions involve sub–64-Kbps channels at 32, 16, and 8 Kbps. New encoding techniques allow voice signals to be encoded at 32 Kbps with similar quality to today's telephone network, and 16-Kbps techniques are also being evaluated. These frugal voice transmission systems have stimulated interest in *subrate switching* at submultiples of 64 Kbps. Subrate switching is especially attractive to private networks, since these techniques would cut recurring leased circuit costs to 50% or 25% of their present levels. Subrate services are also attractive to many circuit-switched data requirements where the larger capacity of a 64-Kbps channel is not required.

Exploratory talks have just begun on new switching technology, variously referred to as burst switching, asynchronous time-division multiplexing, fast packet switching, and mini-packet protocols—but generally lumped (for the moment) under the term "sexy new packet protocols."

Sexy new packet protocols resemble existing packet procedures in the use of statistical multiplexing techniques, but they vary in two significant aspects.

1. Control of individual calls is accomplished by out-band signalling techniques; for example, using the D-channel signalling protocols discussed in sections 3.7.6 through 3.7.11.

2. The functions performed by the packet switch may be selected on a call by call basis. For example, error recovery and flow control techniques for packet data are dispensed with for voice calls but would be available for data calls.

Sexy new packet protocols on broadband interfaces could eliminate the need for any channelization of the aggregate information transfer rate in certain situations.

### 3.5.5  Summary Poem

The following poem was created on July 2, 1983, near the end of one of the CCITT meetings in which Recommendations I.411 and I.412 were developed. These CCITT Recommendations contain the detailed description of reference points, channels, and channel structures previously outlined.

Oh, a channel fit for the user's bits
Is the wonderful channel B.
When transparent it delivers what's sent
And at a very low fee.
At sixty-four k this can be the way
To get user data through.
When nontransparent your voice can be sent
Using coding laws A or mu.

To make a connect so things can be sent
You need this other channel "D."
The signalling goes in a very smooth flow
With packet information "p."[1]
The speed is defined at a later time
With channel structures shown below.
Link level should be procedure LAP D
Which is one that we all love and know.

With higher speeds there is a need

---

1. See section 3.7.

For the larger channel H-zero.
Three hundred eighty-four kilobits (no more)
Is the rate at which this channel goes.
The programme sound encoding is found
Defined by Study Group fifteen.[2]
But the user may find that various kinds
Of submultiplexing solve his needs.

There's lots of fun when controversy comes
On the speed used for H-one.
This channel size cannot be disguised:
Nineteen-twenty kilobits is one.
But some of us say North America's way
Is the path out of this tough fix.
And NTT[3] does also agree
That it should be fifteen thirty-six.

A structure shows how the channels go
Across the user interface.
The basic one that can be done
Uses D at sixteen k.
Two B-channels and one D-channel
Make up the basic channel scheme.
The capability that the user sees
May have one or both B's missing.

The primary rate is really the state
Where things are confusing me.
There seems no end to the structure when
You assemble B, H, and D.
A proposal from France[4] left open the chance
Of a capability in-between:
Six channels wide along beside
A D at perhaps sixteen.

At greater speeds there is the need
For further talk—that is clear.
The Japanese want a secondary slot
But the end of study is quite near.
And thus we'll agree there really should be

---

2. CCITT Study Group XV is developing methods for encoding stereo audio at 384 Kbps with sufficient quality to be used for radio broadcasting.
3. *NTT* stands for Nippon Telegraph and Telephone Public Corporation.
4. Later rejected by CCITT.

Some days in the future when
We'll spend some time in order to refine
This I. four-twelve text again.

# 3.6   Physical Aspects of User-Network Interfaces

Present international standards describe physical interfaces using metallic circuits. The refinement of broadband interfaces will likely stimulate definition of optical interfaces for these faster rates.

## 3.6.1   Connectors

Basic-rate interfaces employ the eight-pin modular plug and jack. This compact connector was attractive to those who had struggled with RS-232 and other "D-series" plugs that protrude from equipment and require screwdrivers for installation.

The center four pins (mandatory) carry transmit and receive signals on electrically balanced wire pairs. See Fig. 3.11. A phantom circuit formed by these two wire pairs and their isolation transformers carries some dc power; the network side of the interface supplies a smaller amount of emergency dc power with reverse polarity during times when no local power is available. The outer four pins (optionally wired) may be used to carry additional power in each direction. When these optional pins are not wired, the ubiquitous U.S. four-pin mini-mod jack may be used; it snaps into an eight-pin socket and connects to the center four pins.

The eight-pin connectors are also being considered for the primary-rate interface. Some people laughed at the idea of portable equipment using primary-rate interfaces; but, as video, teleconference, and high-speed fax equipment shrinks in size, this decision to simplify attaching and detaching primary-rate equipment now seems reasonable. The use of the same connectors and wiring arrangements as those used for the basic-rate interfaces also avoids rewiring the building when primary-rate equipment is deployed. The matter has not been finalized, however. Primary-rate interfaces at 1544 Kbps are modelled after the T1 carrier transmission system, which carries 130 volts dc. It is not certain that Underwriters Laboratories will certify the modular plug and jack at these voltage levels.

Another issue being studied is whether the primary-rate signal interchange circuits must always be terminated. If this is required, then some form of automatic termination must be included in the plug so that termination is reliably provided when the jack is removed.

**Fig. 3.11. Reference configuration for signal transmission and power-feeding in normal operating mode.**

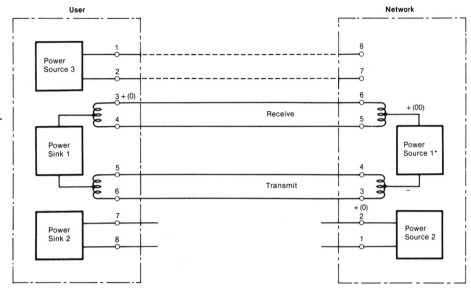

Notes:

*The numbering of leads in this figure does not imply any assumption on pin allocation or physical connectors.*

*Maintenance of polarity on a wired pair cannot be guaranteed in all cases. This must be taken into consideration for terminals drawing power from access leads 7-8.*

*\*Polarity reversed during emergency power situations. Only user equipment that must continue to operate in these situations would draw current from a reverse-polarized source.*

(0) Refers to the polarity of positive pulses
(00) Refers to the polarity of power

## 3.6.2 Wiring Arrangements

The standards permit only one wiring arrangement for a primary-rate interface: point-to-point between two equipments. The equipment on the side of the interface closest to the end user is referred to as the "user side"; the other equipment is referred to as the "network side" of the interface. The terms do not designate ownership. For example, an NT2 PABX is the network side of an interface at reference point S; the TE (or TA) is the user side of the interface at reference point S. Both the NT2 and the TE/TA may be owned by the same organization (e.g., the public ISDN customer). Similarly, at the interface at reference point T, the NT2 becomes the user side; see Fig. 3.12. The terms *user side* and *network side* refer to the entire combination of equipments on each respective side of the interface. This crucial point allows electrical signals, signalling messages, and other information to be processed at arbitrary points within the respective sides. For instance, on the network side of an interface at reference point T, the electrical signals will be terminated in the NT1.

However, the call-control signalling messages will be terminated further into the network; for example, at the central office or even at a distant network control point. Each side of the interface has complete freedom to design the allocation of functions among its hardware as convenient; the standards are not restrictive on this point.

Fig. 3.12.
Network side
versus user side
of interface.

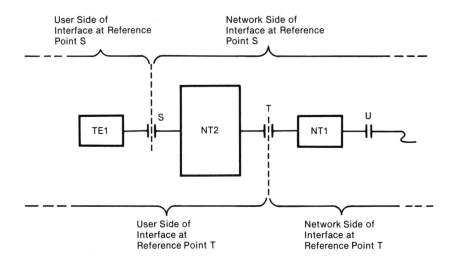

Primary-rate interfaces may be used at any reference point. A common misconception is that only reference point T may operate at the primary rate. A typical PABX installation may employ one or more primary-rate interfaces at reference point T; however, interfaces at reference point S may also operate at the primary rate.

Basic-rate interfaces may also use a point-to-point wiring arrangement, as long as the electrical signal generators and receivers are within 1 kilometer. Alternately, up to eight basic-rate user interfaces may share a passive bus. A passive bus bridges these interfaces on the same set of wiring, without requiring active electronics (although signal amplifiers and regenerators are permitted). In the passive bus configuration, only one network side exists. The eight user interfaces must be clustered within 100 to 200 meters (m) of each other (depending on cable impedance), and the cluster must be located within 500 m of the network side interface. See Figs. 3.13, 3.14, and 3.15.

### 3.6.3   Basic-Rate Interface Multiplexing Methods and Contention Resolution

Since multiple information streams (B-channels and the D-channel) must be transmitted over a single pair of wires on the basic-rate interface, a

Fig. 3.13. Point
to point.

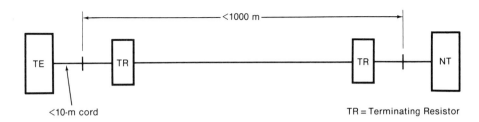

Fig. 3.14. Short
passive bus.

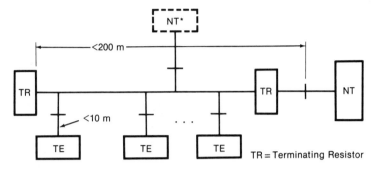

*In principle the NT may be located at any point along the
passive bus. The electrical characteristics in Recommendation I.430,
however, are based on the NT located at one end. The
conditions related to other locations require confirmation.

Fig. 3.15.
Extended
passive bus.

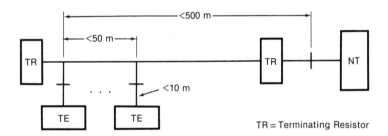

time-division-multiplexing structure must be used to interleave information. Fig. 3.16 presents the multiplexing format for the basic-rate interface.

The passive bus arrangements permitted for the basic-rate interface introduce certain constraints on the design of the interface. Although up to eight user equipments may be bridged onto a passive bus, only one user may transmit on a given B-channel. Authorization to use a B-channel must be received from the network side of the interface prior to use. The D-channel call-control signalling procedures (described in CCITT Recommendations I.450 and I.451) provide the mechanisms for authorizing use of B-channels and for relinquishing B-channels at the conclusion of calls.

Transmissions from each of the user equipments must be carefully

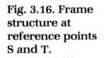

Fig. 3.16. Frame structure at reference points S and T.

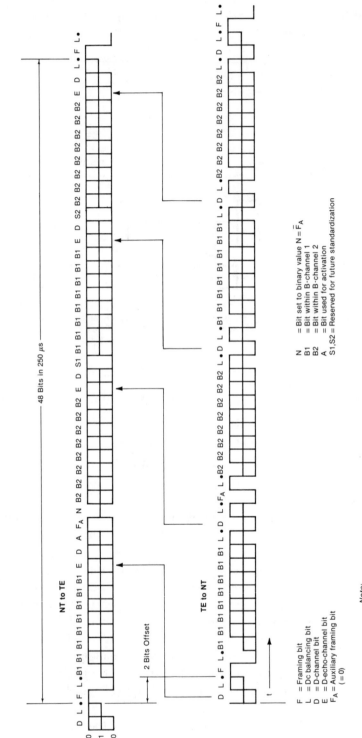

synchronized in order to prevent one equipment's transmission from interfering with another's. Each user equipment slaves its transmission to timing information received from the network side of the interface. All user equipments transmit in synchrony, to ensure that the authorized user equipment transmitting on each B-channel will not be interfered with by the other user equipments on the interface.

Any user equipment can transmit on the D-channel, however; and contention must therefore be resolved. *Contention resolution* relies on the unique electrical characteristics of the interface and on the *D-echo channel.* The network side transmits the D-echo channel, simply repeating each bit heard on the D-channel. The multiplexed frame structure distributes D-channel transmission evenly over time. Each bit transmitted by user equipment on the D-channel is echoed back by the network side before the next D-channel bit may be transmitted. This arrangement permits the following contention resolution principles:

- No user equipment begins transmission on the D-channel until the channel has been idle for a specified number of bits. The *idle channel condition* is defined as a continuous stream of more than seven "1" bits. The D-channel data-link layer protocol prevents this idle pattern from appearing during the middle of a transmission.

- After the specified idle period has passed, any user equipment may start transmitting. After each bit has been sent, the same bit value must be heard as the next D-echo bit. If the same bit value is heard, the user equipment may continue transmission; otherwise, it must immediately stop sending on the D-channel and wait for the idle period before attempting transmission again.

The electrical characteristics of the interface are such that any user equipment transmitting a "0" bit will override user equipments transmitting "1" bits at the same instant. This arrangement ensures that one user equipment will be guaranteed successful completion of its transmission; that is, destructive interference does not occur. In fact, because of the structure of the transmitted data-link layer frames, the contention resolution process boils down to a single transmitting user equipment by the time three octets have been sent.

In order to prevent one user equipment from hogging the D-channel, equipment successfully completing one transmission must wait for a slightly longer idle period (generally one extra bit) before attempting a new transmission. After eventually seeing this longer idle period, the equipment resets to look for the normal idle period.

A further convention guarantees priority for signalling traffic over all other D-channel traffic (X.25 packets, etc.). When user equipment wishes to transmit signalling information, it waits for a shorter idle period (8 bits normally) than for other types of information (10 bits normally).

The time-division-multiplex frame structure of the basic-rate interface and the multipoint configuration create an asymmetrical interface. An integrated circuit implementing this interface must behave differently when operating as the user side of a basic-rate interface than when operating as the network side.

## 3.6.4  Primary-Rate Interface Multiplexing Methods

The time-division-multiplex frame structures used by 1544- and 2048-Kbps primary-rate interfaces follow the corresponding digital transmission system standards closely. Since these interfaces are point to point, no contention resolution mechanism is required and there are no D-echo bits. The primary-rate interfaces are fully symmetrical in this sense; implementations of the physical layer specification may be used on either the network or the user side of the interface. A minor asymmetry exists about timing. The user side of the interface extracts bit timing information from the received primary-rate bit stream, and clocks its transmissions to this derived rate. The network side of the interface is generally expected to obtain timing information from the network's synchronization plan; that is, from a source outside that particular user-network interface.

The obligations to slave timing to the network side can introduce an ambiguity for NT2s (e.g., PABXs) attached to more than one network. See Fig. 3.17. Timing information provided by Network A will not be identical to timing information provided by Network B in the figure. On which basis should the NT2 provide timing to the TEs? If the TE is connected to a call on Network A, but the timing information is derived from Network B, then a disparity between the timing rates will cause bits to be lost from, or spuriously inserted into, the call (*slips*). The dilemma becomes clearer when considering a TE with several simultaneous calls, each through different networks, or a passive bus with TEs connected to different calls routed through different networks. To avoid slips, the NT2 must implement *pleisiochronous* timing techniques, providing an averaged timing to the TEs and using flexible internal buffering to smooth out short-term differences in timing between networks. The size of the internal buffers is chosen by taking into consideration the short-term stability and long-term accuracy of each network's timing and by selecting a sufficiently large size so that slips rarely occur.

Fig. 3.17.
Conflicting
timing
information in
multinetwork
environment.

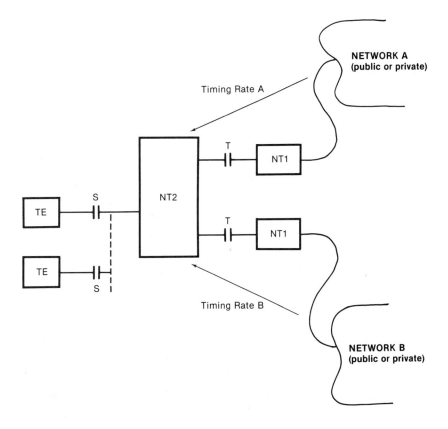

### 3.6.5  Activation and Deactivation

Basic-rate interfaces allow user-side equipment to be partially or fully deactivated when no calls are in progress. Deactivated equipment consumes less power from the interface (less than 25 mW). A bootstrap sequence of signals may be initiated by the network side (e.g., just prior to delivering an incoming call) or by the user side (e.g., just prior to originating an outgoing call) in order to activate the interface.

Primary-rate interfaces do not as yet support automatic activation and deactivation.

### 3.6.6  New Areas of Investigation

In order to increase the economies of scale for ISDN interfaces, only these three interfaces (basic rate and two versions of the primary rate) have been defined. Standards groups have begun exploring higher-speed interfaces. To avoid the dichotomy between the North American–Japanese and European digital-transmission-rate hierarchies, the next larger size of ISDN interface may operate at about 150 Mbps. The exact speed and channelization (if any) of this interface has not been finalized. Such an interface might exploit optical-fiber technology.

### 3.6.7  Summary

- Three ISDN interfaces provide information-carrying capacities through the megabits-per-second range. Each of these interfaces employs time-division multiplexing to channelize this capacity. User information channels operate at 64 Kbps, 384 Kbps, 1536 Kbps, and 1920 Kbps (this latter rate available only in some parts of the world). Each interface provides a separate channel at 16 or 64 Kbps to carry all signalling traffic.

- All of these interfaces support a point-to-point arrangement between user-side and network-side equipment. The basic-rate interface (the smallest in terms of information-carrying capacity) also can support up to eight user-side equipments bridged onto the wiring of a single interface. Under this passive bus multipoint arrangement, each user information channel is never shared simultaneously. A contention resolution mechanism permits sharing of the signalling channel.

- Equipment attached to a basic-rate interface may enter a deactivated, low-power consumption state. An automatic procedure reactivates the interface when required. Primary-rate interface standards do not presently provide activation and deactivation for these higher-speed interfaces.

## 3.7  D-channel Operation across User-Network Interfaces

Within the D-channel, additional protocols specify the signalling necessary to originate, receive, modify, and clear calls. These signalling procedures also provide for special call features. Signalling information is generically categorized as *s information.*

Further, the D-channel protocols allow transmission of packetized user information whenever no signalling information is waiting. Packetized user information is generically categorized as *p information.*

Standards documents also mention low-speed telemetry (*t information*), but telemetry protocols are not detailed. No international or U.S. discussions for telemetry standards are in progress.

A comprehensive discussion of ISDN signalling protocols requires a book in its own. In this limited space, only some very broad highlights of the capabilities of these sophisticated, powerful protocols can be presented.

This particular section focuses on protocols used at the user-network interface. D-channel protocol layering follows the general layering principles of the reference model for open systems interconnection (OSI); see Recommendation X.200 series. The D-channel protocols, however, do not adhere rigidly to these principles because OSI describes a simpler operating environment. For example, OSI principles apply only to point-to-point operation; on a basic-rate interface, the D-channel operates in multipoint mode. In such cases, the development of D-channel protocols has extrapolated from OSI principles.

The D-channel protocols may be used at any interface at reference points S, T, or U. In addition, the D-channel signalling protocols may be used within a private network. For example, consider a portion of a private network consisting of two PABXs, at different geographical locations, and interconnected by dedicated transmission facilities. The D-channel signalling protocols may be used to control all calls between these two PABXs. D-channel signalling protocols are intended to support both user-network signalling as well as private interswitch signalling. Some discussion continues on the extension of the D-channel signalling protocols to also support public network interswitch signalling.

The user-network and interswitch operating environments pose somewhat conflicting requirements on a signalling protocol. For example, on basic-rate interfaces, the passive bus configuration requires the D-channel to operate in a multipoint manner. Calls being delivered to the user side must be offered to multiple user equipments, and several equipments may simultaneously attempt to answer the incoming call. Such a basic-rate passive bus is inherently asymmetric. On the other hand, interswitch signalling is generally completely symmetric. Each switch usually holds equal responsibility for the processing of calls and the general operation of the interface(s) connecting it to the other. D-channel signalling protocols resolve this dichotomy by relegating some asymmetric functions (such as contention resolution) to the physical layer of the interface. Simple replication accommodates other aspects of multipoint operation. For example, multiple instances of the data-link layer protocols operate independently between the network-side equipment and each user-side equipment; each instance operates in a symmetric manner. A point-to-point interface, then, merely represents a degenerate case of the multipoint configuration.

## 3.7.1 Data-Link Layer Protocols

A received D-channel bit stream, extracted by the physical layer from the time-division-multiplexed structure, next encounters D-channel data-link layer processing. The high-level data-link control (HDLC) family of protocols, defined by the International Standards Organization (ISO) during the 1970s, includes those data-link layer protocols used on the D-channel. The D-channel data-link layer protocols operate according to the general principles of the HDLC *balanced class of procedures*, where each equipment retains an equal responsibility for operating the data link. This protocol class was chosen because of robustness, widespread use, and the increasing availability of relatively inexpensive, specialized integrated-circuit implementations.

ISO's HDLC procedures embody three general classes of procedure, plus over a dozen options that combine with each general class to describe a particular protocol variant. Public data networks were among the first to employ a specific HDLC protocol, baptized *LAP* (*link access procedure*) in the 1976 (Yellow Book) version of CCITT Recommendation X.25.

In 1978, an updated version of Recommendation X.25 published in the Grey Book also included *LAPB* (*link access procedure—balanced*), a member of the HDLC balanced class of procedures. LAPB implementations now far outnumber LAP implementations in packet-switching equipment.

For D-channel operation, some further refinements to LAPB produced a data-link layer protocol suitable for ISDN application: *LAPD* (*link access procedure—D*-channel).

### 3.7.2  Multiple Data Links within the D-channel

All HDLC procedures employ transmission bursts called *frames*. Each frame contains one or more octets identifying the equipment receiving or transmitting the frame (data-link-layer *address field*). Multipoint configurations of equipment, such as those found on the basic-rate passive bus, require multiple parallel independent information flows. For example, suppose that three user equipments are attached to a basic-rate passive bus. Signalling information must be exchanged between the network-side equipment and each of the three user equipments, all within the same D-channel. Each user equipment transmits its signalling information to the network side via an independent stream of data-link-layer frames, oblivious to the existence of any other user-side equipment. On the network side of the interface, the network equipment receives three commingled streams of data-link-layer information. The data-link-layer address field allows these streams to be separated and processed independently.

This capability to interleave multiple data links, each from a different source, distinguishes LAPD from LAPB. (Other minor differences in protocol operation between LAPD and LAPB do exist, but these are generally of interest only to the data-link layer protocol design specialist and systems implementers.)

To achieve this multiple parallel data-link capability, LAPD expands on the use of the address field over that of LAPB. The LAPD address field occupies two octets and contains a *terminal endpoint identifier* (TEI) and *service access point identifier* (SAPI).

Each user-side equipment on a particular interface is assigned a unique TEI. TEI assignment occurs either automatically when the equipment first connects to the interface, or manually, or at time of manufacture. In the latter two cases, care must be taken that multiple equipments with the same TEI value do not attach to the same interface. Automatic procedures for TEI assignment exist in order to achieve an important goal in ISDN interface design: the capability of allowing the user to change, add, or delete equipment at will without mandatory prior notification to the network administrators. Without such a principle, the network side of the interface could require a large set of data bases in order to know the exact configuration of user equipment at every user-network interface. In addition, these data bases would in turn require a system to update them, as well as staff people to handle inquiries and notifications of changes. All this amounts to a considerable expense. By making the procedures at the user-network interface as automatic as possible, such expenses can be minimized.

Generally, each piece of user equipment employs only one TEI. Frames transmitted by the user equipment to the network contain the TEI of the user equipment. Similarly, frames transmitted by the network to a particular user equipment contain the TEI of the intended receiving user equipment.

When viewing the interface from the network side and looking towards the user side, one cannot distinguish between the case of two user equipments employing two TEIs and the case of one user equipment employing two TEIs. From a protocol operation standpoint, both interfaces appear identical to the network side. See Fig. 3.18. While probably not the usual case, a single equipment using multiple TEIs could be envisioned. For example, a device acting as a terminal adaptor for several non-ISDN terminals may employ several TEIs, one per adapted terminal, in order to transmit and receive signalling information for each adapted terminal independently. See Fig. 3.19.

The LAPD specification also sets aside a particular TEI value that is used to broadcast information from the network side to all equipments on the user side when delivering an incoming call. Each equipment, therefore, responds both to its particular assigned TEI(s) as well as to the broadcast data link TEI.

Fig. 3.18.
Multiple LAPD
TEIs employed
at a user-
network
interface.

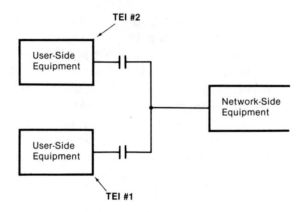

*(A) Two user equipments employing two TEIs.*

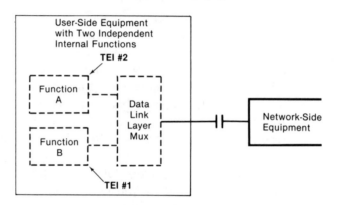

*(B) One user equipment employing two TEIs.*

Fig. 3.19.
Example of
user-side
equipment
employing
multiple TEIs.

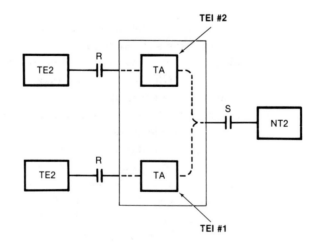

Within the user-side entity designated by a single TEI, the *service access point identifier* (SAPI) provides further discrimination. Particular standardized SAPI values specify particular LAPD capabilities. The LAPD standard does not provide for dynamic assignment of these SAPI values. For example: the SAPI value "0000 00" designates LAPD capabilities suitable for the transport of signalling messages across the interface. The broadcast TEI applies individually within each SAPI. In this example, a frame transmitted with the broadcast TEI and SAPI = 0000 00 would be received by all equipments containing some process (e.g., signalling protocol processing) using the data-link service access point suitable for signalling.

The combination of TEI plus SAPI identifies each independent data link. Fig. 3.20 illustrates five independent data links over a single D-channel, terminating in two equipments on the user side of the interface.

## 3.7.3 LAPD Capabilities

The *frame check sequence*, two octets appended to each LAPD frame, provides error detection. The LAPD receiver discards all frames received in error. LAPD employs timers and sequence numbers to trigger recovery of lost frames.

The *control field* designates the function performed by each LAPD frame transmission. LAPD supports the following functions:

1. Sequential information transfer, using *Information* (I) frames.

2. Flow control of sequential information transfer to avoid exhausting buffers in the receiving equipment. Flow control employs *Receive Ready* (RR) and *Receive Not Ready* (RNR) frames and other special protocols.

3. *Reject* (REJ) frames trigger retransmission of lost information.

4. *Unnumbered Information* (UI) frames provide nonsequenced information transfer, typically used to broadcast information.

5. Link operation start, using *Set Asynchronous Balanced Mode Extended* (SABME) and *Unnumbered Acknowledgment* (UA) frames.

6. Link operation stop, using *Disconnect* (DISC), *Disconnect Mode* (DM), and UA frames.

7. Optional notification of data-link layer parameters via the *Exchange Identification* (XID) frame.

In order to guarantee correct reception of information, sequence numbers identify each *I* frame transmitted and then report those correctly received. Generally, HDLC contains two types of sequence numbers. The first type is modulo 8 numbering, which cycles through values

Fig. 3.20.
Overview
description of
the relation
between SAPI,
TEI, and data-
link connection
endpoint
identifier.

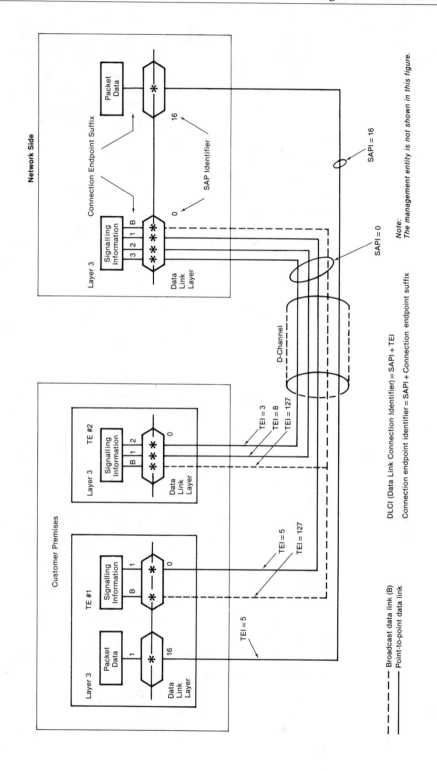

0–7 over and over, never allowing more than seven frames to be transmitted without acknowledgment. The second type of sequence numbers, known as extended numbering, operates modulo 128 and thus cycles through 0–127. In order to increase the economies of scale for LAPD implementations, international agreements reached in 1985 specify only extended (modulo 128) numbering.

Similar goals resulted in the elimination of the "single-frame mode" of operation in LAPD. Earlier documentation of LAPD included the single-frame mode, which provided sequential information flow without sequence numbers. This lack of sequence numbers suggested that a less-expensive LAPD implementation could be developed. However, subsequent study revealed that single-frame mode procedures achieved only a slight reduction in the size of LAPD integrated circuits. Buffer memory dominates the landscape of these chips. The LAPD chip itself forms only a small part of the overall cost of ISDN equipment, thus further diminishing the significance of any economies provided by the single-frame procedures. On the other hand, a requirement to support several incompatible LAPD protocol variations would introduce additional costs in providing universal equipment implementations that could attach to ISDNs anywhere.

The elimination of modulo 8 numbering and single-frame procedures typifies the pressures to remove implementation alternatives in ISDN user-network standards, especially when those alternatives do not offer any significant increase in capability or economy to the network or user.

### 3.7.4  LAPD Description

Along with the usual prose description of the LAPD procedures, LAPD standards provide a detailed description using *specification and description language* (SDL). The formal graphic representation methods of SDL detail a much more precise definition of protocols than can be produced in English text.

Horror stories of equipment incompatibilities caused by ambiguous sections of prose text in protocol descriptions litter recent technological history. A well-scrutinized set of SDL diagrams makes implementation much more straightforward, generally reducing it to a boring job of coding, with no opportunity for creative interpretations. LAPD represents the first application of formal description techniques in a standard data-link layer protocol.

### 3.7.5  Use of LAPD in Other Circumstances

The multiple parallel data-link capabilities of LAPD make it attractive for other applications beyond the D-channel. For example, Fig. 3.21 shows an inexpensive LAPD concentrator multiplexing several X.25 DTEs onto a single B-channel connection. In this example, each LAPD link carries X.25 packet level information for a single X.25 DTE. Flow control imposed by the packet switch on one X.25 DTE does not affect the ability of other DTEs to exchange information. A reduction of costs also occurs because the various X.25 DTEs share a single 64-Kbps connection to the packet switch. The use of LAPB, which allows only one data link on a channel, would have required multiple 64-Kbps connections to the packet switch.

Chip implementations of LAPD will likely stimulate other applications for LAPD by users, both on ISDN interfaces and in other situations.

## 3.7.6  User-Network Signalling Protocols

ISDN user-network interfaces employ out-band signalling techniques. Out-band signalling places all call control information on a signalling connection distinct from the route used by user information during the call. Out-band signalling relieves equipment from the burden of monitoring the user information paths for call control information. The end-user terminal equipment also no longer needs to format user information in a manner that avoids the imitation of signalling (either accidentally or for purposes of public network fraud).

D-channel signalling uses messages to convey signalling events across the interface. The details of this protocol may be found in CCITT Recommendations Q.930 (I.450) and Q.931 (I.451), and in related draft American National Standards; see the references at the end of this chapter. The power of message-oriented signalling lies in its ability to convey richly detailed information. D-channel signalling provides the user equipment with substantially more information about calls than any other network technology makes available. It also gives the user an unprecedented degree of control over the network(s) involved in the provision of each service and connection.

## 3.7.7  Organization of D-channel Call Control

Elementary call control, supplementary services control, and general interface management aspects make up the D-channel signalling procedures. Of these three areas, elementary circuit-switched call control procedures have benefited from the most detailed attention and are therefore the most complete. An excellent, extracted specification of elementary call control procedures is available in draft American National Standards (1986a, 1986b).

Supplementary services control procedures and user-network interface management signalling are just now receiving more focused attention. General agreements on the framework for the more complex supplemen-

Fig. 3.21. Use of
LAPD on a
B-channel.

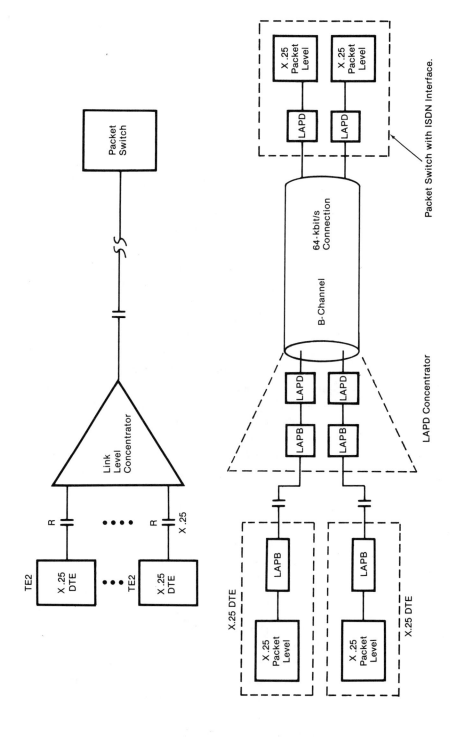

tary services signalling (such as N-way conferencing with off-line consultation among user-specified parties) are expected within the next year. Detailed specifications should follow once the general direction has been established.

Present descriptions of the D-channel call-control protocol intermingle call supervision and local protocol processing (such as confirmation messages). Promising new work recently undertaken would segment the protocol description into distinct activities such as elementary call-control protocol termination, interactive feature functions, channel management, call supervision, and route/resource selection. While the change in description would not affect the actual signalling messages exchanged across the user-network interface, it would appear to simplify the explanation of how signalling events at the user-network interface interwork within interswitch signalling protocols, such as Signalling System Number 7 ISDN user part. The inclusion of new supplementary services descriptions in the protocol may also be simplified.

### 3.7.8  Identification of Activity Being Controlled

The D-channel signalling procedure controls each activity across the user-network interface independently, using a state-oriented procedure. Each received signalling message and each internal event cause a precise change in protocol state, independent from the states of any other activities occurring at the same (or other) interfaces. While prose text outlines most aspects of the D-channel signalling procedure, SDL diagrams provide an unambiguous description.

A *call reference* distinguishes each signalling activity from all others. At the time an activity (call, feature modification, etc.) begins at that interface, the originating equipment assigns a unique call reference. The protocol manages call reference values in a fully symmetrical manner, and no collision (simultaneous attempt by both sides of the interface to use the same call reference for different activities) in call reference assignment can occur. Thus, each call reference points to an independent state machine following the procedures described by the SDL diagrams.

### 3.7.9  D-channel Signalling Message Format

Many earlier protocols used a largely rigid, fixed-message format. The exact contents and organization of each message depended on the particular type of message being transmitted. One item of information rigorously followed another, and each item could be interpreted only on the basis of exact location within the message. Subsequent additions of new information fields, or increases in the length of existing information fields, were difficult to accommodate and usually, if attempted, resulted in incompatible protocol versions.

To avoid these difficulties, D-channel signalling messages employ a modular format. *Information elements*, self-defining building blocks with a length descriptor, are assembled as required to form each message. All messages are constructed from the same pool of information elements. Individual information elements may be absent, present, or even repeated, as required, in any particular type of signalling message. Each information element begins with an octet identifying the type of information contained within the element (e.g., called party number). The length of the remaining part of the information element follows next.

For example, the SETUP message starts the call establishment procedures. If the user wishes to specify that the call transit through certain networks on its way to the destination, the *transit network selection information element* is included in the SETUP message. This information element contains an identification of the network to carry the call.

In order to specify more than one network, the transit network selection information element is simply repeated, each repetition containing a different network identification. If the user does not care about the routing of that particular call (or if that country does not permit network selection, as is frequently true outside the United States), then the transit network selection information element is omitted altogether.

This building block approach, although more complex in initial implementation, greatly enhances the ability to expand messages and procedures gracefully. New information elements can be added in the future, without obsoleting existing equipment. Equipment that does not understand the identifier of a particular information element will read the length field and then skip over to process the next element. A great deal of information can be delivered to the user equipment when an incoming call is offered. The user equipment picks and chooses among this information, taking into consideration some of it and ignoring the rest. The network side of the interface need not be concerned whether or not, for instance, to provide the calling party number to that particular piece of user equipment for display.

## 3.7.10  Service and Compatibility Information

Several information elements used during call establishment and modification phases provide closely related descriptions of:

Bearer service required from the network.

Low-layer protocols to be used during the call.

High-layer protocols to be used between users during the call.

User-user information related to successful call establishment.

The *bearer capability information element* describes the bearer service to be provided during the call. Each of the information transfer and access attributes (as described earlier in section 3.2.2) is defined in this element.

The *low-layer compatibility information element* describes additional aspects about the low-layer protocols (corresponding to the OSI physical, data link, and network layers) to be employed directly between the users during the call, if any. Note that if low-layer protocols are terminated by intervening network switches (as in packet switching), this protocol definition is included in the bearer capability information element. Only descriptions of low-layer protocols to be used directly between the call originator and destination(s), transparent to the ISDNs involved in this call, are contained in the low-layer compatibility information element. This information element allows the call originator and destination to confirm that they will be able to exchange information in a compatible manner during the call.

Similarly, the *high-layer compatibility information element* describes aspects about the high-layer protocols (corresponding to the OSI transport, session, presentation, and application layers) to be employed directly between the users during the call, if any.

The *user-user information element* transports any user-application-to-user-application information necessary to allow successful completion of the call. An example would be the provision of user name and password when calling into a data base application. This information allows the destination to refuse calls from inappropriate sources at an early stage: prior to actually answering the call itself.

The latter three information elements need not be present in every call. A basic speech (telephony) call generally would not require any of these three information elements; they would simply be omitted from the signalling messages.

## 3.7.11  Consolidated Signalling

In certain configurations, a single piece of user equipment may connect to a network with several interfaces. For example, multiple primary-rate interfaces can connect a PABX to the local public network. The D-channel signalling procedures for each of these interfaces can be consolidated onto a single D-channel. The signalling protocol includes a capability to designate both the interface and the channel within that interface that are being controlled by a particular signalling message. The *channel identification information element* provides this ability.

Automatic changeover and audit procedures provide recovery when a D-channel fails and an alternative is available.

### 3.7.12 Summary

The D-channel data-link layer protocol LAPD provides for information from multiple equipments to be independently multiplexed into the D-channel. Within a single piece of equipment, multiple sources of information are also multiplexed onto the D-channel via the same LAPD mechanisms.

LAPD provides sequential information delivery with flow control and error recovery. In addition, information may be broadcast to all other information sinks using the same data-link service access point.

Current LAPD specifications contain a detailed protocol description using SDL. Although a few implementation options still remain within LAPD, none of these options cause incompatibility between equipments that elect different option values.

Signalling, packet data (using X.25 packet level protocol), and possibly low-speed telemetry and interface management messages can each employ independent LAPD links to transport their messages.

The D-channel signalling protocol employs message-oriented out-band signalling techniques to set up, modify, and tear down calls. The protocol is also intended to control special features (e.g., call forwarding), but final details of these protocol aspects remain to be ironed out. The protocol uses a modular message format to simplify future expansion and minimize equipment obsolescence when updated implementations are deployed.

Several different information elements, used together, provide detailed descriptions of the attributes of each call to each involved network and user equipment.

D-channel signalling procedures may be used to control calls and services on one or more interfaces. Signalling between terminals and network, and between two switches within a private network, may both employ the D-channel signalling procedures. Further consideration is being given to using D-channel signalling procedures between public networks.

# 3.8  Public ISDN Architecture

ISDN standards avoid detailed discussion of public network architecture, and almost entirely ignore private network architectures. The silence comes not from a reluctance to tackle thorny issues, but rather from the realization that each public network differs in structure and administration. Furthermore, laws and regulations governing public telecommunications networks vary substantially from country to country. These differences often reflect historical antecedents as well as present technological factors. Thus, technical standards usually provide only the minimum details on network architecture, leaving the actual implementor wide freedom to invent (and to get into trouble!).

This section confines discussion of network architectural aspects to those addressed in public standards. Individual network practices, generic requirements, and other technical references are generally avoided. If you require details on ISDN architecture for specific networks, you should contact the individual network administrator directly.

## 3.8.1  Conceptual Organization

In broad terms, ISDNs involve the transfer of three categories of information:

1.  User information.

2.  Control information.

3.  Management information.

*User information* is, of course, what networking is all about. Control information—which is typified by signalling—is information necessary to control the establishment, modification, and disestablishment of user information paths (calls). *Management information* refers to everything else necessary to successfully operate a network: billing, alarms, maintenance commands, data base updates, etc.

## 3.8.2  Practical Organization

Increasing numbers of standards apply to each of these three information categories. When networks provide transparent information transfer services, users select which (if any) standards apply to the user information sent during calls.

Earlier sections have already dealt with ISDN standard signalling protocols for the exchange of control information between the public network and users. Standard interfaces between public networks have also been defined. For ISDN, the principles of out-band signalling extend to all signalling interfaces. Inter–public-network interfaces employ CCITT *Signalling System Number 7* (SS#7) protocols. Of particular importance are the following parts to SS#7:

*Message Transfer Part* (MTP).

*ISDN User Part* (ISUP).

*Signalling Connection Control Part* (SCCP).

*Transaction Capabilities Part* (TCAP).

*Operations, Maintenance, and Administration Part* (OMAP).

## Signalling Message Transport and Call Control

The MTP and ISUP provide for the exchange of call control information between public ISDNs. MTP provides a general-purpose protocol suite for transporting SS#7 messages. The functions performed by the MTP correspond very roughly to the data-link layer and network layer of the OSI reference model. The ISUP provides call establishment, modification, and clearing procedures for use in controlling calls between public networks. The signalling messages generated by the ISUP are transported by MTP protocols between *network control points* resident in each network. In addition to the MTP and ISUP, the SCCP may be used to manage individual signalling connections between network control points.

Actual public telephony network implementation plans for SS#7 vary considerably. Many European telephony networks expect to deploy initial ISDN systems using extensions to another SS#7 protocol, the *Telephone User Part*, for control of calls; significant deployment of ISUP is not expected during the next several years. Within the United States and Canada, local telephony networks and some interexchange carriers (such as U.S. Sprint) have already committed to the deployment of the ISUP, based, in general, on a recently developed draft American National Standard. AT&T Communications, however, apparently plans to expand on its already significant investment in Common Channel Interoffice Signalling (CCIS, also known as Signalling System Number 6). As mentioned earlier, increasing consideration is being given by some U.S. networks to employing the D-channel call control protocols between public networks, as the final details of this procedure are worked out. In Japan, NTT is considering implementing a modified ISUP that uses the modular message formatting principles of the D-channel signalling protocol. Nontelephony public networks will generally evolve to support ISDN interfaces through other methods that do not initially justify the deployment of these SS#7 call control signalling protocols.

## Management Information Flow

SS#7 TCAP and OMAP protocols provide standard methods for exchanging management information. Early impetus for TCAP implementation resides almost entirely in U.S. local telephone operating companies and non-AT&T interexchange carriers. These telephony network implementors intend to exploit TCAP to provide 800 business telephony calling services as quickly as possible, without relying on AT&T originating-screening-office data bases. OMAP protocol definitions are still generally in a formative stage.

### Other Internal Practical Considerations

Although SS#7 protocol standards officially apply only to the international interconnections between two public networks, protocol designers expect that these standards will be applied to connections between public networks within one country, and between switches within a single public network. For the United States, in particular, standards for interfaces between individual switches have recently assumed a greater importance. The plans of the Bell Operating Companies to procure equipment from a variety of sources mandate much more precise definitions of the interfaces between equipment manufactured by different suppliers.

The use of equipment from several manufacturers within a single public network also places other pressures on network architecture. For example, consider the strategy for deploying a new service feature within a public network. Central office switches become less attractive locations for new service feature intelligence when the network operator must repetitively pay each manufacturer to develop the necessary new software and hardware for each feature implementation. This fact encourages the location of feature intelligence not in the individual network switches but rather in a few more-centralized locations. The switches retain only elementary call supervision and routing functions; remote data bases and control points become involved in handling more complex communications service scenarios.

Centralized administration of these data bases, centralized control of the network, and centralized network maintenance have additional attractions. Economies of scale gained by centralization reduce staffing costs. In the areas of network control and maintenance, a centralized staff is exposed to a much larger cross section of network problems. The likelihood increases that staff have seen a given problem before, and this speeds troubleshooting and repair.

## 3.8.3  Relationship between Networks

In many cases the user will obtain telecommunications services from multiple networks and information service providers. For example, a business person may obtain some services (on-site dialing, reminders) from the local PABX, others from the local telephone operating company's networks, and yet others from remote networks and information service providers. For each of these services, the user must have a means to subscribe, activate, modify, deactivate, and cancel. Many of these actions involve signalling between the user and the service provider.

In an in-band signalling environment, such signalling was relatively easy. The user simply executed some action to get the attention of the appropriate network or information service provider, and then continued with a further exchange of information to instruct the network what to do next. Take as an example the capability available on certain U.S. interexchange carrier telephony networks to re-originate calls. Typically, the user sends an in-band DTMF[5] signal, such as #, to get the attention of the network. The network responds with a new dial tone, and the user then provides the new destination number.

In the out-band signalling environment, the interexchange network will not monitor the call for tones. To get the attention of the remote network, an out-band signalling message must be transmitted from the user to the specified network. The network could then reply with another out-band signalling message, or perhaps break into the connection to provide voice prompts and announcements.

To offer network services to a wide market and for a wide variety of call scenarios, via out-band signalling, requires the addition of a remote-network control point signalling capability. Some initial proposals on how such a capability would be incorporated into ISDN signalling standards are now under discussion. Such a generalized capability can be applied to much more complex scenarios. All of the functions of service registration, activation, modification, deactivation, and cancellation mentioned earlier could be performed through such arrangements.

### 3.8.4 Quality of Service

A multinetwork environment also complicates the engineering of acceptable quality of service. ISDN quality of service planning mimics that for existing multinetwork services.

Before engineering the quality of service, one must define the parameters that represent quality and determine how to measure them systematically. *Hypothetical reference connections* (HRX) lay out the anticipated usual and worst-case call routing scenarios. An *impairment allocation* process then allocates worst case and typical tolerance for each quality of service parameter to each network's portion of the HRX. Each network is then responsible for engineering its equipment and for routing calls across that equipment in such a way as to meet the impairment allocation requirements. Impairment allocation decisions between networks can be quite contentious, since stringent requirements on any network can significantly increase its cost of offering service.

---

5. *DTMF* stands for *dual tone multifrequency* signalling (e.g., TouchTone®).

To verify that network engineering requirements have been met, surveillance equipment systematically tests sample portions of the network for compliance with engineered quality of service parameters.

### 3.8.5 Numbering Plans

A new numbering plan for ISDNs has been agreed on. The ISDN numbering plan (described in CCITT Recommendation I.331) mimics the present international telephony numbering plan (CCITT Recommendation E.163). While international telephone numbers are limited to a maximum length of twelve digits, ISDN numbers may be as long as fifteen digits. *ISDN addresses* may also include up to forty digits of *subaddress* information. Public ISDNs do not process the subaddress but deliver it transparently to the remote end(s) of a call.

To simplify interworking and transition planning, user devices on ISDNs that must interwork with public telephony networks will be assigned ISDN numbers of no more than twelve digits. User devices on ISDNs that must interwork with public data networks will be assigned ISDN numbers of no more than fourteen digits. Fifteen-digit ISDN numbers will be assigned only to user devices that do not expect to interwork with existing public networks for a transitional period of time. At the end of the transition period, other existing public networks will have implemented the necessary interworking methods to accommodate calls containing the longest ISDN numbers. At that time, ISDN numbers up to the fifteen-digit length can be assigned to any ISDN user device. The transition period ends on 1996 December 31 at 23:59:59 Coordinated Universal Time (UTC).

The design and administration of any public numbering plan is closely related to the *dialling plan* and *routing plan*. A dialling plan describes the methods for providing numbers, routing information, and other call data to a network. Routing plans describe how a network processes numbers, service selection information, routing information, network control information, and quality of service constraints in order to select a path for each call. Design flaws in each of these may not become apparent for years or decades, and frequently require expensive fixes when identified. Such design flaws may make it awkward or impossible to add new users or service features to the network, to interconnect networks, or to route calls properly between interconnected networks. Good design requires the careful attention of experts from a variety of operational networks, with due regard to the anticipated business directions of networks and the industry at large. The exact plan for allocation of ISDN numbers among public networks (both local and nonlocal) within the United States is presently under study.

### 3.8.6  Summary

- ISDN technical standards attempt to discuss architectural issues in a nonprejudicial manner; that is, without regard to actual location and implementation of functions.

- Three categories of information—user, control, and management—conceptually define the major categories of internal network information flow.

- Inter-ISDN and intra-ISDN control information may employ certain parts of Signalling System Number 7: MTP, ISUP, and SCCP. SS#7 supports management information via TCAP and OMAP protocols.

- Other business and architectural forces have begun to force intelligence out of individual switches into more centralized data bases and control points.

- Methods for signalling between a user and a nonadjacent network in order to control remotely provided service features are presently under development.

- Hypothetical reference connections and the allocation of performance impairments among sections of such connections are used to characterize the overall quality of service. Individual networks must engineer their systems to meet the allocated impairment limits.

- The new ISDN numbering plan expands on the existing telephony numbering plan. Large numbers and subaddress information are available. Individual networks will define their routing plans. Dialling plans, which describe the operations required at the man-machine interface to the user, will be developed as a separate activity.

## 3.9  Private ISDN Architecture

Private ISDNs share most of the same design considerations as public networks. However, few private networks own and operate all of their required transmission and switching facilities. To a greater or lesser extent, these facilities are shared with public networks or are leased from them. The signalling and management principles behind ISDN provide a greater spectrum of sharing alternatives. Shared public network resources such as switches and data bases can behave as if they were exclusively owned and operated by the private network, even though they may be quite remote physically. Such hybrid networking arrangements may prove economical for many private networks.

# 3.10   Conclusion

This chapter provides but a cursory overview of certain aspects of ISDN. The level of material is analogous to describing American criminal law in as many pages; the reader of such a text may gain an awareness of crime and the law but is not yet ready to practice law. It is hoped that this chapter has provided an insight into general ISDN principles, as well as a framework for future, more technical knowledge of ISDNs. For further detailed information, readers should consult the most recent drafts of CCITT Recommendations (especially series I and Q), American National Standards, and individual network and customer specifications.

At the time of this writing, ISDN remains the paper tiger—attractive in some respects, perhaps fearsome in others, and largely nonexistent. The true dimensions of ISDN will unfold over the next two years. Nippon Telegraph and Telephone continues with its Information Network System trials in Mitaka, Japan. U.S. local telephone operating companies and European telecommunications authorities are now deploying trial systems. The reaction of customers to these trials will shape the evolution of ISDN systems. It is hoped that the technical standards have established a sufficiently flexible technical and architectural framework to permit this evolution to occur gracefully.

Other evolutionary pressures arise from the introduction of new network technologies in the areas of switching, transmission, control, and management. These pressures would principally affect the internal organization of networks, and would consequently impact intra- and internetwork standards. The implications for standards at the man-machine and user-network interfaces should be less severe.

While this chapter focused on technical standards and related aspects to ISDNs, ISDN planning considerations cannot proceed just on this basis. Legal, regulatory, economic, and business aspects have an even larger role in determining if and how networks (both public and private) will incorporate ISDN technologies. The existence of a new technology does not guarantee its introduction or viability in the market. While the high level of interest in integrated services digital networks is promising, networks and their customers have not yet voted with their wallet. Expanding telecommunications and information services are having an increasing impact on the shape of society in developed countries. ISDN technology promises to be one, but not the only, vehicle for conveying these services to the population.

# 3.11   References

### CCITT References

International Telegraph and Telephone Consultative Committee of the International Telecommunication Union (CCITT). 1985a. *Integrated services digital network (ISDN)*. Recommendations of the series I; Red Book vol. III, fascicle III.5. Geneva, Switzerland.

____. 1985b. *Digital transit exchanges in integrated digital networks and mixed analogue-digital networks. Digital local and combined exchanges.* Recommendations Q.501–Q.517; Red Book vol. VI, fascicle VI. Geneva, Switzerland.

____. 1985c. *Specifications of signalling system No. 7.* Recommendations Q.700–Q.795; Red Book vol. VI, fascicles VI.7 and VI.8. Geneva, Switzerland.

____. 1985d. *Functional specification and description language (SDL).* Recommendations Z.100–Z.104; Red Book vol. VI, fascicles VI.10 and VI.11. Geneva, Switzerland.

*Note*: The material on ISDN in the preceding documents reflects agreements reached as of the first half of 1984. Subsequent work has resulted in revision in several areas. Later information may be found in the reports of the International Telegraph and Telephone Consultative Committee's (CCITT) Study Groups II, VII, XI, and XVIII, or by consulting your organization's representatives to CCITT and related national standards groups. If you are contemplating the implementation, specification, or purchase of equipment intended to operate in an ISDN environment, you should consult these representatives and not rely solely on printed information.

International Telegraph and Telephone Consultative Committee of the International Telecommunication Union (CCITT). 1985e. *Data communication networks: Open systems interconnection (OSI), system description techniques.* Recommendations X.200–X.250; Red Book vol. VIII, fascicle VIII.5. Geneva, Switzerland.

### American National Standards References

American National Standards Institute. 1986a. *Draft American National Standard: Minimal subset specification for basic rate interfaces.* Accredited Standards Committee T1D1.2/86-32. Committee T1 Secretariat; c/o Exchange Carrier Standards Association; Parsippany, New Jersey.

____. 1986b. *Draft American National Standard: Minimal subset specification for primary rate interfaces.* Accredited Standards Committee T1D1.2/86-31. Committee T1 Secretariat; c/o Exchange Carrier Standards Association; Parsippany, New Jersey.

# 4

# A BUSINESS VIEW OF COMPUTER-BASED MESSAGING

VINTON G. CERF

This chapter presents a business view of computer-based messaging from the perspective of an engineer engaged in implementing commercial systems for public or private use. It is not intended as a detailed technical guide to the implementation of computer-based messaging systems (but see Chapter 5). Neither is it intended as a definitive history of computer-based messaging. Rather, the objective of this chapter is to define computer-based messaging, to review some of the history behind it in the United States from both a research and a commercial perspective, and to survey many of the aspects of commercial needs that are a factor in the evolution of this technology as a business tool.

## 4.1 Definition of Computer-based Messaging

Probably the most significant aspect of computer-based messaging, the one that distinguishes it from most other forms of electronic communication, is its "deferred" nature. In conventional telephony, it is essential for both parties to be simultaneously linked to make use of the service. *Computer-based messaging is a service or facility that permits the sender to produce and send a message independent of the ability of the receiver to receive it as it is being produced.*

By construction, one can include a rather interesting array of communications facilities and services under the rubric of deferred electronic messaging. For example, the early telegraphy systems, though they required the immediate sender and receiver to be simultaneously linked, were used as deferred communications systems by consumers because messages were originated on paper, transmitted electronically to a receiving station (possibly through several intermediate manual forwarding stations), and then committed to paper once more before being delivered (e.g., as a telegram) to the ultimate recipient.

Obviously, the existence of telephone answering services and, more recently, telephone answering machines, allows users to achieve deferred voice communication. A sufficiently general and abstract definition of computer-based messaging can easily encompass such a facility. However, for purposes of this discussion, the definition of computer-based messaging is restricted to *deferred communication of computer-manipulable information*. This definition need not, however, constrain discussion to text-only communication.

The successor to telegraph services, Telex, improved matters in several ways. For one, the receiving stations did not require attendance at all times, since they could receive automatically and commit to paper, or punched tape, messages sent by originators elsewhere in the worldwide circuit-switched Telex network. Second, the encoding of the message text for transmission was done with a keyboard using the 5-bit Baudot (International Alphabet No. 2) code, rather than through the manual Morse code system of the telegraph.

In the United States, however, the growth of the telephone system and its adoption by business as a tool for real-time communication rapidly overshadowed the then-available deferred communications tools: Telex and telegraphy. Of course, the hard-copy postal service is also a deferred communication tool; but, for purposes of this discussion, it is not considered an example of the computer-based messaging repertoire (but see section 4.8, "Product Scope and Features"). One might argue the point that posting a personal computer diskette, for example, represents deferred communication of computer-manipulable information; but, since the switching, storing, and forwarding is not done via computer, it falls outside the terms of reference for computer-based messaging.

For completeness, some authors include facsimile in the same general category as electronic or computer-based messaging. For the most part, however, facsimile services are point to point, though there are some exceptions. In addition, facsimile service still tends to be paper to paper, rather than computer-manipulable, and thus falls outside our general range of consideration. Teletex services (the successor to Telex, at least in Europe) support an intriguing mixture of text and facsimile representation, as well as text/image composition at Teletex work stations. This 2400-bps service is just beginning to emerge. It still tends to be a point-to-point electronic file transfer service, strongly akin to communicating word processor services, which are similarly organized. All of these classes of service tend to be device-oriented rather than organized around the concept of personal, electronic mailboxes—which are a central feature of computer-based messaging.

For further reading about computer-based messaging, see the *Xerox Network Systems Architecture Manual* (Xerox 1985) and *Electronic Mail and Message Systems* (Kahn, Vezza, and Roth 1981).

# 4.2   History of Computer-based Messaging Systems

In its earliest incarnation, computer-based messaging arose out of the 1960s development of time-sharing systems. These computer-sharing systems had (and have) the property of permitting multiple users to concurrently access machine resources interactively, often from remote locations, by means of dedicated, multidrop or dial-up circuits and, more recently, through packet-switching networks. To minimize interference among the multiple users of such systems, each user typically had his or her own directory of program, text, and data files, which were distinguishable and segregated from those of other users. Of course, many files were commonly accessible (e.g., compilers, text editors, and other tools), while others were accessible only under the control of the file "owner." A common practice in conjunction with such systems was the sharing of files among users. One could create a file and tell other users about it (its name, for example) to permit them to access, manipulate, copy, and possibly modify the data. It became common practice to share files as a means of cooperative communication among users who were not necessarily simultaneously connected to the time-shared computer.

The using community of such facilities typically consisted of programmers working on the time-shared computer, although users of the programs run on these shared facilities soon came to appreciate the convenience of deferred communication through file-sharing. To make the sharing of text information more convenient, conventions for file-naming and data structures evolved. In single, time-shared systems, this evolution meant that users could run message preparation programs that would produce text-file formats compatible with the conventions used by message reading programs. By convention, the message preparation program might "send" a composed message to the recipient by appending a copy of it to a well-known file (something like "mail.txt", for example) in the recipient user's directory of files. The recipient could then read his or her messages with a program that understood the format conventions of the message production programs.

Other variations of this scheme were possible. For instance, the message production programs might simply store a copy of the message as a distinct file, with an agreed-on file name, in the recipient's file directory. Or only a single copy of the message might be created, but a pointer to it might be left in each recipient's directory. Regardless of the details, this method of sharing information became very popular among the users of time-sharing systems.

# 4.3   Early Messaging Systems

In this section, a selection of early systems implementing computer-based messaging is reviewed to provide context for the discussion of more recent commercial offerings. Many of these systems were developed under U.S. Defense Department sponsorship and illustrate the early recognition by the military of the critical importance of strategic and tactical information management. No attempt has been made to be comprehensive, either in the selection of example systems or in the description of those selected. Rather, the choices are intended to illustrate some of the aspects of experimental messaging systems that have colored the development of their commercial counterparts.

## 4.3.1   The AUTODIN System

The military has long had to satisfy its requirements for command, control, and communication by using a variety of telecommunications techniques. In the early 1960s, the U.S. Department of Defense (DoD) contracted with Western Union to develop the AUTODIN (*Auto*matic *Di*gital *N*etwork) system, which continues to serve the worldwide communications needs of DoD (and other departments, as well). The AUTODIN system is a message store-and-forward system, largely aimed at electronic distribution and hard-copy delivery of operational messages. Optical character scanners are used to input hard-copy messages typed in a special font into the electronic store-and-forward network. Delivery of messages can be to hard-copy printers (laser or impact) and to host computers capable of acting as computer-based message system interfaces to AUTODIN.

## 4.3.2   The ARPANET System

When the U.S. *D*efense *A*dvanced *R*esearch *P*rojects *A*gency (DARPA) sponsored the development of the ARPANET (Roberts and Wessler 1970) in the late 1960s, one of the earliest applications of this network of time-shared computers was a computer-based messaging facility that worked by sending text files from one computer to another through the network.

The most important aspects of this application were the format and host computer identification conventions. These conventions made it possible for computers of different types to share text data and for the data to be stored or appended to the appropriate files in the file directory of the intended recipient.

The communications conventions used in the ARPANET were codified by its implementors in the ARPANET documents called "Requests for Comment" (RFCs). In particular, RFC 733 (Crocker, Vittal, and Pogran 1977) and its successor RFC 822 (Crocker 1982) were the basic standards around which all electronic message systems on the network were made to operate. These documents specified the details of mailbox addressing and message header format that each electronic mail transmission had to meet. The twin concepts of *User Agent* and *Message Transport Agent*, now such an essential part of the international electronic mail standards concepts found in the CCITT Recommendation X.400 (CCITT 1984), were born in this research environment.

The User Agents were programs for reading, composing, storing, and retrieving messages. The Message Transport Agents (sometimes called Mail Servers) were the programs that exchanged mail with other cooperating computers in the network. The User Agents generally handled the formatting of message output in accordance with the RFC documentation, so that the Message Transport Agents were able to deal with standard objects with a known structure.

Among the other early electronic message User Agents implemented during the intense research period of the early 1970s were: the BBN Hermes system (Deutsch and Dodds 1979; Mooers 1979, 1982), the SRI International (then Stanford Research Institute) On-Line System (NLS) (Engelbart 1963, 1973, 1978, 1980), and the University of Delaware Multichannel Message Distribution Facility (MMDF) (Crocker, Szurkowski, and Farber 1979). For more information on the ARPANET messaging systems, see Myer and Vittal 1977; Deutsch and Dodds 1979b.

Fundamental to the ARPANET experiments were conventions for mailbox address representation, message header structure, transport protocols, error indications from receiving mail servers, etc. These concepts were refined during the decade from about 1972–1982. They emerged as a second set of standards, after a great deal of experimental experience had been amassed from the first implementation. These concepts included the RFC 822 mentioned earlier and a new mail exchange protocol, Simple Mail Transfer Protocol (SMTP) (Postel 1982a), which was made to operate above a new set of Defense Department standard internetwork communication protocols (Cerf and Cain 1983; Postel, Sunshine, and Cohen 1981a; Postel 1981b, 1981c; and U.S. Department of

Defense 1981a, 1981b). Experiments in the interconnection of dissimilar but cooperating mail systems were carried out, and examples may be found in Cohen and Postel 1983; Crocker, Szurkowski, and Farber 1979; and Shoch 1980.

### 4.3.3 The Network Information Center (NIC)

One other concept of enduring importance uncovered in the DARPA-sponsored network and messaging area was the *Network Information Center* (NIC). This service is operated by SRI International and was an early application of the NLS technology for the ARPANET research and user community. Among the applications of importance were the maintenance of a user directory that could be easily searched by name to find a person's mailbox designator. This idea has emerged in more recent international standardization efforts in the CCITT as part of the Directory Services concept.

As part of its directory, the NIC maintains tables of host computer names, network addresses, protocol and application support capability, operating system type, etc. In addition, the NIC has become the on-line repository for publications related to the ARPANET effort and its successor, the INTERNET research project. It would be hard to overstate the value of such a central information resource in the coordination of a research community of the scale and geographic scope encompassed by the INTERNET environment. A critical element is the machine manipulable/accessible nature of the information contained in the NIC repository. Programs operating in the INTERNET environment can directly access NIC server programs to look up host and mailbox references, find internetwork gateways, etc. Users can transfer copies of on-line documentation using standard file transfer protocols and access methods. Of such stuff, the future of computer-aided business is made.

### 4.3.4 The SRI On-Line System (NLS)

At SRI International, the Augmentation Research Center (ARC), headed by Douglas Englebart, put together a bewildering array of concepts revolving around computer-accessible and -manipulable information in support of what Englebart called "knowledge workers." These ideas, developed during the latter half of the 1960s through the mid-1970s, covered a range of services and features, of which computer-based messaging was but a tiny part of the whole. The On-Line System (NLS), as it was called, included provisions for compound documents and for formal document descriptions to assist in computer-based formatting for hardcopy production, including phototypesetting in multiple-column format; on-line, full-text document storage and retrieval; on-line, automatic referencing of documents (you could be reading a document on-line, point to a reference, and that document would pop up on the screen!); full-screen editing capabilities using a mouse (which was invented as part of the NLS project), and an on-line help subsystem.

It is simply impossible to convey in a paragraph or two the incredible richness of the concepts explored in the NLS system. Many of the ideas that emerged in the Xerox Palo Alto Research Center (PARC) environment had their origin in the NLS effort, which is not surprising since some of the original PARC staff were drawn from the SRI International ARC project.

The NLS/ARC efforts at SRI International ended in the mid-1970s and were moved to Tymshare Corp., where an attempt was made to commercialize the product (now called Augment) and to sell it either as a private system or for shared use, accessed via the TYMNET. The system ran on a DEC PDP-10 (later KL-20) but was transported onto a Foonly processor (made by Foonly Computer Corp.). The Foonly was capable of emulating the DEC KL-10/20 instruction set and used the Tenex operating system. It was less expensive than the DEC system and was used by Tymshare Corp. as a base for commercializing the Augment service. The economics, however, were still not suited to a broad business market.

### 4.3.5 The Xerox PARC Office Automation Experiments

During the course of experimental work on computer-based messaging, one of the ideas attracting special interest was the integration of text and graphics into *compound documents* (multifont text and graphics documents) that could be moved freely from one computer to another, printed, archived, etc. The efforts of Xerox PARC (Pake 1985) along these lines, combined with research there on local area networks (Ethernet in particular), plus personal computers with bit map displays (e.g., the Alto and Dorado systems), and file and laser print servers, produced an extraordinary microcosm of office automation/communication ideas in a usable form. These ideas quickly spread to large fractions of the Xerox Corp. and eventually reached the market in the form of the Xerox Star (Seybold 1981a, 1981b) and later the Xerox 6085 products.

In a remarkably comprehensive volume, Xerox Corp. has published a description of its Xerox Network Systems Architecture (Xerox 1985). This document describes not only its networking philosophy, including the interconnection of its local area Ethernets, but also its application architecture, which encompasses the creation, storage, distribution, exchange, printing, and manipulation of compound documents. A great many of the specifics of the Xerox electronic mail and document management system have found their way into the CCITT X.400 recommendations. Early motivations behind the Xerox architecture are found in Dalal 1981.

Another important concept developed in the Xerox research effort is the Clearinghouse, which dealt with the resolution of names (e.g., mailbox names) to addresses in a network environment (Kluger and Shoch 1985; Oppen and Dalal 1983; and Shoch 1978).

# 4.4   Multimedia Messaging

In a continuing effort, DARPA is supporting research projects in *multimedia communication*, which combines text, graphics, digitized/encoded speech, and imagery into a structured, multimedia document. Among these efforts is a demonstration system called Diamond (Thomas et al. 1985) developed at BBN (Bolt Beranek and Newman) under the direction of R. H. Thomas and H. C. Forsdick. In this advanced messaging system, the objects exchanged have sufficient structure to permit presentation of text, graphics, imagery, and sound under the control of a powerful personal computer developed by BBN for experimental use. The composition and reading of messages created in the Diamond environment is reminiscent of sound film/video editing in its style and accommodation of both spatial and temporal presentations.

Within the DARPA-sponsored research community, an experimental standard for the exchange of multimedia messages has been developed by J. Postel at USC-ISI, and others at cooperating institutions, called Multimedia Mail Transport Protocol (MMTP) (Garcia-Luna 1984; Postel 1980a, 1980b, 1982b, 1982c).

# 4.5   Early Commercial Services

Without attempting to be comprehensive, it seems appropriate to mention several of the well-known commercial computer-based messaging systems, many of which have been offering service since the early 1970s. Some of these, such as the Tymnet Ontyme system, grew out of the time-sharing business base pursued by Tymshare Corp. Like most of its competitors, the Ontyme system supported basic mailbox exchanges between users. Similarly, the GTE Telemail® service, which was an early application of the Telenet network, had its origins in the ARPANET research effort. Telenet was initially developed by Bolt Beranek and Newman (BBN) as a commercial packet service, drawing on the deep experience BBN had acquired in its ARPANET development. BBN had also carried out extensive computer-based messaging research using its Hermes mail system as the principal research tool. Eventually, Hermes service was offered commercially by BBN on the Telenet. The Telemail computer-based mail system was developed separately by Telenet and continues to be one of the major applications on the public network.

A Canadian company, I. P. Sharp, which specializes in time-sharing services and software development for such U.S. and Canadian clients as the U. S. Navy and Bell Northern Research, offers a computer-based electronic messaging facility as part of its time-sharing service support environment.

Most of these systems were centralized, single-host systems, at least at the outset, though many were later extended to support multiple hosts or were embedded in a multiple host environment, such as the DARPA INTERNET in the case of the BBN Hermes system.

# 4.6 Recent Commercial Services

Since about 1980, a variety of commercial messaging services have emerged. Western Union introduced Easylink, a service that features a Telex-like interface to ease the transition from Telex usage to the richer, computer-based Easylink environment. General Electric Information Services Co. (GEISCO) introduced Quik-Comm as part of its panoply of time-shared business services. CompuServe® Corp. began operating a large-scale time-sharing service for residential and business users. Among the services it offers are the Easyplex computer-based messaging system, primarily for residential users, and the Infoplex computer-based messaging system, for business users. Like the GEISCO Quik-Comm system, Infoplex is organized around the concept of closed user groups for each business account. The Source, another large provider of time-sharing services to the public, also began offering a computer-based messaging system. In the early 1980s, Dial-Com, acquired by International Telephone and Telegraph (ITT) and recently sold to British Telecom (BT), added more competition to this marketplace.

MCI Communications Corp. introduced its MCI Mail™ computer-based messaging system in 1983 (Cerf 1984). The primary distinguishing feature of this service was its incorporation of a postal addressing and hard-copy laser printing option for addressees, including courier delivery, if requested. The hard-copy service included the laser reproduction of user signatures and letterheads that had been preregistered with MCI. The only competitor approaching these capabilities was Western Union with its Mailgram® service, and it used uppercase-only impact printers and a fixed, Mailgram letterhead and envelope. MCI Mail also interfaced to the MCI International Telex service, allowing users to send and receive messages via Telex, as did Western Union's Easylink service.

Some of the public systems were also configured for turnkey use by overseas carriers. Among the systems with "clone" configurations are GTE Telemail and BT Dial-Com.

# 4.7   Private Mail Systems

In addition to public computer-based mail systems, a number of computer and word-processing equipment vendors sell messaging software as optional facilities. Representative of these are Digital Equipment Corp.'s (DEC®) All-in-One, DECmail, and VAXmail products; Data General™'s CEO; Hewlett-Packard™'s HP Mail (and more recently, Desk Manager); IBM®'s PROFS and DISOSS systems; Wang™'s Mailway®; and Xerox's 6085 office automation system. Apple Computer Corp. operates an internal mail system called AppleLink™, which is based on the Apple® Macintosh™ personal computer. This service was developed in conjunction with GEISCO and supports the exchange of text, graphics, and spreadsheets among similarly equipped users. AT&T Information Systems introduced an MCI Mail–like internal computer-based mail system in 1985, which it now markets as a public service.

# 4.8   Product Scope and Features

Over time, computer-based messaging systems have evolved to incorporate a variety of services and functions. In the following sections, a number of these features, from the most essential to the most sophisticated, are examined. Commercial systems often make use of special features to achieve product differentiation, but all systems need to support the basic deferred messaging facility to qualify as computer-based messaging systems.

## 4.8.1   Mailbox Names and Directory Services

In its earliest stages, computer-based messaging provided for one basic function: the deferred exchange of computer-manipulable text between user directories on stand-alone or networked time-sharing systems. The first problem confronting users was how to determine a *mailbox name* to which to send the message (document, file, etc.). In the ARPANET system, each computer in the network typically had a directory for each user that was associated with the name under which the user logged into the system. In addition, every host computer had a unique name, assigned with the coordination of the Network Information Center to assure uniqueness. Mailbox names thus had the form:

User@Host

Finding the appropriate mailbox name among the hundreds of computers on the ARPANET would have been impossible, even if one could log into each of them to search the list of their known users. The Network Information Center supplied a central directory of all user names and their mailbox identifiers. This directory could be searched by means of queries using a simple protocol sent across the network. The queries typically included a first and a last name, or possibly only a last name. In reply, the user received mailbox identifiers, postal addresses, and telephone numbers, if these had been registered. Responsibility for registering in the NIC data base was usually undertaken by the administrators of each of the host systems on the ARPANET (later, INTERNET).

The MCI Mail system introduced an interesting variation on this global data-base approach by registering each of its users by *formal name* (the name the user wanted to have presented on all messages sent) and by *user name* (the name under which the user logged in), including information about the user's organizational affiliation and geographic location. Whenever a user composes a message on MCI Mail, the addressee is simply entered as the person's name, for example:

To: J. Robert Harcharik

The system automatically looks up this name in its global directory, and, if a unique addressee is found, the selection is confirmed. If more than one registrant is found with the same name, a list of these is presented, including registered organization and location. The user may select one or may reject them all, as may happen if someone with the same name as the desired addressee is registered but is not the person to whom the message is to be sent. Thus, in this system, directory searching is done automatically as messages are composed. There is also a directory searching command that permits users to verify addressees without composing messages.

Such directory services are important for commercial applications to assist users in finding addressees and in using the service. One need only compare this service to the telephone directory service to appreciate this fact.

When systems become very large or geographically dispersed, it may no longer be feasible to maintain central directory services of the kind offered by the NIC or MCI Mail. For manageability's sake, the assignment of mailbox names and of identifiers needs to be distributed. Xerox Corp., in its experimental work at PARC, pursued the concept of *name domains* to distribute the job of assuring unique names for mailboxes. Of course, the ARPANET "user@host" design distributed this job to each host administrator. What was new in the Xerox idea was the concept of a *name server* called "Clearinghouse" (Oppen and Dalal 1983). Such a name server could be replicated regionally (on administrative or geographical boundaries) and could assist in resolving references to names not known in the originating *domain*.

Like the MCI Mail and NIC facilities, these name servers were available on-line but in addition were configured to know about or to find each other in the event that a query was received that could not be resolved by local data-base information. For example, the mailbox name:

Dalal.pa

could be resolved by any name server by checking with the Clearinghouse system in the Palo Alto ("pa") region.

Although many commercial computer-based messaging systems still use special identifiers rather than personal names to reference mailboxes (e.g., [70000,130] in the CompuServe Easyplex service), the trend seems to be towards personal naming of mailboxes wherever possible. Of course, this introduces the problem of ambiguity since there are so many identical names (try finding all the Charles Browns in MCI Mail!). This problem is typically solved by associating a unique identifier with each mailbox as a "last resort," which can then be referenced by the user when all else fails.

In the ARPANET/INTERNET environment, the concept of naming domains has been formalized by extension of the "user@host" mailbox name structure to the form:

user@host.domain

where the "domain" part can have a substructure as in:

MIMSY.UMD.EDU

In this example, "EDU" is the *top level* domain and refers to the "educational" administrative domain. "UMD" refers to the University of Maryland, and "MIMSY" to a particular host computer in the University of Maryland's administrative domain. The top level domains must be known to most messaging systems, although it is permissible to establish conventions in which messages for unknown domains will be serviced by a well-known host that *does* maintain knowledge of all top level domains. The actual routing of messages need not follow each step of the structured domain name, since the originating host message system may in fact be able to map the qualified domain name directly into an address to which to send the message (Kluger and Shoch 1985; Mockapetris 1983a, 1983b).

### 4.8.2  Lists

The question of directory services and domains plays a critical role in the interconnection of computer-based messaging services and is addressed in section 4.11 on interconnection.

To simplify message addressing, many systems provide for private address lists or books in which a user may enter a nickname for a particular addressee and refer to that name when sending to that recipient. The nickname reference is expanded to be fully qualified before the message is entered into the underlying message transport system for delivery. This feature helps to alleviate some of the pain associated with otherwise ambiguous mailbox name references or with mailbox identifiers, which are hard to remember or lengthy to type. Sellers of computer-based messaging services have learned that the easier the service is to access and operate, the more likely it will be used.

A related facility of considerable value is the *distribution list*. Private distribution lists simplify the task of sending information periodically to the same list of addressees. A more interesting service is a list that can be shared by many users. In the ARPANET environment, such a list typically took the form of a mailbox serviced by a *redistribution agent* (a program that, when it finds a message in the mailbox, resends it to a predefined distribution). For example, sending to:

TCP-IP@SRI-NIC.ARPA

will cause copies of the message to be redistributed to several hundred addressees, all of whom are interested in the DoD standard internet protocols, TCP and IP (U.S. Department of Defense 1981a, 1981b).

In MCI Mail, a similar feature permits users to register "shared" lists on-line and to allow others to refer to them. Access to these lists may be controlled by listing users who are permitted to send through them, or the lists may be available to all users. For example, the XYZ company might register a list, "XYZ Marketing," and allow all its marketing employees to send through it:

To: XYZ Marketing (list)

Great care needs to be taken in implementing such facilities. An obvious mistake is for two lists to refer to each other, since this would create an endless round of redistribution (as has happened now and then on the ARPANET). The MCI Mail redistribution system looks for such recursion during list expansion to protect against this problem.

List facilities are important tools in fostering the use of commercial messaging, since they make it easier to support common interest groups that can readily exchange messages with each other. In the business of messaging, anything that creates a critical mass of users for an application contributes to the viability of the product. Creating critical mass has become the essential goal of messaging service providers (see section 4.10 on infrastructure).

Distribution lists represent one method for achieving a kind of selective dissemination of information, long the holy grail of information management systems and a buzzword of the 1960s. One can see evidence of this in the research community where hundreds of such distribution lists are used daily for numerous purposes (e.g., to broadcast queries to a selected group of interested participants, to raise issues for discussion with a selected group, etc.). What is potentially misleading about this type of utilization in the research community is the "free good" phenomenon. The cost of these services is frequently hidden from the users, since it is often paid for by a central funding authority rather than from a researcher's individual budget. Consequently, the intensity of use may be higher than it would be if the service were offered in a traditional commercial environment. Nevertheless, this kind of widespread, interorganizational information exchange represents the essence of freedom of communication: the bedrock of our social, economic, and scientific development. Clearly, the technology of business communication is aimed at enhancements to this essential freedom; however, it remains to be seen whether the economics are aligned with the demonstrated high utility.

One rather interesting phenomenon has evolved in the research environment: the *moderated distribution list*. In this variation, a person acting as moderator receives all contributions to the list and, after editing, publishes periodic summaries or edited versions of submissions. The benefit of this strategy is that the information actually distributed tends to stay on subject, and occasional episodes of "flaming" (uncontrolled, self-serving, sometimes libelous rhetoric) can be filtered out.

### 4.8.3 Bulletin Boards

A communication facility related to the shared distribution list is the *bulletin board*. There are two essential variations:

1. Everyone can post and read.

2. One can post; any can read.

In the first variation, the service acts rather like a community bulletin board at the neighborhood grocery store. Typically, one finds personal computer implementations of this service, sometimes called Community or Computer Bulletin Board Servers (CBBS), in addition to commercial, time-shared versions, among the vendors of messaging service. In the second variation, only the owner of the bulletin board can post to it. This facilitates the moderated type of publication, since the bulletin board owner (or "operator") determines what is actually placed on display.

The bulletin board mechanisms differ in at least one important way from the distribution lists. Rather than placing new items directly in users' mailboxes, requiring recipients to wade through them when they arrive, the bulletin board keeps all messages related to that bulletin board in a common place. The contents can be scanned at will by interested readers.

The MCI Mail Bulletin Boards are of the second variety and come with one additional feature: a control list to limit access to the board to those users authorized by the board owner. The bulletin board owner can also specify unlimited access for a given board, if desired.

### 4.8.4 Conferencing Systems

Bulletin boards and distribution lists are crude manifestations of a much more elaborate facility for on-line communication called *conferencing systems*. In these systems, most of which are housed on a single, shared mainframe, minicomputer, or personal computer, each conference has a declared subject. Conferences may be hierarchically organized, and new conferences may be spawned under the control of the parent conference administrator. Every submission (message) sent to a conference is cataloged as to date/time/originator and is given a unique identifier to facilitate scanning. Conferees may connect to any conference for which they are registered and can review unseen entries or go back over entries already reviewed. The system typically keeps track of which entries a participant has seen. Side conversations (interpersonal messages) are often supported, so that not all submissions are made available to all participants. The conference administrator often can edit the conference record, control the list of authorized participants, etc.

Although still in its infancy as a commercial technology, conferencing has a significant potential in the business marketplace. The most significant barrier to its growth is the upward scaling of the technology, which would allow it to support thousands, or hundreds of thousands, of participants using tens to hundreds, to thousands of interlinked computers engaged in managing multiple conferences spanning many computers at once. The maintenance of a distributed, global data base containing the conference record is the main challenge, especially given the need for very rapid dissemination of conference submissions to all appropriate computers in the system. Regretfully, there is neither time nor space in this chapter to do justice to this emerging technology. It is the author's opinion that computer-based messaging and conferencing services are, for the most part, complementary. They are, however, competing for the attention of the same business community that is still not sure what critical benefits either of these technologies may have to offer to the business world.

### 4.8.5   Automatic Answering Service

Another variation on the concept of mailboxes served by programs rather than by people is a service for automatically answering incoming messages when the recipient expects to be out of contact for an extended period (such as a vacation). The idea was to set up the receiving mail system to publish to the sender a short message (e.g., "Thank you for your message. I am on vacation right now and won't be able to respond until November 23."). In at least one such implementation, there was no checking to see if such a message had already been sent at least once to the originator. When two users of this facility went on vacation, set up the automatic answering service, and then notified each other by a message that they were going on vacation, the ensuing automatic barrage of mail rapidly filled all available storage and the system collapsed. A worse variation occurred when one of the originators was a redistribution list that redistributed the automatic reply back to the redistribution list, which redistributed the reply to the automatic reply system, which sent a reply back.... The entire distribution list had received hundreds of copies of this exchange until someone disabled the automatic reply feature.

The idea of associating programs rather than people with the receipt of messages and charging the programs with the responsibility for processing them leads to some very powerful mechanisms of potential interest to business. Among the most visible examples of such facilities are the Print Servers developed in connection with the Xerox Network Architecture. Messages in appropriately coded form could be sent to mailboxes served by programs that automatically interpreted the message content and produced high-quality laser-printed copies of the messages received. Special instructions as to fonts, number of copies, interoffice or postal addresses, and so on could be included. Rapid distribution to a printer near the recipient could be achieved by proper selection of the appropriate mailbox name. The MCI Mail system used this method to automatically route electronic message copies to hard-copy laser print servers located in major U.S. and European cities.

### 4.8.6  Document Representation

An important element of changing message presentation from electronic to paper is the format of the result. Experience with experimental and commercial messaging services seems to show that users are far more sensitive to the appearance of printed messages and documents than their electronic counterparts. Word processing systems reflect this concern by providing a number of formatting cues and features in their representation of documents. This concept extends to distinguishing between the internal, still-manipulable form of messages (often called *revisable* form) and the presentation form (often called *final* form) of messages. Some messaging systems, such as the IBM PROFS service, explicitly permit user control over the format (revisable or final form) of messages sent.

For messaging systems users, standardization of document representation could hold some important benefits. Chief among them would be the improvement of opportunities for document exchange and of the chances for hard-copy representations that reflect the intent of the writer. While some attempts have been made to agree on document representation standards (such as the U.S. Navy initiative on Document Interchange Format), there remains a multiplicity of representations, and some businesses are founded on the conversion of one format to another. Examples of representations include all the native forms of word processing systems, including personal computer-based systems, the Xerox Interpress format, Adobe PostScript, and text-based document description languages such as NLS (see section 4.3.4), Unilogic Scribe, and Hewlett-Packard Text and Document Processing (TDP) language, to name just a few.

An extension of the notion of structured text documents is the *compound document*, which is able to accommodate both text and graphics or bit-map imagery within a single message. Such representations open up opportunities for electronic publishing, which include combined text and graphics. One important standard for such representations, called *North American Presentation Level Protocol Standard* (NAPLPS), has emerged in the context of Videotext services. This standard, however, is limited to fairly low-resolution displays and is not currently suited or applied to full-page, combined text and graphics documents. The Interpress and PostScript representations support multifont text and combined graphics that can be displayed either on soft-copy bit-map screens or in hard-copy form using high-resolution laser or dot-matrix printers.

### 4.8.7   The MOSIS System

Another ambitious project relying on careful definition and adherence to message structure is the *Metal Oxide on Silicon Implementation Service* (MOSIS) (Lewicki et al. 1984; USC Information Sciences Institute 1984). In the MOSIS system, several computer-based messaging systems are utilized as the communication link level of an application that permits users to submit very large scale integrated circuit designs for analysis, functional verification, and automatic implementation (chip production)—all via purely electronic message exchange, except for the last, actual fabrication step.

The MOSIS system accepts messages from the ARPANET and public electronic mail system sources and interprets them according to their format and content. Since the circuit designs may be very long, the MOSIS system accommodates the transmission of multiple electronic mail messages, which are sequence numbered for reassembly. The MOSIS application is a prime example of computer-aided design and manufacturing. For this particular case, application software, operating on messages received from electronic messaging services, preprocesses the chip designs for consistency with the fabrication design rules. The software then automatically lays out multiple circuit designs per chip and multiple chips per wafer, produces the fabrication mask description files, documents the required bonding of pins to circuit pads, and tracks administrative information for each fabrication project.

### 4.8.8   News Clipping Service

The concept of automatic processing of messages can be applied to a number of applications. One idea of some interest is a news clipping service. The subscriber sends an electronic request for copies of news items containing information referring to some particular set of keywords. The receiving program updates a profile for the requestor and analyzes incoming news (from Associated Press, United Press International, Reuters, or other news wire service) for matching references. Messages containing the articles of interest to the consumer are copied and sent as ordinary computer-based message traffic. Experiments with this idea were carried out successfully at Stanford University in the 1970s, and some commercial messaging services do offer a news clipping service.

### 4.8.9  Electronic Data Exchange

Variations on the electronic exchange of structured messages have extended to binary files (executable programs, spreadsheets, etc.) and to compound objects that include digitized voice and music as well as text and graphics. The utility of such structured objects is likely to be realized only when there is widespread agreement on one, or perhaps a few, compound object representations that have a high probability of being processible upon receipt.

A related area of increasing product interest revolves around standardizing the representation of high-volume commercial correspondence, such as purchase orders, purchase requisitions, bills of lading, invoices, payments, and so on. At least one *American National Standards Institute* (ANSI) committee, X.12, has been working in this area for several years. In addition, specific industries, notably the transportation, banking, and grocery industries, have developed ad hoc document representations intended to support direct, computer-to-computer exchange of orders, confirmations, and the like. This area of product development falls under the general rubric of *Electronic Data Interchange* (EDI). Opportunities exist for capturing information in a friendly fashion from originators and providing it in formatted, bulk fashion to receivers for further processing. Implementation strategies range from special PC software packages to support by means of shared mainframes.

Successful application of computer-based messaging in EDI applications will not be merely a matter of collecting and properly formatting information. A major problem to be resolved is identification of sender and recipient. Some standards make use of the Dun and Bradstreet identifiers for this purpose. Clearly, in the long term, international agreements on addressing and identification will be required to support efficient routing and delivery of EDI messages among the correspondents.

### 4.8.10  Cryptographic Keys

As business correspondence migrates to an electronic basis, there will be an increasing need for verifying or assuring the authenticity of the originator and the associated message. A number of electronic signature methods, based on public key cryptographic methods (Israel and Linden 1982; Kent 1981; Needham and Schroeder 1978), have been proposed or implemented to address this need. As message traffic is used more and more for formal and binding business communication (including electronic funds transfer), the need for privacy and authentication will increase opportunities to apply this technology and more conventional cryptographic methods. One should not minimize, however, the challenge of managing and distributing cryptographic keys. The public key systems, which, on the surface, simplify this problem, are heavily dependent on the strength of algorithms that have not yet stood the test of time. Full utility of such techniques, as for structured documents, is very much a function of commonality. It is still too early to predict which methods will gain broad enough support to form the fundamental basis for widespread message authentication.

## 4.8.11   Error Protection

The support of machine-to-machine communication by means of computer-based messaging brings with it another requirement that may not be obvious to the casual user. Most public messaging services are reached by dial-up access through the voice telephone network. Despite the relatively high quality of U.S. telephone facilities, random line and switch noise can corrupt the information transfer. When this occurs during a simple interactive session (terminal to time-shared computer), it is often easy to recover the information by reentering the corrupted information or by redisplaying it. When two computers are trying to exchange information in accordance with an agreed-on format and procedure, it may be more difficult, if not impossible, to detect such errors and recover from them, without agreeing on an error detection and correction procedure ahead of time.

Several initiatives in this direction are under way; examples include the X.PC protocol, developed by Tymnet Corp.; the Microcom Network Protocol (MNP) developed by Microcom Corp.; and public domain protocols, such as X.Modem and Kermit (which are file transfer oriented). Synchronous protocols, such as the CCITT standard High Level Data Link Control (HDLC) protocol and proprietary standards such as IBM's Synchronous Data Link Control (SDLC) protocol and its older Bisynchronous protocol, are also aimed at reliability on the access links to messaging services. Over time, the evolution of Integrated Services Digital Networks (ISDN) among the carriers may focus on a common standard for achieving link reliability, but until this happy day, service providers must anticipate the need to support one or more reliable link protocols. Even with the use of long-haul packet network techniques, the immediate access link connecting a subscriber computer (PC or mainframe) will need some protection. The alternative is to exercise this protection from end to end between the originating computer and the service provider. The disadvantage of the latter approach is that errors on the local access line have to be recovered across the long-haul network and the user may be charged for all such retransmissions. Of course, the concomitant advantage is end-to-end error control.

To the extent that anyone can speculate about trends in a business still in its infancy, it appears that there is a predictable migration away from the time-shared computer-based messaging paradigm to personal computer-based services. Along with this, it appears likely that economics will encourage large computer-based service users to install private systems wherever feasible (see the next section). Such a trend does not necessarily herald the end of the public service provider, but rather changes the role to interconnection of private, domestic, and international messaging services (see section 4.11 on interconnection).

# 4.9   Economics

A great many parameters enter into the equation for determining the optimum technology for any particular computer-based messaging application. It is even misleading to suggest that there is a *single* equation or that an *optimum* can be determined or implemented.

Some examples of factors that influence the best match between system and requirement for a particular application are accessibility, service price structure (or capital cost, if acquisition of a private system is contemplated), number and geographic dispersion of users, message volumes, length, multiplicity of addressees, types (text, binary, compound,...), response time requirements (delivery speed), off-network delivery requirements (Telex, hard copy, external message system, etc.), equipment types that must be accommodated, and so on.

## 4.9.1   Public versus Private Systems

At the very low end of the scale (few users, small numbers of messages), it may prove cost effective to acquire a personal computer (or supermicro) and a number of terminals and modems to create a private business bulletin board system. In a very constrained case (all users in the same building serviced by a digital PBX or local area network), it may even be feasible to dispense with modems. Solutions of this type may quickly become inadequate, however, if any of the following conditions exist: the total number of users outgrows the capacity of the system, the accessing delay becomes too long, the duration of sessions becomes excessive, or the volume of message traffic exceeds the store-and-forward or archival capacity of the PC.

If users of the system are widely dispersed and must rely on direct distance dialling, wide-area telephone service, or packet network service, then access costs, not acquisition costs, may be the dominant cost factor. A potential, hidden cost is the operation of a private system. This cost might be more readily absorbed if the host system is used for a variety of other applications and a support staff is available to carry out the additional tasks required to operate the private messaging facility.

The choice between dedicated private and shared public computer-based messaging service may hinge on the initial incremental cost of implementing the service. A client with an existing base of terminals or personal computers already equipped to support communication may find that the incremental cost of adding a computer-based messaging service is lowest if the service is acquired from a public service vendor. This is especially likely if geographic dispersion is large and initial message volumes are small.

If the usage tends toward interoffice communication and if there are concentrations of users in a few places, the average message cost may be reduced if one or more private, dedicated systems are acquired and operated. However, this may not be the case if the public service is able to offer sufficient price discounts in exchange for volume. In general, computer-based messaging costs are strongly influenced by volume: the larger the volume, the lower the incremental cost, assuming, of course, that the system has sufficient capacity to reduce per-message costs significantly with increasing message volume.

Anecdotal information suggests that in a typical office messaging application, some 70% of all messages are intraoffice and the rest are interoffice or intercompany. As volumes increase, it becomes more attractive to acquire and operate a private system at fixed acquisition, maintenance, and operation cost up to the limit of its capacity and to utilize public services for intersystem and intercompany communication.

Public messaging systems exhibit a variety of pricing elements ranging from connect-time to message length, addressee count, and volume discounts. Some services use a postal model in which there are charges only for sending messages. Others charge for resource utilization regardless of its type (e.g., message composition, sending, reading, storage, retrieval). Special services, such as return receipt notification, often carry special charges.

It is not always a simple matter to predict the cost of messaging services. This fact alone sometimes makes it difficult to decide on the best mix of public and private services needed to create a corporate, computer-based messaging system.

For public systems, as for private ones, communications access and usage charges may represent a substantial cost per message. Higher-speed modems may reduce access and usage charges sufficiently to justify their added acquisition cost. These same arguments may apply when considering dedicated versus switched access to public service or between public and private services. Sufficient volumes may offset the fixed cost of dedicated access and yield lower incremental communication cost per message.

### 4.9.2 Cost Comparisons

Some applications of computer-based messaging may displace other costs. For example, interoffice telephone calls frequently result in "telephone tag." This carousel of calling can sometimes go on so long that, when the parties finally do connect, neither remembers the purpose of the original call! Deferred messaging services can reduce incidents of this type, displacing the cost of fruitless calls. It is worth noting that such calls bear at least two costs: the cost of the call itself, assuming it is completed, and the cost of the caller's time in making repeated, unsuccessful calls. The vendors of telephone answering systems attempt to deal with this problem, but the mechanics of trying to have a deferred conversation in this way are awkward.

Examples of cost comparisons are often offered by public and private message service providers. One popular comparison is the cost of a letter prepared for postal delivery in the conventional way with the cost of one prepared for electronic delivery. The dominant cost is usually the cost of message preparation. The "transmission" costs for domestic postal delivery range from \$.22 per message to several dollars, depending on length. Some private mail systems reportedly achieve costs as low as \$.05 per message. In any event, the attempt is made to show that transmission costs are small compared with message preparation costs, and that message preparation is less expensive if the message is to be sent electronically because it needs less editing and formatting.

It is often a mistake, however, to compare on the basis of cost alone the efficacy of computer-based messaging with more conventional, usually paper-based or telephone, messaging. A major advantage of computer-based message delivery is the time and location independence of sending and receiving. The sender need not have any idea of the current geographic location of the recipient. Unlike paper and telephone messaging, the recipient need not pick up messages from a specific geographic location. Instead, with suitable equipment, the recipient can read and respond to messages regardless of his or her own location or that of correspondents.

Although scaling of volume may bring some attractive benefits in the form of reduced incremental message costs, it may have some unexpected costs for the individual recipient. Current systems (private and public) are not well organized to support significant volumes of messages delivered to an individual user mailbox. The problem is not so much the cost of storing quantities of messages until they are read but, rather, the time cost to the recipient of wading through 50 to 100 or more messages in a day. Even in systems that support variable priority marking on messages and ordering by highest priority first, reading through and processing large numbers of messages can still be very time-consuming. Until strategies are developed to assist users in automatic sorting of messages according to originator, topic, priority marking, and actual message content, many users will rely on third-party filtering by secretaries or other assistants. If the latter course is adopted, the supporting message system should provide a way to deny access to confidential messages to all but the intended recipient through encryption or passwords or by means of a separate, private mailbox. The interested reader should refer to Malone et al. 1985 and 1986 for some very recent results in the management of message routing and retrieval.

In cases where access costs are a major factor in the per-message cost of computer-based messaging, a means of archiving copies of messages and documents may be an effective way to reduce the cost of sending repeated copies of messages. Many public and private messaging systems support private filing of messages, which later can be forwarded in response to requests for copies. Depending on storage charges and the application, storing messages and forwarding them upon request may be the most economical choice. In terms of convenience, not having to resend a long document is generally superior. Some systems deal with this problem by supporting automatic and unattended document uploading under control of the source personal computer and the host messaging system. File server systems in local area office automation applications, which provide archival facilities and simplify document forwarding, can deal with some of these problems.

One potential economic barrier to widespread use of computer-based messaging in a business application is the provision of equipment to access and use the service. In the research environment out of which the technology emerged, terminals and mainframes were already in place and used to support other applications. As a consequence, the computer-based messaging application had only an incremental cost to implement. Even in that case, a number of users who normally made no use of computer equipment had to be outfitted with terminals and, in some cases, modems to access the mainframe servers.

In the business world, word processors and, increasingly, personal computers are becoming more widely used. Not all of this equipment is outfitted with the software and communication gear needed to access computer-based messaging services. Indeed, most users of this gear are skilled in only a narrow range of its uses. As a result, the provision of proper equipment, software, and training can present business with a costly hurdle to overcome before it can benefit from these new services. This topic is explored further in the next section on infrastructure.

# 4.10 Infrastructure

*Infrastructure* is the collection of facilities, services, and practices that are required to maintain an economy or some aspect of an economy. For example, the U.S. automobile industry is heavily dependent on the infrastructure provided by the road system, the network of automobile service stations, licensing, law, and safety enforcement practices, to say nothing of the petroleum refining industry and its supporting structure.

Infrastructure is pervasive. Another example may be found in the electrical appliance industry. It could not exist without the power generation and distribution infrastructure, which includes the power distribution system network, building wiring code standards, power delivery standards (60-cycle, 120-volt, etc.), and power generation systems and their operation and maintenance.

The penetration of computer-based messaging services into the U.S. economy requires the concomitant evolution of an infrastructure of standards, common training, equipment availability, and ubiquitous service access and interconnection. These concepts are explored in this section.

## 4.10.1 Critical Mass

Computer-based messaging services are virtually useless unless a *critical mass* of correspondents can be assembled for a particular application. The crucial term here is "critical." It is one thing to bring many correspondents together in a single system but quite another to have a sufficient number for any particular application. The absolute number of users required to achieve critical mass may be quite small for a particular application. For example, in a small company of twenty to thirty people, an intraoffice messaging service may need to include only company employees or perhaps just all employees in a given department. The key idea is that all, or nearly all, who have a common interest or objective, which the messaging service can help them achieve, must be included in the service or it will not achieve its potential. An essential property of such services is that they make it easier for the group to communicate. This is almost always true only if there are few if any exceptions: virtually everyone must be reachable through the system.

Computer-based messaging services that have a large user population may still not have achieved critical mass for any particular application.

Businesses seeking to create a market for this service must begin by identifying applications for which a critical mass can be readily achieved.
This may mean that the user group must already have equipment and
software suitable for accessing the service. Alternatively, the buyer must
be prepared to pay the cost of acquisition, installation, and training to
achieve this condition.

In systems like MCI Mail, which support hard-copy postal and courier
distribution service and access to Telex, it often happens that the critical
mass of on-line users is as small as a single person publishing a newsletter, since the recipients (the other part of the critical mass) will be
reached by hard copy. In other cases, a few people need to be on-line to
the messaging service and the rest may already be a part of the Telex
service. The choice of services and connectivity to existing communications systems can strongly influence opportunities for creating critical
mass usage for particular applications.

## 4.10.2  Public Attitudes

A second element of infrastructure is the widespread recognition of personal computers and word processors as communication instruments.
This is still not a common view. In a sense, this may be the most significant hurdle to overcome: creation of widespread understanding of, and
appreciation for, the utility of computer-based communication. The for-
mation of this intellectual infrastructure could take a generation, like the
slow accumulation of sedimentary layers over geological time periods.
There are perhaps two different layers to develop: familiarity with, and
appreciation of, personal computers and similar equipment, and recognition that such equipment can play a crucial role in communications.

In the United States, the telephone is nearly as ubiquitous as the television. One does not have to take classes in the use of the telephone, since it
is readily available in most homes, and demonstrations in its use occur several times a day in most households. Children are taught almost by osmo-
sis. Personal computers and word processors are beginning to penetrate
the educational systems at all levels. Children in elementary school today
will enter the business world with a familiarity and possibly a well-
developed dependence on this type of equipment in stand-alone applications. Less common today is experience in the use of personal computers
in a communications environment. This is still a moderately unusual usage, although popular services such as the Source, Compuserve, and Dow
Jones News/Retrieval Service are making headway in the residential world.
The critical point is that the creation of this intellectual infrastructure is a
slow process that may take considerable time to mature.

Even when the equipment and communications barriers have been scaled and the user appreciates the potential of computer-based messaging, there is a habit barrier to overcome. Most computer-based message systems do not automatically deliver incoming messages to the recipient's equipment. Part of the reason for this is that the equipment is often multiplexed among many uses and, when not being used on-line to access computer-based messaging, is applied to other tasks. As a result, users must regularly log into the messaging service to retrieve their messages and send any new ones. This is a habit that is hard to form until the service becomes an essential part of the user's business (or residential) routine.

The messaging habit problem may be partly overcome through the use of increasingly smart personal computer software that can automatically dial-up messaging services, pull any unread messages, send any awaiting processing, sort the incoming messages into different levels of importance, etc. Nonetheless, the incoming messages will be of no utility if the recipient does not look at them on a fairly regular schedule. In an effort to stimulate this habit, MCI Mail developed an ALERT service, which permits message senders to request that MCI telephone the recipient to advise that an important message is waiting. Other service providers have included paging services to achieve a similar objective. Some systems support forced delivery, as does Telex, to reduce the dependence on the recipient to collect awaiting messages.

### 4.10.3 User Interfaces

Another area in which infrastructure development is needed revolves around the user interface to the messaging service. Most services today use screen or scroll mode command user interfaces. Some, such as the Applelink application, are oriented around the icon-mouse model of the Apple Macintosh, the Xerox Star and the Xerox 6085 systems, to name three examples out of many. Simplifying the user interfaces may help to foster the spread of the intellectual infrastructure. One should not imagine, however, that windows, icons, and mice represent a magical panacea. Mouse-style interfaces can be fairly clumsy in situations where expert users might have used more efficient command-driven interfaces with type-ahead.

The world communications infrastructure is undergoing a major revision in the form of *Integrated Services Digital Networking* (ISDN). In effect, communications providers are working towards provision of digital connectivity, worldwide, over the course of the next ten to fifteen years. This effort, involving international standardization of equipment, signalling methods, data communication protocols, and services, will form the basis for business and residential computer-based service provision at rates (64,000 bps) 25 to 50 times those commonly available today

(1200–2400 bps). ISDN demonstrations and trials of both business and residential digital services are already taking place. In all of these, voice services are digitized at the origination and carried in a multiplexed form with a variety of data services. The resulting digital streams can be demultiplexed and routed to a variety of destinations, some via packet-switching services, some on digital switched circuits, and others via dedicated point-to-point routes.

The communications, business equipment, and intellectual infrastructures will evolve over time. Costs will fall as these facilities find larger communities of users willing to pay for the equipment and services. Agreements on equipment interface standards, compound message representations, communication protocols, and addressing standards will be essential to realize the benefits of this technology. Many of these agreements are already in progress, but they will not happen by accident. Considerable investment is still needed on all fronts to achieve the necessary agreements; to develop and build the equipment, software, and systems; and to educate the potential user community.

# 4.11   Interconnection

For the same reasons that the interconnection of the worldwide telephone system is essential to its utility, the interconnection of computer-based messaging systems (private and public) is an essential step in the creation of its critical infrastructure. This is not to say that every computer-based messaging system must be able to access all others—there are good reasons for the creation and operation of closed systems (for example in classified applications)—but, for the most part, the greatest utility is achieved by the greatest ubiquity.

The providers of computer-based messaging systems and services are engaged in the standardization of formats and procedures for the exchange of messages among systems. This effort, sponsored by the CCITT (*Comite Consultatif Internationale Telephonique et Telegraphique*), has produced a series of recommendations under the general rubric of X.400 (CCITT 1984), which specify the details of an architecture for computer-based message exchange. The objective of these Mail Handling Systems (MHS) recommendations goes beyond making possible international, intersystem personal messaging. The recommendations are aimed at general-purpose exchange of digital objects, although the earliest applications contemplated are, in fact, interpersonal messaging. Included in the recommendations are procedures for interconnecting messaging systems with the Telex and Teletex services, as well as incorporating interoperability with facsimile and mixed text and facsimile documents. For further details of the CCITT X.400 Recommendations, please refer to Chapter 5.

At the time of this writing, many of the technical strategies for inter-connection have been worked out, although many details remain to be fully specified, and considerable implementation and testing among systems will be needed to gain confidence and full understanding of the requirements.

## 4.11.1  Directory Services

The two principal problems remaining to be solved to achieve general guidelines for interconnection of messaging systems involve accounting and directory services. Of these, the issue of directory services has received the most attention so far, but the issue of accounting is even more critical from the business point of view. With respect to directory services, the problem is twofold:

1.  Message system registration service for intersystem routing.
2.  Directory services to find user mailbox addresses.

What is needed, first, is a facility for unique naming of public and private messaging services. This service might be hierarchical in nature, starting with an international organization such as the CCITT that could assign standard country names, within which national-level assignment could be managed by each country. This type of hierarchy may conflict with multinational private-message-system naming conventions, but so far it has been the practice of the CCITT not to deal with private naming problems. Perhaps in this context the concepts can be extended so that multinational organizations might register systems internationally.

It is desirable to permit private systems to be linked to more than one public messaging service but to retain a unique identity. The less desir-able alternative is to assign more than one name to a private system, depending on which public system(s) links to it. The public system must maintain knowledge of other public systems and of those private systems that it serves directly. Automatic means are needed to provide updates to all public systems as new public (and private?) services are linked into the global network. Since it is unlikely that a central facility can maintain a data base that includes all private systems, the messaging system nam-ing conventions will need to provide clues as to which public system(s) serves the private system to which a message is to be sent. At worst, the naming conventions will need to refer to the country in which the public and private systems are registered. In addition, some on-line means must be found for messaging systems that can resolve names into addresses and addresses into routes (at least to the next messaging system, which can then forward the message towards the destination). One need only consider the voice telephone system to appreciate the extent of the conventions needed to support directory services and automatic routing from source to destination messaging system.

The MCI Mail system has taken a modest step towards this goal by supporting the registration of messaging systems other than MCI Mail in a routing table, which is maintained centrally by MCI's operational staff. Provision has been made to support indirect routing of messages destined for registered messaging services through other directly connected systems. Messaging systems directly connected to MCI Mail are called "Remote Electronic Mail Systems" (REMS), and those that can be reached through the REMS are called "Very Remote Electronic Mail Systems" (VREMS). The following example shows the addressing conventions used by MCI Mail to permit its users to send messages to other message services:

```
To: Alexander Trevor (EMS)
    EMS: COMPUSERVE
    MBX: [70000,130]
```

This is an example of a message addressed to a user of Compuserve's Easyplex service. The EMS name, COMPUSERVE, tells MCI Mail to forward the message to a mailbox associated with the Compuserve system. MCI Mail does not validate the mailbox identifier (MBX: [70000,130]), but this is validated during the message exchange with Compuserve when that system picks up waiting messages and forwards any for MCI Mail (or for systems reachable through MCI Mail).

In the event that a messaging service is reachable through another service, as might be the case for a private mail system served by GTE's Telemail, then the VREMS concept would be invoked. MCI Mail would register the private system in its local tables as a VREMS that is dependent on Telemail for access. An MCI Mail user would send the message to the private system, say, XYZMAIL, in the following fashion:

```
To: Joe User (EMS)
    EMS: XYZMAIL
    MBX: JUSER@HOST2
```

In this case, XYZMAIL would be registered as a VREMS, and Telemail as a REMS, assuming there is a direct link between MCI Mail and

Telemail. The message would be left in the Telemail (REMS) mailbox for pickup and forwarding by that system. Of course, the most likely scenario is that all such messages would be converted into X.400 formats and the exchange accomplished by that means.

Examples of other system interconnections of importance to the business world include links to Telex and its successor, Teletex, as well as interfaces into the hard-copy environment of the domestic and foreign postal and courier services. Integrating access to these services into the computer-based messaging paradigm provides a uniform facility for originating messages, documents, programs, spreadsheets, and other digitally encoded information and sending it to recipients in a variety of environments. The utility of such interconnections is, in part, dependent on the ability of the originating and receiving systems to understand the structure of the objects exchanged and to convert the structure to representations appropriate to the delivery medium. In some instances, the delivery medium may not support an adequate representation (for example, digitized voice could not be readily converted to printed form, short of doing an artificially intelligent voice recognition task!). In those cases, it is essential for the target system to understand the limitations of conversion and to provide notification back to the originator if the message proves undeliverable. An alternative is to deliver that which is deliverable and advise the recipient that some material has been elided.

Besides the problem of routing among interconnected private and public messaging services, the other directory service problem must be addressed: resolving the mailbox address of a named recipient. There appear to be several means by which this may happen. The most obvious is that the target party may provide his or her mailbox address on a business card, in other correspondence, on the telephone, or in a face-to-face meeting. Another possibility is that directory services will be available by telephone, similar to those used in the telephone system. A third possibility is that on-line directory services will be accessible by means of the public data networks or, when available, the integrated services digital network. Yet another alternative is to provide a means for probing well-known directory servers by sending electronic messages to well-known mailboxes. The directory servers would examine the message contents that contain a reference to the desired party and would provide information in return. Of course, a variety of interesting questions arises as to the nature of the information required in the probe, the method by which multiple servers can be probed, if necessary, etc. Static directories in the form of books may also prove useful, just as they are in the telephone network.

## 4.11.2  Accounting

As interesting and complex as the directory services and technical interconnection issues may be, a far more difficult challenge awaits the serious attention of messaging service providers: accounting, interexchange payment, and reconciliation. Here the territory is much less familiar, except for the fact that similarities exist between message system interexchange and the linking of public packet networks, the Telex systems, and the public telephone systems. The most serious challenge is the variation in services provided by messaging systems and the wide range of pricing algorithms employed.

The first step will be to recognize that a distinction needs to be made between the pricing rates of services to subscribers and the accounting rates to be used between public systems. These need not be directly related. Furthermore, the accounting elements must be agreed on, at least bilaterally, in addition to rates. Dealing with transit traffic in which a pair of messaging systems exchange messages through a third system will be necessary to achieve the broadest possible connectivity among systems, although this will require sophisticated data collection and accounting practices to keep the reconciliations straight.

One mistake to avoid, if possible, is that of relying on the target system to provide rated accounting data that can be fed directly into subscriber invoices. One can easily see that the need to accept rated accounting files from $N$ systems before it is possible to produce an invoice for a subscriber creates an $N$-squared file exchange effort. If accounting rates and elements can be agreed on ahead of time, then originating systems will be able to predict with reasonable accuracy the cost of forwarding messages to other services and decouple the rating of subscriber's traffic from the reconciliation and accounting rates needed to settle with other public services. This will not be a simple matter because of the wide range of services that may be accessible through other systems. Examples include information services such as the news clipping service described earlier. Careful segregation of services billed through interexchange and those that might have to be billed directly by the serving system could simplify the reconciliation practice but leave users with multiple invoices.

In this author's opinion, the accounting requirements will not become clear until concrete examples between messaging services are worked out and then fed into the standardization process to guide its more practical aspects. Until this problem is resolved satisfactorily, interconnection cannot become a general reality, although some bilateral or even multilateral agreements will undoubtedly emerge in ad hoc ways over the next year or two.

## 4.12 Summary

This chapter has attempted to offer a view of messaging system technology from the standpoint of its adoption in the business and residential environment. Despite the long history of its use in private situations or in special research environments, the technology has not achieved widespread use in the general business context. The principal barriers are the evolution of an intellectual, equipment, software, accounting, and communication services infrastructure.

There is intense interest among the potential service providers and the software and equipment vendors to spark the growth of these services. Potential users are bombarded with "office automation" concepts, but lack of interoperability and interconnection of the many potential applications among the offered services makes it very difficult to assemble the array of offered facilities into a coherent support environment. These problems are not insuperable, and the high rate of standards development suggests that the business sector and, later, the residential sector can expect increasingly useful and broadly interoperable services and systems from the vendor community.

## 4.13 References

Birrell, A. D., R. Levin, R. M. Needham, and M. D. Schroeder. 1983. Grapevine: Two papers and a report. CSL-83-12. Palo Alto: Xerox Corp., Palo Alto Research Center.

*Note:* The preceding report includes the following:

Birrell, A. D. n.d. The Grapevine interface.

Birrell, A. D., R. Levin, R. M. Needham, and M. D. Schroeder. 1982. Grapevine: An exercise in distributed computing. *Communications of the ACM* 25, no. 4 (April).

Schroeder, M. D., A. D. Birrell, and R. M. Needham. 1984. Experience with Grapevine: The growth of a distributed system. *ACM Transactions on Computer Systems* 2, no. 1 (February).

CCITT. 1981–1984. Recommendation X.400. Final report on the work of Study Group VII during the study period 1981–1984. Document AP VIII-66-E, Report R38. Part III.10, Recommendations X.400 to X.410.

Cerf, V. G. 1984. The MCI Mail architecture. *Proceedings of Networks 84*. Online Conferences, Ltd., Wembly, England (July). (Also available as DPO Document 99 from MCI Digital Information Services Co.; 2000 M Street, NW, Suite 300; Washington D.C. 20036.)

Cerf, V. G., and E. Cain. 1983. The DoD INTERNET architecture model. *Computer Networks* 7, no. 5 (October).

Cohen, D., and J. B. Postel. 1983. Gateways, bridges, and tunnels in computer mail. From *Local networks: Distributed office and factory systems. Proceedings of Localnet 83.* New York.

Crocker, D. 1982. Standard for the format of ARPA INTERNET text messages. RFC-822 (August). Network Information Center, SRI International.

Crocker, D., J. Vittal, and K. Pogran. 1977. Standard for the format of ARPA network text messages. RFC-733 (November). Network Information Center, SRI International.

Crocker, D., E. Szurkowski, and D. Farber. 1979. An internetwork memo distribution capability—MMDF. *Sixth Data Communications Symposium, ACM/IEEE* (November).

Dalal, Y. K. 1981. The information outlet: A new tool for office organization. *Proceedings of the Online Conference on Local Networks and Distributed Office Systems* (May).

Deutsch, D. P., and D. W. Dodds. 1979. Hermes system overview. BBN report no. 4115. Bolt Beranek and Newman, Cambridge, Massachusetts.

———. 1979b. Message systems on the ARPANET: An evolution. Communications techniques seminar (March 20). Princeton University.

Engelbart, D. C. 1963. A conceptual framework for the augmentation of man's intellect. From *Vistas in information handling,* edited by Howerton and Weeks. Washington, D.C.: Spartan Books.

———. 1973. Coordinated information services for a discipline- or mission-oriented community. *Proceedings of the Second Annual Computer Communications Conference.* San Jose, California. (Also, *Proceedings of the NATO Conference on Computer Networks,* University of Sussex, England, September 1974, published in *Computer Communication Networks,* edited by R. L. Grimsdale and F. F. Kuo, Noordhoff, Leyden, 1975.)

———. 1978. Toward integrated, evolutionary office automation systems. *Proceedings of the Joint Engineering Management Conference* (October): 63–68.

———. 1980. Evolving the organization of the future: A point of view. From *Emerging office systems. Proceedings of the Stanford International Symposium on Office Automation* (March), edited by R. Landau, J. Bair, and J. Siegman. Ablex Publications Corp., Norwood, New Jersey.

Garcia-Luna, J., A. Poggio, and D. Elliott. 1984. Research into multimedia message system architecture—Final report. MMM-26, SRI International Project no. 5363 (February).

Israel, J. E., and T. A. Linden. 1982. *Authentication in star and network systems.* Palo Alto: Xerox Corp., Office Systems Division.

Kahn, R. E., Albert Vezza, and Alexander D. Roth, eds. 1981. *Electronic mail and message systems.* Reston, Virginia: AFIPS Press.

Kent, S. T. 1981. Trends for data security and authentication in electronic mail and message systems. From *Electronic mail and message systems: Technical and policy perspectives.* Reston, Virginia: AFIPS Press.

Kluger, L., and J. F. Shoch. 1985. Names, addresses and routes. *UNIX Review* 4, no. 1 (January).

Lewicki, G., D. Cohen, P. Losleben, and D. Trotter. 1984. MOSIS: Present and future. *Proceedings of the MIT Conference on Advanced Research in VLSI* (January). Cambridge, Massachusetts. (Also available as USC-ISI reprint ISI/RS-83-122, March 1984.)

Malone, T. W., S. A. Brobst, K. R. Grant, and M. D. Cohen. 1985. Toward intelligent message routing systems. CISR WP no. 129 (August). Center for Information Systems Research, MIT, Sloan School.

Malone, T. W., K. R. Grant, and F. A. Turbak. 1986. The information lens: An intelligent system for information sharing in organizations. *Proceedings of the CHI 86 Conference on Human Factors in Computing Systems* (April).

Mockapetris, P. 1983a. Domain names—Concepts and facilities. Request for Comments 882 (November). Network Information Center, SRI International.

____. 1983b. Domain names—Implementation and specification. Request for Comments 883 (November). Network Information Center, SRI International.

Mooers, C. D. 1979. Functional description of the Hermes message system. BBN report no. 4122 (October). Bolt Beranek and Newman, Cambridge, Massachusetts.

____. 1982. Hermes guide. BBN report no. 4995. Bolt Beranek and Newman, Cambridge, Massachusetts.

Myer, T. H., and J. J. Vittal. 1977. Message technology in the ARPANET. *Proceedings of the National Telecommunications Conference* (December). Los Angeles, California.

Needham, R. M., and M. D. Schroeder. 1978. Using encryption for authentication in large networks of computers. *Communications of the ACM* 21, no. 12 (December).

Oppen, D. C., and Y. K. Dalal. 1983. The Clearinghouse: A decentralized agent for locating named objects in a distributed environment. *ACM Transactions on Office Information Systems* 1, no. 3 (July).

Pake, G. D. 1985. Research at Xerox PARC, a founder's assessment. *IEEE Spectrum* 22, no.10 (October).

Postel, J. B. 1980a. Internet message protocol. RFC-759 (August). USC/Information Sciences Institute.

____. 1980b. A structured format for transmission of multimedia documents. RFC-767 (August). USC/Information Sciences Institute.

____. 1981a. Internet protocol—DARPA Internet program protocol specification. RFC-791 (September). USC/Information Sciences Institute. (Available from the Network Information Center, SRI International.)

____. 1981b. Transmission control protocol—DARPA Internet program protocol specification. RFC-793 (September). USC/Information Sciences Institute.

____. 1982a. Simple mail transfer protocol. RFC-821 (August). USC/Information Sciences Institute.

____. 1982b. Internet multimedia mail transfer protocol (revised). MMM-11 (March). USC/Information Sciences Institute.

____. 1982c. Internet multimedia mail document format (revised). MMM-12 (March). USC/Information Sciences Institute.

Postel, J. B., C. A. Sunshine, and D. Cohen. 1981. The ARPA Internet protocol. *Computer Networks* 5, no. 4 (July).

Roberts, L. G., and B. D. Wessler. 1970. Computer network development to achieve resource sharing. *AFIPS Conference Proceedings* 36:543.

*Seybold Report*. 1981a. Xerox's Star. *Seybold Report* 10, no. 16 (April).

____. 1981b. The Xerox Star: A professional workstation. *Seybold Report on Word Processing* 4, no. 5 (May).

Shoch, J. F. 1978. Internetwork naming, addressing and routing. *Seventeenth IEEE Computer Society International Conference* (COMPCON) (September).

Shoch, J. F., D. Cohen, and E. A. Taft. 1980. Mutual encapsulation of Internet protocols. Trends and applications: 1980. *Computer Network Protocols* (May). IEEE catalog no. 80CH1529-7C. (Also in *Computer Networks* 5, no. 4, July 1981.)

Thomas, R. H., H. C. Forsdick, T. R. Crowley, G. G. Robertson, R. W. Schaaf, R. S. Tomlinson, and V. M. Travers. 1985. Diamond: A multimedia message system built upon a distributed architecture. *Computer* (December): 65–78.

U.S. Department of Defense. 1981a. Transmission control protocol. MIL-STD-1778. Naval Publications and Forms Center, Code 3015, 5801 Tabor Avenue, Philadelphia, Pennsylvania 19120.

____. 1981b. Internet protocol. MIL-STD-1777. Naval Publications and Forms Center, Code 3015, 5801 Tabor Avenue, Philadelphia, Pennsylvania 19120.

USC ISI. 1984. The MOSIS system (what it is and how to use it). ISI/TM-84-128. USC Information Sciences Institute, 4676 Admiralty Way, Marina del Rey, California 90292-6695.

Xerox Corp. 1985. *Xerox network systems architecture general information manual*. XNXG 068504. Palo Alto, CA: Xerox Corp., Office Systems Division (2300 Geng Road, Palo Alto, California 94303).

# 5

# ELECTRONIC MAIL SYSTEMS

This ~~chapter~~ *paper* ~~discusses~~ *presents* the <u>functions</u>, <u>architectures</u>, and standards associated with today's electronic mail systems. For the purposes of this chapter, an *electronic mail system* is a set of one or more computers, networks, terminals, and other input/output devices that collectively aid its users in the preparation, dissemination, and management of units of communication called *messages*. Although messages are most often composed of text, they may contain other media, such as facsimile images or digitized voice, either singly or in combination.

The rest of this chapter is divided into five major sections. It begins with an overview of what an electronic mail system does for its users. Next, a model for electronic mail systems is explained. The model describes the different components of an electronic mail system and how they interact. The model provides the basis for the discussion in the remaining three sections of the chapter. These sections explore different architectures for electronic mail systems and their trade-offs, discuss the ways users are identified by electronic mail systems, and describe a set of standards for electronic mail systems.

## 5.1  What an Electronic Mail System Does

*An electronic mail system combines word processing, data base management, and telecommunications facilities to provide an integrated office automation tool for its users.* Its functions can be grouped into three categories:

1. *Message Creation.* This category refers to the facilities used to create messages and to specify to whom they are to be sent. For example,

in the course of creating a new message, a user may perform the following: invoke a data-base facility to refer to, or extract information from, an earlier message; employ word processing functions to capture, edit, and format the new message; and then call on a second data base to determine or confirm the proper electronic mail addresses of the message's intended recipients. Draft messages may be stored in a data base before they are sent to anyone.

2. *Message Transfer.* In this category are the facilities used to convey a message from its sender to its recipients. This process can be as simple as adding an entry to a data base. At the other extreme, a message may traverse several communications networks or even have its data converted from one format to another (from text to facsimile, for example) before it is delivered.

3. *Postdelivery Processing.* This category comprises the facilities used to store, retrieve, and examine incoming messages. In addition to these data-base–oriented functions, other, more specialized, processing may be performed. For example, a user may request that incoming messages automatically be forwarded to someone else. Conversely, a return-receipt may be generated once the message has been read, if that action has been requested by the message's originator.

Several characteristics distinguish electronic mail systems from most other communications tools. Perhaps the most important difference is that *messages are sent to particular individuals,* not terminal devices. This means that, provided there is appropriate access to the electronic mail system, a user can move across the hall, across town, or across the continent and still receive mail without having to inform his or her correspondents about the change in location.

Another distinguishing trait of an electronic mail system is that the *transmission of messages is asynchronous.* There is no requirement that the originating and destination systems be available at the same time. Seconds, hours, or even days may pass between the time a message is sent and when it is delivered. Furthermore, a single message may be delivered to each of its recipients at different times. This is possible because electronic mail systems use *store-and-forward* techniques to transfer messages among their users.

Finally, it is important to remember that in an electronic mail system *messages are structured.* The information in a message can be interpreted, and acted on, by an electronic mail system. This is the basis for many of the application-specific functions provided by electronic mail systems.

The following sections describe some capabilities provided by today's electronic mail systems.

### 5.1.1 Message Creation

Every electronic mail system provides its users with a way to enter text. This capability may be as simple as recording whatever a user types, or it may feature a wide array of functions, including text editing, filing, and formatting. Some electronic mail systems will accept text that has been prepared using external facilities, such as word processors. Other office automation systems also can be interfaced to an electronic mail system. An electronic mail system may provide its users with the ability to include facsimile images, spreadsheets, fully formatted documents, or any other information that might be found in the office computer as part of a message.

Electronic mail systems allow users to include handling instructions and other additional information with a message. A subject, cross-references to other messages, a list of recipients for potential replies, and a request for a return receipt are examples of this.

Electronic mail systems provide special facilities to help users reply to incoming messages or to pass them on to new recipients. There is usually a special command for answering a message. When asked to, an electronic mail system can do all the bookkeeping necessary for a response, such as figuring out who should get the reply and cross-referencing the original message in the new one. There are usually facilities for forwarding received messages intact to new recipients or for extracting portions from old messages as part of making a new one.

Electronic mail systems provide a variety of tools to help address messages to their proper recipients. Since many electronic mail systems use special names or addresses to identify their users, some systems provide personal or system-wide directory services. Some systems allow users to define distribution lists that are stored and referenced by name. Some allow nicknames to be defined and used instead of the formal electronic mail address. Some systems can validate a user's electronic mail address before it is used. Finally, people may be able to browse through a directory of users or to query it like a data base.

Sometimes the task of writing a message is a long one and must be interrupted. In certain cases, a user may wish to "sleep" on a message before sending it. That is why many electronic mail systems also allow users to temporarily store outgoing messages before deciding to send them.

### 5.1.2 Message Transfer

When told to deliver a message, most electronic mail systems will try several times before giving up. If delivery should fail, the person who sent the message is told about the problem. Upon request, some electronic mail systems will also inform the sender when delivery is successful.

Some systems allow the sender to give a time by which a message must be delivered. On the other hand, users may be able to ask that a message *not* be delivered until a certain time. Some electronic mail systems, especially ones that are geographically distributed, allow users to assign priorities to messages. Higher priority messages would be transferred before lower priority messages, possibly at an extra charge.

Some electronic mail systems allow the sender of a message to designate an alternate recipient in case it is impossible to deliver the message to the original addressee. Other systems provide a similar function in the form of a dead-letter office. Some electronic mail systems allow their users to redirect their incoming mail. For example, someone going on vacation might ask that messages be delivered to a secretary instead. In that case, there might also be facilities allowing the sender of a message to mark it "private" or to ask that it not be redirected.

Finally, some electronic mail systems are joined with other communications facilities, such as Telex or Teletex. Because each communications system has its own rules for the kind of information it can carry and the way that information is to be represented, messages may have to be converted from one format to another when they cross the boundary between systems. Even inside a single electronic mail system, users may have access to different types of terminals and other input/output devices. For example, one user may have a graphics terminal, while another has a less sophisticated device that can handle only text. If the first user tries to send a message containing a picture to the second user, the second user may not be able to see the picture, or delivery might fail. An electronic mail system might keep track of which users can receive which kinds of information, so that these types of problems can be avoided. In addition, an electronic mail system may be able to translate one kind of information into another. For example, a page of text would be transformed into a page of facsimile. These capabilities are not widely implemented now, but they are becoming more common as electronic mail systems are joined with each other and with other communications systems.

### 5.1.3   Postdelivery Processing

Electronic mail systems tell their users when there are new messages waiting to be read. This is always done at the beginning of a session. Some systems also inform their users about new mail during a session. The notification usually includes the sender of the message, its subject, and its length. After reading an incoming message, a user has the option of saving or deleting it. (The default varies from system to system.) Messages may be saved for the length of a session or for days, months, or even years.

Reading a message can trigger some special processing. For example, a return receipt may be sent. A recipient may be able to control certain aspects of the message's appearance, such as which parts of a message are displayed, the order in which they are displayed, or line widths.

Many systems allow users to organize their saved messages into electronic file folders. Because the message is in a computer instead of on paper, it can be in several folders at the same time. In addition, most electronic mail systems permit users to retrieve messages based on a variety of attributes, such as who sent it, when it was received, or what the subject was.

# 5.2 Model for Electronic Mail Systems

Some special terminology has been developed to name the parts of electronic mail systems and to describe how they interact. A *functional model* for electronic mail systems provides logical abstractions that clarify the operation of electronic mail systems and define when responsibility for a message changes hands. A *layered model* for electronic mail systems shows how their protocols fit in the seven-layered Basic Reference Model for Open Systems Interconnection (the OSI reference model). These models were originally developed by the International Federation for Information Processing's Working Group 6.5 and were further enhanced by the CCITT working group that developed the X.400 series of recommendations.

Certain words, such as "submit" and "deliver," have special meanings when used in the electronic mail system literature. These meanings are defined in terms of the two models.

## 5.2.1 Functional Model for Electronic Mail Systems

The functional model for electronic mail systems deals with units of communication called *messages* and with the abstract functional entities that act on them. Since the model is independent of any particular implementation, it serves as a reference point and provides a common vocabulary for describing all electronic mail systems.

The functional model treats a collection of interconnected mail systems as a single logical entity called a *message handling system* (see Fig. 5.1). A message handling system is said to be composed of a number of *User Agents* (UAs) and a *Message Transfer System* (MTS). A User Agent is a logical entity that acts on behalf of a single user. It provides all functions related to message creation and postdelivery processing. The Message Transfer System transmits messages between User Agents. The

MTS may perform routing and temporary storage functions, among others, in order to do its work.

**Fig. 5 .1.**
**Message**
**handling**
**system.**

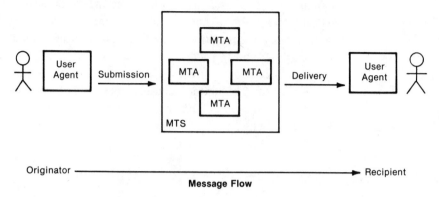

The transfer of responsibility for a message from a User Agent to the Message Transfer System is called *submission*. The User Agent that submits a message on behalf of its user is called the *originating* User Agent; its user is the message's *originator*. Conversely, the transfer of responsibility for a message from the Message Transfer System to a User Agent is called *delivery*. The User Agent that takes delivery of a message acts on behalf of the message's *recipient*. A single message can be sent to multiple recipients, but it can have only one originator.

Delivery of a message does not imply that its recipient has seen it or even knows that it exists. *Receipt* is a term that denotes the action or actions by which a recipient becomes aware of a message and what it says. One interpretation of this is that receipt has occurred after the body of the "message content" has been displayed to the user for the first time.

The Message Transfer System is composed of one or more *Message Transfer Agents*, or *MTAs*. Sometimes the same MTA that accepts submission of a message also performs delivery. In other cases, it is necessary to *relay* a message from one MTA to another until it reaches an MTA that can perform delivery. For example, if only some MTAs have access to the proper long-distance communications paths, a message addressed to a distant UA might be relayed in several stages. The message would go from the first MTA to one nearby that could perform long-distance communications, from that second MTA to another located close to the final destination, and finally to the MTA that would deliver the message. Using relays also eliminates the need to have all UAs and MTAs available on a 24-hour basis. This store-and-forward action makes it feasible to treat electronic mail components like any other office equipment that gets turned off at night.

The notion of responsibility is very important in the functional model. The concept of a *management domain,* or *MD,* is used to divide the message system into distinct parts for which different organizations or individuals are responsible. A very general definition of management domain would say that a management domain contains either one or more MTAs, one or more UAs, or a combination of MTAs and UAs. The commonly accepted definition (as adopted by standards organizations) is a bit more specific. It requires that a management domain contain at least one MTA and zero or more UAs. As shown in Fig. 5.2, management domains can be hierarchically divided into subdomains in order to further delegate responsibility.

**Fig. 5.2 Management domains.**

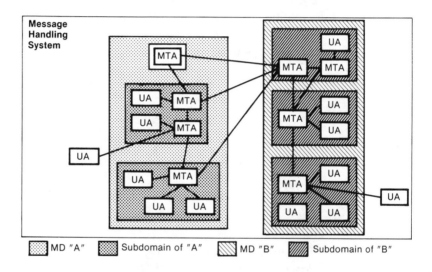

As has already been noted, a message is the basic unit of communication in the functional model. A message is divided into two parts, a *content* and an *envelope* (see Fig. 5.3). The content is information that is dealt with by User Agents. The envelope is processed by the MTS.

**Fig. 5.3. Message structure.**

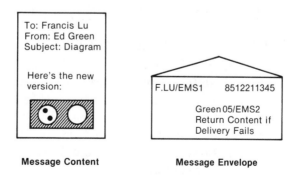

The message content is often modelled as a memorandum, divided into a *heading* and a *body* (see Fig. 5.4). The body contains the primary information intended for the message recipient. It may be composed of text, facsimile images, digitized voice, or any other binary-coded medium, either alone or in combination. The heading is divided into a number of distinct *headers*, each of which conveys some particular information about the message. A header may tell whom the message is from, for whom it is intended, what the message is about, or where replies should go. A header may also suggest how the message should be processed after it is delivered, indicate how to refer to the message in future messages, or contain references to other messages. Most, if not all, of the information in a message's heading is made available to its recipient by his or her UA. Like the content, the message heading may include any type of binary-coded information.

Fig. 5.4.
Structure of
message
content.

When a message is being transferred between UAs, its content is enclosed in another layer of information called an envelope. The envelope gives instructions regarding the delivery of the message and carries a record of what has happened to it so far. Some of the information that makes up the envelope may be redundant with the message content. This is because MTAs do not read or alter the message content unless the originating UA specifically requests that they do so. The exact nature of the message envelope and content is discussed later in this chapter.

## 5.2.2   A Layered Model for Electronic Mail Systems

The purpose of the layered model for electronic mail systems is to identify the different protocols used by electronic mail systems and to show how they are related to each other and the rest of the telecommunications environment. It accomplishes this by adopting the principles of the OSI reference model, as defined by the International Organization for Standardization (ISO) and the International Telephone and Telegraph Consultative Committee (CCITT).

The notion of layering embodies a few powerful concepts. Each protocol layer (layer *n*) provides a set of services to its user in the next higher layer (layer *n+1*) by building on the services provided to it by the next lower layer (layer *n−1*). Layers are defined in such a way that similar functions are grouped together and interactions across layer boundaries are minimized. In addition, layers are defined so that different protocols may be used within a layer without affecting the layer service definition. This is sometimes called *layer independence.* Layer independence means that the implementation details of a layer are considered hidden from its users. This makes it possible, for example, for the same transport protocol to be used on packet-switched as well as circuit-switched networks.

The interactions that occur between a layer and its users as services are requested and supplied are specified by abstract *service primitives.* A service primitive describes each service as a series of one or more interactions. For each interaction, the information that passes between the layer and one of its users is specified. Service primitives do not specify any implementation-dependent details of the interface between a layer and its users, such as how information is to be represented or the mechanism by which it is to be passed.

The OSI reference model defines seven protocol layers, ranging from the physical layer (layer 1) on the bottom to the application layer (layer 7) on top. Readers interested in a complete description of the OSI reference model should obtain a copy of ISO 7498 or CCITT Recommendation X.209, *Basic Reference Model for Open Systems Interconnection.*

The layered model of an electronic mail system is based on the functional model. The message handling system is said to be divided into two protocol layers, the *User Agent Layer (UAL)* and the *Message Transfer Layer (MTL).* The UAL and MTL are not new layers to be added to the OSI reference model. The term "layer" is used here simply to indicate that the principles of layering have been applied. In terms of the OSI reference model, the protocols used within the UAL and the MTL are all application layer protocols.

The UAL is the upper of the two layers. It contains a number of protocol entities called *UA Entities* (UAEs). Each UAE embodies the protocol-related functions of a single UA. UAEs cooperate to provide the services of the UAL. An example of a UAL service is receipt notification, in which a message's originator is notified after a particular recipient has read the message.

The MTL is the lower of the two layers. It contains two kinds of protocol entities, *MTA Entities* (MTAEs) and *Submission and Delivery Entities* (SDEs). Each MTAE embodies the protocol-related functions of a single MTA. MTAEs cooperate to provide the services of the MTL. An example of an MTL service is delivery notification, in which the originator of a message is notified after the message has been successfully delivered to a particular recipient's UA. The purpose of an SDE is to make the services of the MTL available to a UAE. An SDE is required in any system that contains a UAE but does not have an MTAE.

In summary, the users of the message handling system are the users of the UAL, which contains as many UAEs as there are users; UAEs are the users of the MTL, which contains a number of MTAEs and SDEs; and MTAEs and SDEs are all users of the presentation layer (layer 6 of the OSI reference model).

The layered model uses three kinds of protocols, $P_1$, $P_3$, and $P_c$. $P_1$ is the *Message Transfer Protocol*. It is used when relaying messages between MTAEs. $P_3$ is the *Submission and Delivery Protocol*. It is used between an SDE and an MTAE in order to provide a UAE with access to the MTL. $P_c$ defines the syntax and semantics of a message content being transferred. Later in this chapter there is a discussion of the CCITT P2 protocol (Recommendation X.420), which is an example of $P_c$. Fig. 5.5 illustrates UAEs, MTAEs, SDEs, and their protocols.

Fig. 5.5. UAEs, MTAEs, SDEs, and their protocols.

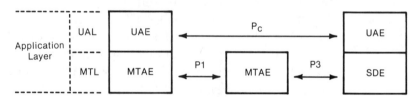

It is important to remember that $P_c$ is not a protocol between a host and a terminal. A terminal, or any input/output device that provides access to a UA, is considered part of that UA. The means by which a terminal communicates with the rest of its UA is not a formal part of the model. However, a class of *Interactive Terminal to System Protocols* ($P_t$) is identified for that purpose. ($P_t$ has also been called $P_4$ in some early documents on this subject.)

# 5.3   Architectures for Electronic Mail Systems

The architecture of an electronic mail system can be described from several viewpoints. Its *physical mapping* allocates the logical entities of the functional and layered models among a number of computer systems. Its *protocol set* describes how information is communicated from one part of the electronic mail system to another. Finally, its *implementation algorithms* tell how each part of the electronic mail system does its job. This section explains each of these three areas by giving some representative examples and discussing their design trade-offs.

## 5.3.1 Physical Mappings

The physical mapping of an electronic mail system relates the various logical entities, such as User Agents and Message Transfer Agents, to the system's software and hardware architecture. Two entities are said to be *coresident* if they are located in the same processing system. It is also possible to implement a single logical entity in a distributed manner, such that it resides in a number of processing systems at once. Some electronic mail systems employ the same set of physical mappings throughout. Others are heterogeneous. Fig. 5.6 shows some typical physical mappings.

Fig. 5.6. Typical physical mappings for an electronic mail system.

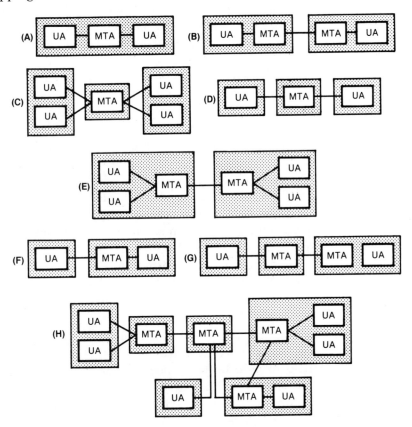

Examples A through E in Fig. 5.6 are fairly simple. The first one shows an entire electronic mail system implemented on a single processing system. This is how many electronic mail systems are implemented. Users access the system via local terminals or employ a communications network, such as the telephone system or a public data network, to connect their terminals to their User Agents. In some cases, a number of processors are teamed to provide greater reliability or enhanced performance. Although there may be a number of internal communications interfaces, this type of system implements neither $P_1$ nor $P_3$ protocols. It could be argued, however, that some $P_c$ is implemented between the User Agents, since they are exchanging information and cooperating as peers, even though they are in the same system.

The second example, B, in the figure illustrates how an electronic mail system for networked personal workstations could be designed. Each workstation has sufficient storage to temporarily buffer incoming and outgoing messages, in addition to more permanently saving draft and received messages. The workstations also have enough memory and the computational capability required to act as MTAs as well as UAs. (An MTA may be required to implement seven layers' worth of communications protocols.) Systems such as this one implement the $P_1$ and $P_c$ protocols.

Examples C and D are similar. In the first case, the processing systems on the left and right have sufficient capability to support several User Agents; in the second case, there is only one User Agent in each end-system. A system such as C might use shared minicomputers to implement User Agents. On the other hand, a system such as D could use personal computers for the same purpose. Given that there could be more than one MTA included in the system, either of these architectures could be used to provide a distributed, private electronic mail system for a corporation or other organization. Systems built according to these architectures are very modular, making expansion to accommodate new users or locations relatively easy and economical. Since the UAEs and MTAEs are in different processing systems, each end-system must contain an SDE in addition to its complement of UAEs. Systems such as C and D implement the $P_c$ and $P_3$ protocols. If there are multiple MTAs present, the $P_1$ protocol is also required.

The architecture of E shows what would happen if two of the systems illustrated in A were joined together. As in A, the processors used would have to be fairly powerful, since they must support all the functions of UAs and MTAs. This architecture may be chosen over A when the user population is concentrated in a number of geographically dispersed locations, since placing a mainframe near each concentration of users would likely reduce communications cost and overhead. Architecture E is also used when mixed-use mainframe computers are networked together. Systems of this type implement the $P_c$ and $P_1$ protocols.

Illustrations F through H show some heterogeneous architectures.

The advantage of a heterogeneous architecture is the ability to tailor a single electronic mail system to meet the requirements of different user populations by operating on a diverse hardware base. While F and G are both fairly simple examples, H begins to hint at the diversity that is possible within a single system. Further, it should be noted that the paths shown connecting the entities in this example may each be complex communications systems, not necessarily simple direct lines.

Especially in the case of mixed-use computers, it should be noted that a single electronic mail system may include several different kinds of UAs and MTAs. In fact, there may be more than one type of UA present in the same computer. In that situation, there is often a common internal interface between the UAEs and the SDE or MTAE. For example, outgoing messages would be placed in files with special names and formats, with an analogous method used for incoming messages. Common procedure calls or use of an interprocess communications facility would serve the same purpose.

The use of a file or files to hold incoming messages is extremely common. The terms *mailbox* and *inbox* are often used to denote this file or set of files. While the concept of a mailbox is appealing because it seems so intuitive, it is also very ambiguous. One cannot always say with certainty if it is the UA or the MTA that is responsible for a mailbox. It can be implemented either way, or the responsibility can be shared. That is why "mailbox" does not appear in the functional model and is why delivery is defined simply as a transfer of responsibility, not as the placement of a message in a special storage area.

Just as there can be more than one UA in a processing system, it is also possible to implement a single UA that is distributed among several systems. For example, consider the case in which a number of personal workstations are located on a local area network, along with a file server and a processor that serves as a gateway to external networks. The MTA resides in the gateway machine. When an incoming message arrives at the MTA, it asks the file server to store it in a special file associated with the intended recipient's UA. This special file cannot ordinarily be read or altered by any user other than the recipient; once the incoming message has been stored there, the MTA cannot call it back or write over it. The recipient's UA periodically checks the special file to see if new mail has arrived.

In this example it is clear that responsibility for the message passes from the MTA to the UA once an incoming message has been stored by the file server. Storage by the file server means that delivery has taken place and that the special file in which the incoming message is stored is part of the User Agent. So, in this case, the UA is distributed across its user's workstation and the file server machine. (In fact, the file server supports a similar piece of each UA on the local area network.) Similarly, it is also possible to distribute the MTA or to distribute UAs along other lines, such as placing their user interface ("front-end") functions in intelligent terminals with their "back-end" functions elsewhere.

## 5.3.2   Protocol Sets

There are nearly as many different electronic mail protocols as there are electronic mail systems. All electronic mails systems implement a $P_c$ protocol, and those that operate over multiple systems implement a $P_1$ protocol, a $P_3$ protocol, or both. This section discusses the major problems that must be solved by protocols in the UAL and the MTL and explores some common approaches to them.

### The User Agent Layer

While certain User Agent functions, such as word processing, are provided to a user by his or her UA working alone, other services, such as sending a return receipt once a recipient knows about a message, require cooperation among UAs. This cooperation can take a number of forms. Sometimes, as in the case of a return receipt, a recipient's UA performs special functions at the express request of a message's originator. Most often, however, UAs cooperate by making information available to their users. For example, a message recipient's UA cooperates with the originator's UA when a list of the message's secondary (*Cc:*) recipients is shown to the recipient. Cooperation is required in this case because the originator's UA had to represent that information in a way that the recipient's UA could understand.

The purpose of any $P_c$ protocol is to facilitate cooperation among User Agents. A $P_c$ protocol does this by accomplishing two tasks. First, it defines a set of *cooperating UA services* that will be supported by the protocol. The definition of each service gives its purpose and says what information must accompany a request for the service and what each UA must do to help fulfill the request. Second, a $P_c$ protocol defines what information must be transferred from one UA to another in order to provide the cooperating UA services. The encoding rules for the information in the message content may also be defined by a $P_c$ protocol, although, strictly speaking, the representation of information is a function belonging to the presentation layer. The encoding rules give the *syntax* of the message content; the service definitions give its *semantics*.

The majority of communications protocols are designed to allow two entities to have an electronic "conversation." There is a beginning phase, during which the ground rules for the conversation are agreed to, a data transfer phase, in which information is exchanged, and a terminating phase, in which the two entities agree to end their conversation, and then do so. Over the course of the conversation, the entities may also exchange status information or invoke some of the ground rules in order to keep everything running smoothly. In some cases, the entities take turns talking. In others, one entity may make a number of statements before it must wait for a response from the other. This sort of protocol is said to be *connection-oriented*.

A $P_c$ protocol is not connection-oriented. There is no conversation, with a beginning, a middle, and an end. The operation of a $P_c$ protocol is more akin to sending telegrams. One User Agent sends a $P_c$ data unit (a message content) to a second UA, makes note of the fact, and attends to other tasks. Some time later, perhaps after both UAs have sent and received a number of additional $P_c$ data units (including communicating with each other and with third parties), the second UA may respond to the first $P_c$ data unit by sending a new $P_c$ data unit to the first UA. This method of operation reflects the asynchronous nature of electronic mail systems. As a result, a user can send an arbitrary number of messages before receiving responses to any, even if responses had been requested.

$P_c$ protocols are differentiated by the sets of services they support and by the way their $P_c$ data units are represented. Certain cooperating User Agent services are almost universally implemented. Among these are indication of primary and secondary recipients (*To:* and *Cc:*), authorizing users indication and originator indication (*From:* and *Sender:*), and subject indication (*Subject:* or *Re:*). Other services, such as replying message indication (*In-reply-to:*), blind copy recipient indication (*Bcc:*), and receipt notification (*Return-receipt-to:*) are also widely implemented, but with varying definitions.

An implementation of replying message indication depends on whether the particular electronic mail system supports the notion of a *message-identifier*. The value of a message-identifier is intended to unambiguously identify all copies of the same message content. Systems that assign message-identifiers to new messages also may use them when formulating replies. This is done by using the message-identifier from the old message as the value of the replying message indication in the new message. On the other hand, systems that do not implement message-identifiers often generate a text string based on some other information from the original message, which then serves as the value of the replying message indication (e.g., "Fernando Garcia's message of P2 January 1987").

In the case of blind copy recipient indication, the differences concern which recipients of a message are told about each member of its *Bcc:* list. In most systems, recipients who are not included in the *Bcc:* list get a version of the message content that does not include a blind copy recipient indication. (The message's authorizing user or originator may be given copies that contain the entire *Bcc:* list.) A member of the *Bcc:* list is sent a message that has either the entire *Bcc:* list or a *Bcc:* list containing that recipient alone.

Implementations of receipt notification vary in two areas. Few electronic mail systems implement the same notion of "receipt." At one extreme, some systems define receipt as occurring when a recipient is first notified that the message has arrived. At the other extreme, receipt is defined as occurring only after the recipient has seen or heard the entire body of the message content. Once receipt has occurred, some systems automatically send a receipt notification back to the originator's User Agent; other systems involve the recipient in the process by asking permission, for example. Some systems that automatically send return receipts specifically warn recipients about messages for which return receipts have been requested. This allows the recipient to choose to discard such a message without ever reading it. In that case, no return receipt would be sent.

The choice of the "right" definitions for cooperating User Agent services is very often a matter of organizational policy. However, when different electronic mail systems are connected together, differences in service definitions can have adverse effects. In some cases, errors will be caused. For example, a system that expects a message-identifier as the value of a replying message indication may not be able to handle a message that contains a free-form text string instead. The problems caused by differences in user expectations are even more serious. For example, if the originator of a message understands that getting a return receipt back means that the message has been read and confirmed by the recipient, and the recipient's system automatically generates a return receipt after telling its user that the message is waiting to be read, the originator may be led to believe that a message has been read and understood when it has not been examined at all. Standards for interconnecting electronic mail systems supply common definitions for cooperating User Agent services in order to avoid this sort of problem.

The data representation scheme used to provide cooperating User Agent services must be *machine-readable*. In other words, the bits and bytes that make up a message content must follow a regular format that can be interpreted by a computer. If this were not the case, a UA would not be able to extract and act on the information that it finds in a message content. It would be able to pass the entire content to its user verbatim, but it could perform none of the cooperating User Agent services that require specialized processing. In addition to allowing a User Agent to identify and understand the different indications and notifications that, together with the body, make up the message content, some data representation schemes also include information to control the *rendition* of the message content. This dictates various aspects of the appearance of the message, especially the treatment of text. Examples of different rendition characteristics include paper size and orientation, choice of font, line and page breaks, and so on.

While most electronic mail systems support very similar sets of cooperating User Agent services, there are significant differences in their approaches to data representation. Some systems use text-based encoding schemes. Others use byte-oriented binary encoding. The following discussion explores each of these methods by describing a simple example of each kind of encoding scheme, representative of actual systems, and examining the trade-offs between them.

### An Example of Text-based Encoding

This section describes a text-based encoding scheme that uses punctuation, linear white space, and line breaks to divide a message content into its major constituent parts. (Linear white space consists of nonprinting characters, such as spaces and tabs, used to separate printing characters, such as alphanumerics and punctuation marks, on the same line.) Each indication or notification is usually represented by one or more lines of text, called *headers*. The collection of all headers present in a message content is called its *heading*. The body of the message content appears after the heading, separated from it by one or more empty lines.

Each header is composed of two parts, a *header-label* and a *header-content*. The header-label identifies the purpose of the header, for example, conveying the subject indication. The header-content supplies the information required to fulfill the purpose of the header. In the case of subject indication, the header-content would consist of a few words that describe the information in the body of the message content. In this example, the header-label is represented by using alphanumeric characters plus hyphens and is always terminated by a colon. It begins at the first character on a line of text. The header-label and the header-content are separated by a linear white space. The following are examples of valid header-labels:

```
To:
From:
In-reply-to:
Subject:
```

Each header-content is composed of printing characters and linear white space and is terminated with a carriage-return character followed by a linefeed character. The text found in the header-content can be free-form or highly structured. Structuring is necessary when specific types of data, such as dates or electronic mail addresses, are to be represented in a form that can be understood by a computer. For example, a date and time stamp might be represented in the form

<DD>-<MMM>-<YY> <linear-white-space> <hh>:<mm>:<ss>
<part-of-day>-<ZZZ>

where

| | |
|---|---|
| <DD> | is two digits representing the day. |
| <MMM> | is three alphabetic characters representing the month. |
| <YY> | is two digits representing the year. |
| <hh> | is two digits representing hours, using a 12-hour clock. |
| <mm> | is two digits representing minutes. |
| <ss> | is two digits representing seconds. |
| <part-of-day> | is either the character string "AM" or the character string "PM." |
| <ZZZ> | is three characters designating the time zone. |

03-MAR-82 12:36:00PM-EST is an example of a date and time stamp using these rules.

Representing dates and times in this form allows UAs to interpret the information in the header-content. For example, consider a cooperating User Agent service called Expiry Date Indication, which allows an originator to indicate a date and time after which the message is to be considered invalid. A recipient's UA might encounter a message containing the following header:

Expiry-Date: 30-Sep-87 05:00:00PM-PDT

which would mean that the message's originator will consider the message to be invalid after five in the afternoon, Pacific daylight time, on September 30, 1987. If the date in the header-contents were not in a machine-readable form, the only processing that the recipient's UA could perform on the Expiry-Date header would be to pass the header-contents to its user. Because the information is appropriately structured, the recipient's UA can understand it and act on it; for example, automatically deleting or archiving the message after it had expired. Furthermore, the UA becomes able to help its user manage his or her store of received messages by allowing messages to be retrieved or organized based on their expiry-dates.

The names and addresses by which users are known to electronic mail systems are another example of structured information that can be found in headers. For the purposes of this very simple example, a user's name will be represented by a series of alphabetic characters. A list of users' names will be represented by a series of individual names, separated by a comma character and a linear white space.

Given these rules, a message could be represented as follows (the ¬ stands for the end-of-line sequence):

```
To: Maria, George, Andre¬
Cc: Kim, Mariko, Fred¬
From: Sean¬
Reply-to: Fred¬
Reply-by: 23-JUN-87 09:15:00AM-CDT¬
Message-ID: 998AXX000037¬
Subject: Company picnic¬
¬
Don't forget that the company picnic is only a few weeks away!¬
Please tell Fred if you plan to come. He will also want to know¬
how many adults and children will be in your group.¬
¬
Last year's outing was a great success. This year it should be¬
even better. Koko the clown will be there to entertain the¬
children, and there will be a number of sailboats available for use¬
by adults.¬
```

In this example, the *To:* and *Cc:* headers convey the information associated with indication of primary recipients and indication of secondary recipients. The *From:* header contains the authorizing user indication. The *Reply-to:* and *Reply-by:* headers together convey the information in a reply request indication, which tells the recipient's UA that any replies should be directed to Fred and that they are expected to be received by 9:15 on the morning of June 23. The *Message-ID:* header contains the unique message-identifier assigned to the message content. The *Subject:* header conveys the subject indication, summarizing the information in the message body. The following example shows how the header-contents in this message might be reflected in a reply message:

```
To: Fred¬
From: Mariko¬
```

In-reply-to: 998AXX000037⌐
Message-ID: 998GHK1010425⌐
Subject: Company picnic⌐
⌐
I plan to attend the company picnic with my husband and our two⌐
older children. Will there be supervision for very young⌐
children? My youngest is only 12 months old. I will bring him if⌐
there will be someone to look after the babies. Otherwise I will⌐
arrange for a sitter.⌐

This encoding scheme is simple and moderately powerful. However, it does have some limitations, especially if headers are to be formatted by people using text editors and if messages are to be displayed in the same format as they are received. For example, since a header must fit on one line, the width of a page or terminal screen places a practical limit on its length (about 70 characters). Anything longer would run off the edge of the page or screen when displayed "as is." Text-based encoding schemes overcome this limitation by providing for multiline headers. For example, the header-contents of contiguous headers with the same header-label could be treated as one, or a header-content could be continued by beginning the following line with a linear white space.

The format used for dates is another drawback to this encoding scheme. It is very rigid, and may seem unnatural to users if they must follow it when typing. One solution is for User Agents to allow a variety of forms for entering dates and then to convert what the user types to the "standard" format. Another solution is to allow a variety of formats for dates and times to be used in messages. Either approach increases the complexity of the software.

The simplicity of the format used for user names limits the usefulness of the example's encoding scheme. The very nature of a large, distributed user population would make it difficult for both administrators and users to keep track of the assigned names. Furthermore, the format makes no provision for the use of distribution lists or interworking with the conventional mail or Telex service. These features may be required by some users. The role and the structure of names are discussed at length later in this chapter.

Finally, it should be noted that this encoding scheme lacks a set of conventions for header-labels. There must be definitions for which headers are to be used to support which cooperating User Agent services. There must also be definitions for the values of their header-labels and for the bounds, if any, on the values of the corresponding header-contents. These conventions would also specify the matching rules for header-labels; for example, comparisons might be upper- or lowercase insensitive (Subject:, SUBJECT:, subject:, and SuBJEct: would all be considered the same), or there might even be alternate header-label values used to denote the same header (for example, Subject: and Re:).

In general, the advantages of a text-based encoding scheme like the one in the example are the simplicity of the software required to create and display messages and the ease of extensibility. An outgoing message in a text-based encoding scheme such as this one could be prepared by using a general-purpose text editor or word-processing program. (The person typing the message would have to be very careful to follow the formatting rules, though.) An incoming message could be printed as if it were an ordinary text stream. Finally, text-based encoding schemes are easily extensible. While a recipient's User Agent might not be able to determine the specific meaning of an unfamiliar header, it would always be able to display it to its user.

The major disadvantages of a text-based encoding scheme are the overhead and expense required to portray nontextual binary-encoded information, such as facsimile or digitized voice; the relative difficulty (when compared with binary encoding) of parsing the headers; and its ties to human language (English in the examples given here).

There are a number of ways to represent binary information using ASCII or other character sets. For example, successive groups of 4 bits could each be represented as a hexadecimal character ("0" through "9," "A" through "F"). This mapping is extremely easy to compute, but the resulting representation is twice as large as the original. Other possible mappings utilize space more efficiently but require more computation or include control characters in their output. Even small amounts of computation add up quickly, since an ordinary facsimile image can contain millions of bits. Control characters can interfere with certain data communication protocols, which assign special meaning to them. (Historically, control characters were intended for just that purpose.)

Parsing a very simple text-based encoded scheme, such as the example given previously, is fairly straightforward. Assuming each message is stored alone in a file, the logic for the routine might be as follows:

```
begin
    Open the file to its beginning
    repeat
        begin (read each header and process it)
            Read each character until the next colon or end-of-line or
                end-of-file
```

```
                    If we stopped at an end-of-file then
                    begin
                        Process the error
                    end
                  else if we stopped at an end-of-line then
                    begin
                        if the line was empty then remember that
                        else
                            begin
                                Process the error
                            end
                    end
                  else (we stopped at a colon, so this is a legitimate)
                       (header)
                    begin
                        Match the characters that we've just read
                            against our list of known header-labels
                        If we recognize the header-label then
                    begin
                        Read the rest of the line
                            (that's the header-content)
                        Process it according to the rules for
                            the particular header
                            (which may include printing all or)
                            (part of the header)
                    end
                        else (deal with unfamiliar headers)
                            begin
                                Print the header-label and a colon
                                Read and print the rest of the line
                            end
                    end
                end
              until an empty line  (this marks the beginning of the message)
                                   (content's body)
                  or an error is encountered
              Print the rest of the file  (in this example, the content body is)
                                          (unstructured text)
            Close the file
        end
```

This procedure is very simple because the encoding scheme is very simple. If multiline headers were allowed, it would become somewhat more complex. The procedure is deceptive, however, because it hides the amount of work that may be required to parse header-contents. Any structured header-content must be parsed with a routine whose complexity reflects the nature of the structuring.

The problem addressed by any encoding scheme for message contents goes beyond identifying and delimiting each header. For each header, there must also be a means of identifying and delimiting each of the different parts of its header-contents. To further complicate the matter, there may be hierarchy or nesting present in the information to be represented. This must also be accounted for by the encoding scheme. In addition, user friendliness or other functional requirements—combined with the desire to permit headers to be entered by, and displayed to, users in the same form in which they are sent and received by User Agents—may dictate that several distinct, alternative forms for the same part of a header-contents be permitted. User names and addresses are examples of information that can have a very complex structure and a variety of permitted forms.

Text characters can take on special meanings inside a header-contents. Linear white spaces, punctuation characters, parentheses, and square, angle, and curly braces are all commonly used as delimiters. Position, choice of delimiter, choice of structure (which can imply a choice of data type), or explicit labeling may be used to identify each part of the header-contents. The same character might have different meanings, depending on the context in which it is found.

As the number of different kinds of headers grows, so does the number of routines needed to parse them. Depending on the complexity of the encoding scheme, these routines can grow large and intricate. The cost of developing the parsing software and the amount of memory and storage space that it occupies put text-based encoding schemes at a disadvantage when compared with binary ones.

How human language can be a factor in a text-based encoding scheme is illustrated in the previous example. English words are used as header-labels. English words can also be used to identify the various parts of a message's header-contents. This makes perfect sense if the plain text portions of the message, such as the subject indication or the message body, are in English too. If not, unless the message recipient is bilingual, it would be necessary to translate header-labels and other English words used as identifiers into the language used in the rest of the message before they are communicated to the user. It is not difficult to write software that recognizes several sets of words as identifiers, each set belonging to a different language. However, a requirement for this sort of software would imply that it is not feasible to simply display the message in exactly the form in which it was received, and this would eliminate a major advantage of text-based encoding schemes. Language-dependency is often not a factor in an electronic mail system used within a single organization. However, it becomes an important consideration in multilingual communities or in message exchanges among different electronic mail systems.

### An Example of Binary Encoding

The byte-oriented binary encoding schemes implemented by electronic mail systems use binary values to identify and delimit the parts of a message content. As with text-based encoding, a message content is divided into a heading and a body. The differences are the ways in which the parts of the message content are delimited and identified.

Like most other application software, electronic mail systems that employ binary encoding use byte-oriented schemes. That means that all information is represented in units of bytes, most often 8 bits long. A text character might take up 1 byte, an integer value 4 bytes, and so on. Another possibility would be to use bit-oriented binary encoding. This is not commonly done. The choice of byte-orientation reflects the fact that the majority of today's computer architectures are designed around bytes, not bits.

Here is an example of a simple binary encoding scheme for message contents. Each header is composed of an identifier that is 1 byte long, followed by a length code that is also 1 byte long, followed by the header-contents. The identifier tells the purpose of the header. The length code, which is interpreted as an unsigned 8-bit integer, says how many bytes are in the header-content. Each header-content is composed of a series of one or more character strings, date and time stamps, message-identifiers, or user names. Each of these pieces of information is also represented using a 1-byte identifier and a 1-byte length code, followed by the bytes that give the actual data. The message body is represented in the same way, except that it uses a 2-byte length code.

Given these rules, the first of the two messages used in the example of text-based encoding might be represented as follows:

$F1_{16}16_{16}$ $04_{16}05_{16}$Maria $04_{16}06_{16}$George $04_{16}05_{16}$Andre
$F2_{16}13_{16}$ $04_{16}03_{16}$Kim $04_{16}06_{16}$Mariko $04_{16}04_{16}$Fred
$F3_{16}06_{16}$ $04_{16}04_{16}$Sean
$F4_{16}06_{16}$ $04_{16}04_{16}$Fred
$F5_{16}09_{16}$ $02_{16}07_{16}570617100F00F7_{16}$
$F6_{16}10_{16}$ $03_{16}OE_{16}$ $01_{16}0C_{16}998AXX000037$

$F7_{16}10_{16}$ $01_{16}OE_{16}$Company picnic

$FF_{16}0182_{16}$ $01_{16}B4_{16}$ Don't forget that the company picnic is only a few weeks away! Please tell Fred if you plan to come. He will also want to know how many adults and children will be in your group.

$01_{16}CA_{16}$Last year's outing was a great success. This year it should be even better. Koko the clown will be there to entertain the children, and there will be a number of sailboats available for use by adults.

Key:
$01_{16}$ = identifier for character string
$02_{16}$ = identifier for date and time stamp
$03_{16}$ = identifier for message-identifier
$04_{16}$ = identifier for user name
$F1_{16}$ = identifier for primary recipients header
$F2_{16}$ = identifier for secondary recipients header
$F3_{16}$ = identifier for authorizing user header
$F4_{16}$ = identifier for reply-to header
$F5_{16}$ = identifier for reply-by header
$F6_{16}$ = identifier for message-content-identification header
$F7_{16}$ = identifier for subject header
$FF_{16}$ = identifier for body

Encoding of date and time stamps:
Byte 1 = year (starting from 1900)
Byte 2 = month (1 is January)
Byte 3 = day of month
Byte 4 = hours (24-hour clock, using Greenwich mean time)
Byte 5 = minutes
Byte 6 = seconds
Byte 7 = local time zone given as an offset from Greenwich mean time (first four bits F means minus, 0 means plus, last four bits give the number of hours)

*Note:* This encoding of date and time stamps is only an example constructed to correspond with the example of text-based encoding. An actual implementation would be more likely to use a representation based on the computer's internal representation of dates and times or to follow a standard for the representation of dates and times.

This example illustrates the two characteristics that distinguish many binary encoding schemes from text-based encoding schemes. Each unit of information is delimited by use of a byte-count conveyed by the length code, and each unit of information is identified using an explicit binary code.

The explicit length codes in the example are used to delimit each piece of information to be represented. Compared with text-based encoding, this makes finding the end of one data object (and, therefore, the beginning of the next) very easy. The computer just skips over the intervening bytes without having to examine them at all. However, the specific way that lengths are represented does have some shortcomings. The length of each header-content or character string is limited to 255 bytes, and the length of the body is limited to 65,535 bytes. The size of any piece of information must be known before it can be transferred to another system. Also, since there are different rules for representing the lengths of different data objects, the encoding scheme cannot be extended in a backwards-compatible manner.

The absolute limitations imposed by this particular encoding scheme on the sizes of data objects can be overcome in a variety of ways. One way is to make the length codes take up a greater number of bytes. There would still be limitations, though, and, as the sizes of length codes increased, the representation of small objects would become relatively inefficient. Another way is to use self-extending length codes, eliminating practical absolute limits on the sizes of data objects and minimizing the amount of overhead that they impose, at the cost of a small amount of computation. A third way is to represent a large data object as a chain of smaller segments of known size, with a flag marking the last one. This also sidesteps the potential problems posed by very large pieces of data. Certain data objects, such as facsimile images, are so large that some User Agents may not be able to buffer the entire data object at one time, making it impossible for the User Agent to know the final size of the object before it must start transmitting it. The price for the use of chaining is that a User Agent may have to read a segment's length and then skip to the next segment several times before reaching the beginning of the next data object.

Actual binary encoding scheme implementations may use all or any of these techniques. The X.400 series of recommendations for Message Handling Systems uses a combination of self-extending length codes and chaining. Some systems also use fixed-length representations, where a given data object is always the same size and no byte count need appear in the message content. Fixed-length schemes are very efficient, but at the cost of flexibility and extensibility.

*Tagging* is another important aspect of binary encoding schemes. Tagging occurs when a piece of data of a particular type is marked as belonging to another, more specialized type. The example of a message-identifier illustrates this notion. In this case, a message-identifier is made out of a character string. This is shown explicitly in the following encoding of a message-identifier:

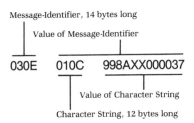

The character string is used as the value of the message-identifier. This type of tagging is called *explicit* tagging because the identifier and the length of the character string are included in the final representation. A second method, called *implicit* tagging, uses only the value of the character string:

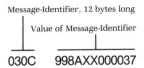

Both methods of encoding mean the same thing: 998AXX000037 is a message-identifier. Explicit tagging allows a variety of underlying data types to be used for the same purpose. For example, integers could be used as well as character strings for the purpose of constructing message-identifiers. Implicit tagging permits no such choice but does reduce the number of bytes needed to represent the desired information. In the example, user names are encoded using implicit tagging.

There are two major advantages to a byte-oriented binary encoding scheme. It is very easy to represent structured information, and the routines required to parse messages can be very simple and small. Binary encoding makes it easy to include nontextual information in a message, and this capability increases the overall usefulness of the electronic mail system.

Because binary encoding schemes can make it easy to identify and delimit individual pieces of data, representing and processing nested or structured data is also simplified. In the example, the primary-recipients header contains a series of user names. Because each user name is a separate data element, and because it can be skipped over without the need to examine its value, operations such as counting the number of names in the header or extracting one of them, are easily performed by a User Agent. In addition, this same aspect of binary encoding eliminates design problems caused by conflicts between the ways that a certain type of data and the larger structure in which it is to be imbedded are represented, and the limitations that this may impose on users. For example, some electronic mail systems that use text-based encoding do not allow the space character to appear as part of a user name because the space character is used for other purposes. This problem can easily be avoided in systems that use binary encoding.

The routines required to parse binary-encoded messages are smaller and faster than their text-oriented counterparts. The following routine illustrates this point. It is functionally equivalent to the one used for parsing the example of text-based encoding discussed earlier.

```
begin
     Open the file to its beginning
     repeat
         begin (read each header and process it)
             Read the first byte
             If we stopped at an end-of-file then
                 begin
                     Remember that
                 end
             else if the byte we just read is a valid identifier for a
                         header then
                 begin
                     Read the next byte  (that's the length of)
                                         (the header contents)
                     Pass the bytes that make up the header contents
                         to the appropriate routine
                 end
             else if the byte we just read is FF₁₆ then
                 begin (we've found the message body)
                     Read the next two bytes  (that's the length of)
                                              (the message body)
                     Pass the bytes that make up the message body
                         to the appropriate routine
                 end
             else  (the byte doesn't identify anything that we expect)
                 (to find)
```

```
                    begin
                    process the error
                    end
            end
          until an end-of-file or an error is encountered
          Close the file
        end
```

The only comparison made by the routine that deals with binary encoding is to check to see if the identifiers that it reads are valid for the context in which they are found. In contrast, the text-oriented routine must look for colons and end-of-line sequences, as well as compare text strings in order to validate header-labels. The differences between the routines that parse more highly structured data would be even greater. A binary-oriented routine would be quite similar to the one just given; a text-oriented version would be considerably more complex because it would have to look for, and match against, an even larger collection of different individual characters and text strings in order to delimit and identify each piece of data.

Binary encoding gives mail system designers a chance to achieve both user friendliness and ease of computer processing. This is shown in the comparison of the two methods for representing date and time stamps. The text-based encoding scheme compromises between machine readability and user friendliness. The binary encoding scheme uses a representation that would be very easy for a computer to comprehend. Date and time stamps represented this way would be reformatted when they are shown to a user. When a user inputs a date or time, it would be translated into the binary format. This small amount of extra work is well worth it because User Agents would be able to compare two dates very easily, using only a single arithmetic operation.

Perhaps the most important benefit afforded by binary encoding schemes is that nontextual information can easily be accommodated in messages. Little, if any, storage or processing overhead is required. This can enormously increase the usefulness of the electronic mail medium, since it becomes feasible to include facsimile images, spreadsheets (in processible form), charts, graphs, and other data used in office processing systems, directly in the messages that discuss them.

There are some disadvantages associated with binary encoding schemes. Because of their very nature, it is not feasible to use an ordinary text editor to create the final form of an outgoing message, nor can incoming messages simply be printed as is. In addition, because any byte value may occur in messages, it is necessary to use transparent protocols when transmitting them. Not all communications systems implement transparent protocols throughout. Finally, binary encoding schemes are fragile. An error in a data object's identifier or length code would render it, and any data that follows it, incomprehensible. Thus, error-free data transmission and storage are required if a binary encoding scheme is to be used.

## The Message Transfer Layer

Unlike the User Agent Layer, which contains only the $P_c$ protocol, the Message Transfer Layer contains two protocols, $P_1$ and $P_3$. $P_3$ makes MTL services available to UAEs that are located in systems without an MTAE. It also carries the information necessary to provide access control between UAEs and MTAEs. $P_1$ performs the rest of the work necessary to provide the MTL services.

$P_3$ protocols are really remote procedure call protocols. When a UAE requests an MTL service from its SDE, the SDE passes the service parameters to an MTAE in the form of arguments to a remote procedure call. The results of the remote procedure call are returned to the SDE, which passes them back to the UAE as the parameters of the corresponding service confirmation. A similar sequence of events takes place when an MTAE uses $P_3$ to pass a service indication to a UAE. Because $P_3$ is primarily concerned with passing parameters between UAEs and MTAEs, while the specific functions relating to Message Transfer are carried out by $P_1$, this section will concentrate on $P_1$ protocols.

A $P_1$ protocol is used to transfer a message from its originator's MTAE to the next MTAE, and so on, until the message reaches, and is delivered by, its recipient's MTAE, or until transfer fails. Then, if the appropriate service has been requested, $P_1$ is used to transfer a notification of delivery or nondelivery back to the originator's MTAE. $P_1$ protocols must carry all of the information that is part of the end-to-end services visible to the originating and receiving UAEs. $P_1$ protocols must also carry additional information (such as the route a message has taken so far) that is required by intermediate MTAEs.

Like $P_c$ protocols, $P_1$ protocols vary in the services they offer and in the encoding schemes they use. $P_1$ protocols also differ in their structure and in the approaches they take to routing messages.

### Protocol Services

Most $P_1$ protocols offer only a few basic services. Messages are delivered to their recipients' UAEs. If delivery should fail, a notification including the reason for failure is returned to the originator's UAE, which passes it to the user. This notification may contain all or part of the original message content, or a reference to it. A message may be addressed to more than one recipient. However, delivery to any one recipient is independent of the others. Some copies may be delivered successfully, but others may fail; and the copies that are delivered may arrive at different times.

In addition to notifying users when the delivery of a message fails, some systems allow their users to request positive acknowledgment of a message's delivery. This is called *delivery notification*. Delivery notification is not the same as return receipt. Delivery notification for a message means only that it is in the possession of its recipient's UA. *Receipt notification* means that the recipient has been made aware of the message, and may have read it. Even if delivery succeeds, receipt may never occur. Therefore, users may wish to request both delivery notification (a $P_1$ service) and receipt notification (a $P_c$ service) for the same message.

Some $P_1$ protocols allow UAEs to specify an earliest time for delivery, a latest time for delivery, or both. Similarly, some $P_1$ protocols permit UAEs to assign a relative priority to messages. Higher priority messages would be processed before lower priority messages. It is possible to meaningfully combine delivery time and priority services. For example, a message may contain information that is not very important but ceases to be useful at all after a certain time. The originating UAE might request low priority transfer for that message and might also stipulate that delivery be aborted if it does not take place by a given time.

The Message Transfer System is often regarded as more trustworthy than User Agents, which may be independently supplied and are more prone to tampering. A recipient's UAE often relies on the MTL to accurately identify which UAE submitted a message. Some $P_1$ protocols authenticate additional information, such as the time the message was submitted or the complete list of recipients supplied as part of the submission process.

Even in normal circumstances, the recipient list known to the $P_1$ protocol may differ from what appears in the message content. First of all, the message content may differentiate between classes of recipients, such as primary, secondary, and blind copy recipients. This would not be reflected in the information available to the $P_1$ protocol. Second, consider what happens when a message is submitted for delivery to a number of recipients, all mentioned in the message content, and one copy fails to be delivered. If the originator resubmits the same message content for delivery to the one recipient who didn't get it the first time, the information in the message content and the information available to the $P_1$ protocol will not agree. This very reasonable discrepancy illustrates why there is overlap between the information in a message content and the information on the message's envelope.

Instead of returning a nondelivery notification, some $P_1$ protocols will attempt delivery to an alternate recipient if delivery to the originally intended recipient fails. In some cases, the alternate recipient is designated by the originator's UAE when the message is submitted. Other systems support the notion of a "dead-letter office."

Conversely, some $P_1$ protocols allow potential recipients to redirect their incoming messages for a time. A similar service may be supplied by the $P_c$ protocol. In that situation, a successfully delivered message would be automatically forwarded to some other UA. The important difference between these two approaches is that in the first method the message's originator may receive a delivery notification saying that a message was redirected by $P_1$ and where it was ultimately delivered, but in the second method he or she cannot tell what happened to a message that was delivered and then automatically forwarded elsewhere.

Some versions of $P_1$ allow users to validate electronic mail addresses before actually sending a message. This provides an extra measure of assurance to the user and alleviates the need for alternate recipients or a dead-letter office to handle critical messages.

Finally, as electronic mail systems are joined with each other and with other sorts of communications systems, some $P_1$ protocols are beginning to support encoded-information-type conversion. This facility translates information among different formats, such as ordinary text, formattable documents, Telex, Teletex, Videotex, digitized voice, or facsimile. Not all conversions are feasible, and some may involve a loss of information. For example, the formatting commands embedded in a formattable document would be lost if the document were transformed into simple text. Other conversions may be deceptive because the result of the conversion may imply information not present in the original. For example, when ordinary text is transformed into a facsimile image, a font style and a size must be chosen, even though neither was indicated in the source information. To help users avoid the need for conversions, some electronic mail systems can tell message originators about the rendition capabilities available to potential recipients.

### Protocol Encoding

$P_1$ protocols may use either binary or text-based encodings. The trade-offs between these two approaches are similar to those of the $P_c$ protocol. Binary encoding brings greater flexibility and improved performance, but it is not human-readable and requires the use of reliable protocols at the lower layers. Text-based encoding is more robust and can be read directly by people, but it lacks the flexibility of binary encoding and is somewhat more difficult to process.

The choice of an encoding scheme for $P_1$ is often linked to the choice of a $P_c$ encoding scheme. Most, if not all, electronic mail systems that use binary encoding for $P_c$ also use it for $P_1$. Many electronic mail systems that use text-based encoding for $P_c$ also use text-based encoding for $P_1$. Some situations may call for a text-based $P_c$ and a binary $P_1$. On the other hand, combining a binary $P_c$ and a text-based $P_1$ seems inadvisable. The problems of representing arbitrary binary information in a text-based encoding scheme were touched on earlier in this chapter.

### *Protocol Structure*

Even though there are end-to-end acknowledgments, a $P_1$ protocol is best thought of as taking place between pairs of adjacent MTAEs. A $P_1$ protocol works as follows:

1. The MTAE in possession of a message determines if it is responsible for delivering it.

2. If it is, it attempts to deliver the message. (Any resulting delivery notification or nondelivery notification would be handled in approximately the same way that an ordinary message would.)

3. If it is not responsible, it picks a second MTAE that is "closer" to the message's destination.

4. A data connection is established with the second MTAE.

5. The message envelope and the message content are transferred to the second MTAE.

6. If all is well, the second MTAE accepts responsibility for the message, and the $P_1$ protocol transaction is terminated.

7. Go back to step 1.

There are several different ways to carry out step 5. One approach is to transmit the message envelope and the message content as a single unit. Another is to transmit the message envelope first, wait for approval from the second MTAE, and then transmit the message content. A third method is to transmit the message envelope first, one parameter at a time, waiting for a positive acknowledgment after each parameter. For example, the recipient list might be transmitted first, then the handling instructions, and so on.

These three methods trade early problem detection against simplicity and flexibility. The first approach is simplest from the viewpoint of the protocol implementation, but the effort of transmitting the message content is wasted whenever the second MTAE must reject the message based on some information in the message envelope. The second approach has a slightly more complex implementation, but the message content will never be transmitted unnecessarily. The third approach allows transfer to be aborted even before the entire message envelope is transmitted, but there is a catch. In this case, the nature of the $P_1$ protocol is determined by the definition of the message envelope. If the envelope definition were modified, the $P_1$ protocol would probably have to be changed too. Furthermore, MTAEs whose $P_1$ protocols follow either the first or the second approach can accept and pass along unfamiliar envelope parameters so that they can be interpreted later by other MTAEs. MTAEs whose $P_1$ protocols follow the third approach may have no way to do that.

### Routing Messages

The path that a message takes through the Message Transfer System is called its *route*. A route is a series of one or more MTAs. In order for messages submitted by one UA to be delivered to another, there must be at least one valid route between the two UAs. (Certain routes may not be permitted because of operational or policy constraints.) A message's route may be determined entirely at its point of origination, or it may be computed in separate steps as it passes through the MTS. The first case is called *source routing*, and the second is called *incremental routing*.

Source routing minimizes the work that must be done by intermediate MTAs. It has the potential to permit originating systems or their users to exercise control over the choice of route. Source routing requires that the originator's UA or MTA be able to determine which MTA can deliver a particular message. It is most practical in small electronic mail systems, where information about all UAs, all MTAs, and the connectivity between them can be kept at each originating node or in a central repository.

Incremental routing minimizes the work that must be done by originating UAs and MTAs. Because each step of the route is determined just before it is taken, MTAs that are busy or temporarily out of service can be avoided. This improves the overall performance and robustness of the MTS. Incremental routing requires that each MTA be able to determine an appropriate next MTA, given a message's destination.

Electronic mail systems vary in how they treat messages addressed to multiple recipients. In some systems, the originating UA or MTA makes a copy of the message for each recipient. Then the copies are routed independently of each other. Other systems determine the complete or incremental route for each destination first. Copies of the message are made only when the routes diverge. This practice, called *mail-bagging*, is illustrated in Fig. 5.7.

Fig. 5.7. Mail-
bagging.

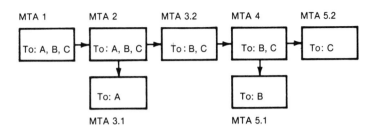

| DESTINATION | ROUTE TAKEN |
|:-----------:|:-----------:|
| A | 1-2-3.1 |
| B | 1-2-3.2-4-5.1 |
| C | 1-2-3.2-4-5.2 |

Mail-bagging economizes on data transmission time and costs. It also reduces the amount of temporary storage that each MTA must have available to buffer messages in its possession. These factors can be very significant in electronic mail systems that process a large number of messages. Routing decisions may keep mail-bagging in mind. Implementing mail-bagging adds slightly to the complexity of the $P_1$ protocol and the MTA software, but is usually well worth it.

### 5.3.3   Other Design Decisions

This section surveys some aspects of User Agent and Message Transfer Agent architecture that are not directly related to physical distribution or protocols.

#### User Interface

The interface between a human user and an electronic mail system is critical to the system's ease of use and commercial viability. The *user interface* is an example of a User Agent function that is provided independently by an individual User Agent and, therefore, is not reflected in a $P_c$ protocol. Like other applications, electronic mail systems use a variety of styles in their user interfaces, such as menus or prompts. Whatever style is used, the software that implements the human interface often represents a significant fraction of the total software in a User Agent.

#### Message Storage

*Message storage* is another aspect of User Agent design that is not governed by a $P_c$ protocol. Without considering the particulars of data-base software and design, three facets of message storage distinguish electronic mail systems. First is the general approach taken to storing each user's messages. Some electronic mail systems store one copy of each message in a cache and give each of its recipients access to it. A reference count of the

number of recipients who have access to the message is kept. As each recipient "discards" the message, the reference count is decremented. When
the reference count reaches zero, the message is deleted from the cache.
Other systems store separate copies of a message for each of its recipients.
In this second approach, messages can be stored separately or together in
the same computer file. Message caching saves on storage overhead. The
alternative approach economizes on computation.

Another design decision deals with how the messages that belong to
a single user are organized. Many electronic mail systems allow users to
group related messages in electronic analogues of folders. A single message may appear in a number of different folders. Some systems implement folders as separate computer files, each with its own copy of the
relevant messages. Other systems implement folders using references or
indexes to the user's message store. The trade-offs here are the same as
for message caching.

Finally, some electronic mail systems use separate computer files to
segregate new, incoming messages from messages that have already
been seen by their recipients. This may be done for reasons of access
control or convenience. For example, consider a group of distributed
User Agents all located on the same local area network. Incoming messages for the User Agents would be stored at a gateway node where a
Message Transfer Agent resides. The rest of each User Agent would be
implemented on a personal computer or workstation that would be powered off when not in use. As each User Agent is activated by its user, it
would retrieve the incoming messages that have been stored at the gateway node.

### Message Transfer System Topology

The Message Transfer System works by relaying messages among the
originating User Agents and the recipients' User Agents. If all the hosts
that make up an electronic mail system were available all the time, and if
all nodes implemented the same protocols and had the same capabilities,
and if administrative and political policy played no part in electronic mail
systems, then no more than one relay would ever be needed to transfer a
message from its originating MTA to its recipient's MTA. However, this is
not the case. Requirements for special functions, such as protocol conversion or medium-term buffering can determine the route taken by a
message. Policy may dictate that access to the MTAs in a management
domain be carefully controlled. As a result of these factors, the topology
of a Message Transfer System can be hierarchical, with certain MTAs,
such as those that implement access control or protocol conversion, acting as gateways to others.

### Implementing Distribution Lists

A distribution list allows an originator to send a message to a group of recipients without having to enumerate each one. There are three major approaches to implementing distribution lists:

1. *Store a distribution list in a user's UA*. Only that user has access to the distribution list. The distribution list is expanded before the message is submitted.

2. *Store the distribution list in a special UA called an Expander*. All users may send messages to an Expander. As its name implies, when an Expander receives a message, it resubmits it to each member of the distribution list. An Expander may keep a copy of each message it receives, or it may perform other processing related to its duties.

3. *Store the distribution list in a directory that is external to any UA*. Expansion is performed before the message is submitted, or it is performed by an MTA. Access to the contents of a distribution list may be restricted or open to all.

In addition to dealing with issues such as storage location and access control, electronic mail systems face some special problems when distribution lists are nested and the lists refer to each other by name. First, steps must be taken to detect potential loops that may be formed. Furthermore, there are a variety of approaches (including doing nothing) that can be taken to suppressing potential duplicate entries.

Distribution lists interact with other aspects of electronic mail systems. For example, a message that was sent to a distribution list as its only recipient may generate a number of nondelivery notifications, each referring to an intended recipient that is potentially unknown to the originator's User Agent. If a message is sent to a distribution list whose ultimate size (number of members) is not known or knowable by the originating User Agent or Message Transfer Agent, how can it be determined that all copies of the message have been successfully delivered? These factors make the implementation of distribution lists one of the trickier parts of an electronic mail system.

## 5.4  Naming and Addressing

Throughout this chapter there have been many references to the names and addresses used to identify users of an electronic mail system. This section explores what these names and addresses are like, how they are assigned, and how they are associated with each other.

Names, addresses, and routes are closely related concepts. It is important to distinguish among them. Most simply put, a *name* says what an object is, an *address* says where it is, and a *route* says how to get there. An address is a special form of a name because it identifies an object by telling where it is. A route is a special form of an address because it locates an object by giving directions for getting there. Fig. 5.8 represents these relationships graphically.

**Fig. 5.8. Venn diagram showing relationships of names, addresses, and routes.**

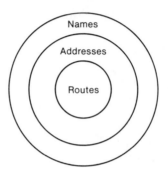

Names, addresses, and routes play different roles in an electronic mail system. User Agents are named after their users, so their names frequently describe people. For example, User Agents' names may be based on their user's names. Since addressses are concerned with locations, a User Agent's address frequently includes information identifying the MTA that delivers messages to that UA. As discussed earlier in this chapter, a route identifies the series of MTAs forming the path that a message takes through the Message Transfer System.

A User Agent's name, address, and a route to it from a given starting point can all be different. When it is necessary to locate a User Agent, its name is *mapped* into its address. This mapping can change with time. In a similar fashion, a route is determined, based on a User Agent's address. The route chosen between User Agents' addresses may also change with time.

It would be possible to use only addresses to refer to User Agents, and some electronic mail systems do. User Agent names can provide an extra measure of user friendliness. They also provide indirection. That means that even if a User Agent's address must change, its name would remain constant. Users of the system could remain totally unaware of the change.

It would also be possible to use only routes to locate User Agents. Again, some electronic mail systems do. User Agent addresses are absolute and independent of network connectivity. In other words, the meaning of a User Agent's address does not vary with time or where it is interpreted. Routes can become invalid or lead to different places as network connectivity changes. Furthermore, a route linking two User Agents is only useful to those User Agents. It does not tell any third parties how to communicate with them.

User Agent names may denote individuals, roles, or distribution lists. Most users of electronic mail systems are individuals in either a business or a private setting. A *role* is a particular position that may be filled by a variety of people. The dispatcher at a taxi service is an example of a role. A *distribution list* is a list of potential message recipients. The purpose of a distribution list is to allow a user to send a message to a group of people without having to enumerate each member of the group. A distribution list can contain other distribution lists.

In order for an O/R (Originator/Recipient) Name to be used successfully, it must be meaningful to each User Agent or Message Transfer Agent that acts on it. The portion of a communications environment in which a name is meaningful is called the *domain* of that name. A *naming domain* is a portion of a communications environment over which a set of names is defined. Naming domains can overlap. They can also be subdivided. An electronic mail system might consist of a single naming domain, or it might be divided into several.

Naming domains usually follow administrative or organizational boundaries. Each naming domain is administered to avoid duplicating names. A *naming authority* is an administrative entity that assigns names. A naming authority must make sure that the same name is not assigned to two different objects. As an electronic mail system grows, its administrators may delegate authority to assign names for different segments of the user population. This authority can be further delegated as the situation requires.

Naming domains may be related to management domains, but they are not necessarily the same. Similarly, the people who administer a naming domain may or may not be responsible for the operation of the portion of the directory in which the names are stored.

The following sections discuss in greater detail the nature of User Agent names, User Agent addresses, and the directory.

## 5.4.1 User Agent Names

A *naming convention* is a method for making up names. It defines the structures of names and places bounds on their values. Naming authorities must follow a naming convention when assigning names. The set of names that follow a given naming convention is called a *name space*. Each electronic mail system has a naming convention. All connected electronic mail systems that follow the same naming convention form a single naming domain and share the same name space.

The simplest naming convention is based on primitive names. A *primitive name* is a label that cannot be broken into meaningful parts. "Fred," "Europe," and "Taurus" are examples of primitive names.

A naming convention that uses primitive names to identify all objects works best if the number of objects to be named is small. Otherwise, names become increasingly arbitrary and difficult to manage. For example, it might make sense to use people's given names as User Agent names in a very small electronic mail system, but that would not work in a large system. There would be multiple people with the same given name, and any way of resolving the conflict (e.g., assigning "John1," "John2," "John3"... as User Agent names) would deviate from the original intention and would be hard for people to remember correctly. Because the name space for such a naming convention would have no structure, any division of the name space among potential subdomains would also be arbitrary.

*Multipart names* are built out of two or more primitive names. Most electronic mail systems use naming conventions based on multipart names. A *hierarchical name* is a multipart name whose components are hierarchically related. If there is no hierarchical relationship among the components of a name, it is said to be *flat*. Primitive names are flat, too. The name space generated by a hierarchical naming convention is hierarchical. A similar relationship holds true for flat naming conventions.

Hierarchical naming conventions are very convenient if naming authority is to be delegated. The hierarchy of administrative entities that assign names can be reflected in the names themselves, simplifying the management of the name space. For example, consider an electronic mail system used by the XYZ Corporation, a very large, geographically distributed company. Naming authority for User Agents might be divided between the research division and the production division. The research division might further delegate naming authority among its various departments. The production division might delegate naming authority by plant. "XYZ/Research/Chemistry/JRoberts" could be the name that belongs to the UA of a research chemist who works for the XYZ Corporation. On the other hand, "XYZ/Production/Pittsburgh/JRoberts" could be the name that belongs to the UA of a different chemist who works for the XYZ Corporation at a plant located in Pittsburgh.

Many electronic mail systems use tree-structured naming conventions similar to the one described in the previous paragraph. Tree-structured naming conventions are very flexible. There can be as many levels in the tree as necessary. Not all branches need be the same length.

For example, one department in the research division of the XYZ could add an additional level for the employee's laboratory.

In mathematics, a *directed graph* is similar to a tree, but its intermediate nodes and leaves can have more than one parent. An *acyclic directed graph* is a directed graph that has no loops. That is, no path leaving a given node will return to that node, either directly or indirectly. A tree is actually a special case of an acyclic directed graph. Fig. 5.9 shows this relationship.

Fig. 5.9.
Comparison of
trees and acyclic
directed graphs.

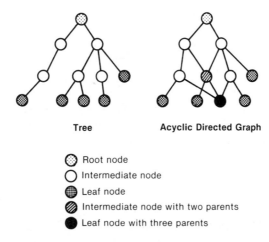

Tree                              Acyclic Directed Graph

⊛ Root node
◯ Intermediate node
⊕ Leaf node
⊘ Intermediate node with two parents
● Leaf node with three parents

In both tree-structured and graph-structured naming conventions, the object to be named is represented by a leaf node. The intermediate nodes represent layers in the naming hierarchy. Names are associated with the arcs between the nodes. The ordered set of names of the arcs forming the path from the root to a leaf is the name of the object represented by the leaf.

Graph-structured naming conventions are more flexible than tree-structured ones. In a tree-structured naming convention, there is only one path from the root of the tree to a given leaf node, so the object represented by the leaf has only one name. In a graph-structured naming convention, there can be many different paths leading from the root to a given leaf node. This means that an object represented by a leaf node can have several different names. For example, in Fig. 5.9 the object represented by the black leaf node would have four different names. Each name would be associated with one of the four different paths that could be taken from the root to that leaf. The ability to have multiple names for the same object is very useful. One application is to make one or more parts of a name optional. This is done by creating arcs that bypass one or more nodes on the graph, as in Fig. 5.10.

Fig. 5.10. Use of
an arc to bypass
one or more
nodes.

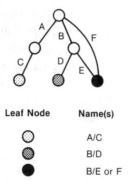

| Leaf Node | Name(s) |
|-----------|---------|
| | A/C |
| | B/D |
| | B/E or F |

Graph-structured naming conventions also permit the same object to have completely different names. Cases in which the same UA might logically belong in two different subdomains of the same naming domain are easily accommodated. An example of this is a consultant who is directly employed by the ABC Consulting Company and works on site for their client, the PQR Company. The ABC Company and the PQR Company both are naming authorities for their own subdomains of an encompassing naming domain. Giving the consultant's UA two names, one for each subdomain, would make it easier for people in either organization to correctly specify the consultant as the recipient of a message, since they would be following the naming convention used by their particular organizations. Unfortunately, joint administration of a single subdomain by two separate naming authorities is not always simple. There must be a high degree of cooperation and coordination. In cases where this would not be feasible, another solution is to designate a node in one naming domain as an alias for another node that belongs to the other naming domain. The path to the alias node would be replaced by the value of the alias, and the name would be reevaluated. For example, assume that Multinational, Inc., which is headquartered in Ontario, Canada and has a branch in Belgium, employs someone named Jean Roberts at its Belgian office. The naming domain for Multinational, Inc., is administered from its headquarters. An alias, Belgium/Multinational, Inc., is established to make the naming scheme more intuitive for people who wish to correspond with employees of the company's branch office. The name "Belgium/Multinational, Inc./Jean Roberts" would be evaluated as follows:

1.  Starting at the root, the arc labeled "Belgium" is traversed.

2.  Next, the arc labeled "Multinational, Inc." is traversed.

3.  The node that has been reached is marked as an alias for "Canada/Ontario/Multinational, Inc." A new name, "Canada/Ontario/Multina-

tional, Inc./Jean Roberts" is formed by using the information found at the node and the portion of the original name that had not yet been evaluated. Return to the root node.

4. The arc labeled "Canada" is traversed.

5. Then the arc labeled "Ontario" is traversed.

6. Next, the arc labeled "Multinational, Inc." is traversed.

7. Finally, the arc labeled "Jean Roberts" is traversed. Given that the name evaluation took place as part of a name-to-address mapping operation, the final node would contain a User Agent address.

This process is illustrated in Fig. 5.11.

**Fig. 5.11. Example of the use of aliases in naming schemes.**

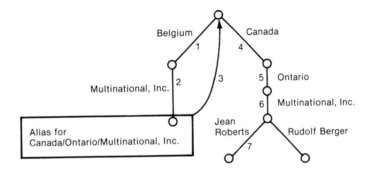

User friendliness is a very important element in the design of a naming convention. User Agent names should be easy for people to remember and type. They should also be easily deduced, given information that might normally be known about the users with whom they are associated. One way to achieve this goal is to construct names out of attributes. An *attribute* is an assertion about the object being named. It is composed of a type-value pair. Personal-name, company-name, postal-address, and telephone-number are all attributes that might be used in User Agent names.

Names composed of attributes can be flat or hierarchical. If they are hierarchical, each attribute would be associated with an arc of the tree or graph. Sometimes the correct hierarchical relationship between two attributes is not obvious. For example, in some cases a city-name attribute would logically be "higher" than a company-name attribute. In large, geographically dispersed firms, the reverse might be true. In the interest of user friendliness, an electronic mail system might sidestep this problem by allowing users to specify attributes in any order. Unfortunately, this introduces a new problem. Now the naming tree or graph must be coordinated to eliminate the possibility that two names could be composed of the same attributes, but in different orders.

In Fig. 5.12, two Acme companies are located in Massachusetts. One, which runs a chain of garages, has locations in Worcester and Boston. The other, which deals in glass, has its only office in Worcester. There are two people in Worcester by the name of Anna Cappelli. One works at the Acme Garage, and the other works at the Acme Glass Company. Following the paths from the beginning of the fragment, their UAs' names would be [(Company = "Acme"), (City = "Worcester"), (Personal-Name = "Anna Cappelli")] and [(City = "Worcester"), (Company = "Acme"), (Personal-Name = "Anna Cappelli")], respectively. The only difference between the two User Agents' names is the order in which the attributes appear. This distinction would be confusing to most people. Electronic mail system designers and naming domain administrators must always be careful to avoid this sort of problem.

Fig. 5.12. Use of ordered attributes to distinguish User Agents' names.

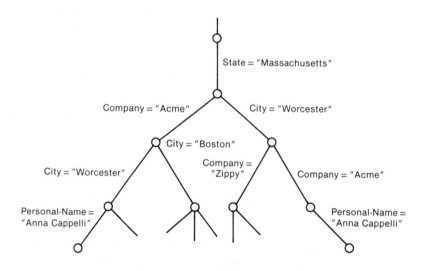

The use of attributes to form User Agent names is a relatively recent development. Because most electronic mail systems are used in a business environment, the most attention has been paid to attributes that make sense in a business setting. The following business-oriented attributes are representative of those that have been implemented or proposed:

| | |
|---|---|
| ***Organization-Name*** | The name of a company, association, or governmental agency. |
| ***Organizational-Unit*** | The name of a division, department, or other part of the given organization. |

| | |
|---|---|
| *Country* | The name of the country in which the organization is located. In multinational organizations, this may be the country in which the user's office is located. |
| *Postal-Address* | The region, town, and street name and number at which the user's office is located. (Postal codes and boxes might also appear here.) |
| *Position* | The user's job title or role. |
| *Personal-Name* | The user's surname, given name(s), initials, and/or generational qualifier (e.g., "Jr."). |
| *Telephone-Number* | The telephone number at the user's office. |

As the number of "residential" users increases, appropriate additional naming attributes may be adopted for use by electronic mail systems.

Not all sets of attributes are names. For a set of attributes to be a name, it must unambiguously identify a single object. For example, in an American company with 10,000 employees, there are probably many people named "John Smith." It would be necessary to specify more information than an organizational-name and a personal-name to distinguish among them. A *base attribute set* is a set of attributes that is guaranteed by the naming authority to identify no more than one object. A base attribute set for business-oriented User Agent names might consist of Country, Organization-Name, Position, and Personal-Name attributes.

## 5.4.2   User Agent Addresses

A *User Agent Address* is a User Agent name that gives the location of a User Agent. The same rules for other User Agent names also apply to User Agent addresses. User Agent addresses differ from User Agent names in the information that is used to build them.

In a communications environment, a location might be the name of a network, a subnetwork, a host node, a network node, a port on a node, or a terminal. This kind of information is frequently found in User Agent addresses. If several User Agents can be located in the same host node, additional information might be included to distinguish among them.

Many User Agent addresses are hierarchical. The most popular form for a User Agent address has two parts, a host specification and a user specification. The host may be identified by using symbolic or numeric network addresses or by a name that must be mapped into a network address. Some electronic mail systems identify the host using a route composed of hosts that are willing to act as relay points. The user may be identified by a log-in identifier or other information known to the host. Together, these pieces of information specify a User Agent's location in a form that is easily dealt with by communications software.

In cases where a User Agent is implemented in a terminal, an X.121 address or a terminal address is sufficient as a User Agent address. If only one User Agent is implemented on a host, only the host need be specified.

## 5.4.3   The Directory

The mapping function between User Agent names and User Agent addresses is provided by a logical entity called a *directory*. The directory may also be used to implement distribution lists or to store additional information not directly related to the mapping of User Agent name to User Agent address. For example, there may be "yellow pages" information that is of interest to the user population, or there may be registration information that is accessed by the administrators of the electronic mail system. The directory may be implemented in a central location, or it may be distributed. It may also be redundant. Just as different parts of an electronic mail system can be supplied and operated by different organizations and individuals, responsibility for the implementation and operation of the directory can be dispersed.

Directories are a comparatively recent development in the evolution of electronic mail systems. The following sections give an overview of their operation.

### Directory Model

The directory model helps explain the structure and operation of the directory. It does this by describing relationships and interactions among directory users, Directory User Agents, and Directory System Agents. Users of the directory can be people, User Agents, Message Transfer Agents, or other computer applications or processes.

A *Directory User Agent* (DUA) is a logical entity that acts on behalf of the user in the context of the directory. It is nearly analogous to a User Agent in the model for electronic mail systems. The exception is that, unlike UAs, DUAs do *not* communicate with each other.

A *Directory System Agent* (DSA) stores directory information, such as the mapping between names and addresses. A DSA provides directory information to a DUA or to another DSA in response to a query. A DSA may use information that it has, or it may request help from other DSAs. Alternatively, the DSA may respond to a query by suggesting a different DSA that may be better able to provide the needed information. This kind of response is called a *hint*. The use of hints illustrates another aspect of the directory model; namely, that a DUA (or a DSA) can choose the DSA to which it directs a query. DSAs work cooperatively to provide the services of the *Directory System*. Taken as a whole, the information stored and processed by the Directory System is called the *directory*.

A DSA receives queries from DUAs and from other DSAs. A DSA must eventually return a response for every query that it receives. If the DSA has the requested information, the response indicates success and the desired information is included. If the DSA does not have the right information, it can return an error response or take one of four actions:

1. It can return a hint.

2. It can pass the query request to another DSA. This is the DSA that would have been mentioned in a hint. This type of action is called *chaining*.

3. It can multi-cast the query request to a number of DSAs. This is done when it is not clear which DSA is more likely to have the required information.

4. It can split the original query into a number of new queries, sending each new query to one DSA. This action is taken when the DSA believes that the required information is distributed among several other DSAs. These interactions are illustrated in Fig. 5.13.

In addition to phrasing queries, users may add, delete, or modify the information in the directory. These operations are requested via the users' DUAs. Suitable access controls may be implemented to control who may examine or change sensitive or critical information.

Just as an electronic mail system can be divided into management domains, a Directory System can be divided into a number of *directory management domains* (DMDs). Each directory management domain contains at least one DSA and zero or more DUAs.

### Relationship to Naming Domains and Naming Conventions

The directory model just described is independent of naming domains and naming conventions. Logically, the portion of the directory stored in the DSAs of a given directory management domain may coincide with a single naming domain, contain a subdomain of a naming domain, or over-

Fig. 5.13.
Operations
performed by a
DSA.

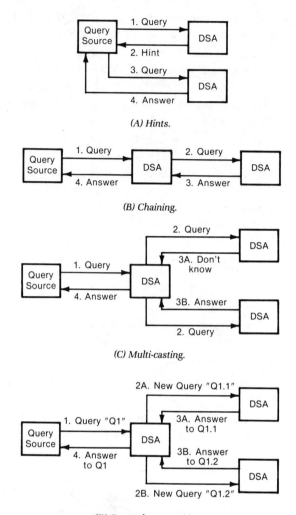

(A) Hints.

(B) Chaining.

(C) Multi-casting.

(D) Query decomposition.

lap several naming domains or their subdomains. User Agent names and associated information from a single naming domain or subdomain may be stored in multiple directory management domains.

The organization responsible for a directory management domain may or may not be a naming authority. Naming authorities may also be responsible for directory management domains, but they don't have to be.

The choice of a naming convention may suggest ways to distribute directory information among the DSAs that would facilitate the efficient operation of the directory. For example, given a hierarchical naming convention, directory information might be distributed in a pattern that reflects the hierarchical structure of the name space. Also, the choice of naming convention and of other information to be stored with a name may suggest the implementation of particular forms of queries. However, there is no required relationship between the naming convention and the structure of the directory.

### Relationship to Management Domains

Directory management domains are logically independent of electronic mail system management domains. An organization may be responsible for a directory management domain, a management domain, both, or neither. Unless policy or practicality dictates otherwise, a user, User Agent, or Message Transfer Agent may freely choose the DUA to which it directs a query.

## 5.5 Standards

A standard for electronic mail systems specifies rules that each system must follow to exchange messages with other systems. In order for a standard to be successful, vendors and users must feel that it is desirable. From the technical point of view, a successful standard must specify rules for everything that requires cooperation between two systems. In the case of electronic mail systems, that means the $P_c$, $P_1$, and, if needed, $P_3$ protocols. There must be sufficient flexibility to allow reasonable variation between products. On the other hand, there should not be so many options that the ability to interwork is compromised.

Although large, networked electronic mail systems have been in operation since the early 1970s, widespread commercial use of electronic mail systems is a more recent phenomenon. Thus, it is only in the past few years that much attention has been paid to the standardization of commercial electronic mail systems. Until then, the only standards that had been developed were for research networks or for proprietary products.

The International Telephone and Telegraph Consultative Committee (CCITT) developed a set of eight recommendations (which have the effect of being standards) for electronic mail systems during its 1980 through 1984 study period. These recommendations are known collectively as the X.400 series. The International Organization for Standardization (ISO) and the European Computer Manufacturers' Association (ECMA) have followed suit and are developing their own versions of the CCITT Recommendations. Since the CCITT is concerned primarily with communications across national boundaries, its recommendations do not deal with aspects and details of protocols that could reasonably vary from country to country or user to user. These matters are being addressed by the ISO and ECMA.

Because the X.400 series is the predominant force behind current efforts to standardize electronic mail systems, the rest of this section is devoted to an overview of this series.

The eight recommendations that make up the X.400 series are as follows:

| | |
|---|---|
| X.400 | *Message Handling Systems: System Model-Service Elements.* Defines the message handling system model, discusses naming and addressing, and defines services provided by the User Agent and Message Transfer Layers. |
| X.401 | *Message Handling Systems: Basic Service Elements and Optional User Facilities.* Divides services into required and optional groups. |
| X.408 | *Message Handling Systems: Encoded Information Type Conversion Rules.* Discusses conversion between Telex, International Alphabet No. 5, Teletex, Group 3 Facsimile, Text Interchange Format 0, Videotex, digitized voice, Simple Formattable Document, and Text Interchange Format 1 codes and formats. |
| X.409 | *Message Handling Systems: Presentation Transfer Syntax and Notation.* Defines a binary encoding scheme and an associated human-readable notation. |
| X.410 | *Message Handling Systems: Remote Operations and Reliable Transfer Server.* |

|  | Describes general mechanisms for remote operations and bulk data transfer using the session layer. |
| X.411 | *Message Handling Systems: Message Transfer Layer.* Specifies $P_1$ and $P_3$ protocols. |
| X.420 | *Message Handling Systems: Interpersonal Messaging User Agent Layer.* Specifies a $P_c$ protocol. |
| X.430 | *Message Handling Systems: Access Protocol for Teletex Terminals.* Describes mechanisms and protocols for interworking with Teletex. (Conversion of encoded information formats is specified in X.408.) |

The model and the vocabulary used in the X.400 Recommendations closely parallel those used in this chapter. The term "message handling systems" is used instead of "electronic mail systems." The phrase "Interpersonal Messaging User Agent" is used to denote User Agents concerned with communications between people. (Other types of User Agent are possible. For example, a different kind of User Agent might be used to allow people to order products from a computerized store.)

The following sections discuss the X.400 series. The first section describes the layer services. The second explores the protocols in terms of their encoding scheme and the particular mechanisms they employ. Finally, the provisions made for interworking with other telematic services are discussed briefly.

## 5.5.1 Services

Layer services are divided into three categories: basic, essential, and additional. *Basic* services are intrinsic to the operation of the system. They must always be implemented. *Essential* services are required to be implemented and are invoked at the option of the user. *Additional* services are not required to be implemented. Each Message Transfer Layer service is placed in one of these categories. In the case of the User Agent Layer, optional services (those that are not basic) receive two designations, one for origination by UAs, the other for reception by UAs. Many of the services that are additional for originating UAs are essential for reception by UAs. For example, UAs are not required to allow users to mark a message "private"; but, if a UA receives a message that is marked private, it must handle it in an appropriate manner.

## Message Transfer Layer Services

The following Message Transfer Layer services are classified basic:

| | |
|---|---|
| *Message Identification* | Provides the User Agent with a unique reference identifier for each message that it submits. This identifier may be used later in conjunction with other services. |
| *Nondelivery Notification* | Informs the User Agent that a message it had submitted could not be delivered. |
| *Registered Encoded Information Types* | Allows a User Agent to specify a set of encoded information types that can be delivered to it. |
| *Original Encoded Information Types Indication* | Permits a User Agent to indicate the encoded information types present in a message being submitted. |
| *Converted Indication* | Informs a User Agent about any encoded information type conversions that may have been performed on a message being delivered. |
| *Submission Time Stamp Indication* | Informs both the originating and the recipient User Agents of the date and the time at which a message was submitted. |
| *Delivery Time Stamp Indication* | Informs the recipient User Agent of the date and the time at which a message was delivered. |

The following Message Transfer Layer services are classified essential:

| | |
|---|---|
| *Alternate Recipient Allowed* | Enables delivery to a dead-letter office if the originally supplied address was in the correct form but no User Agent was identified. |
| **Conversion Prohibition** | Forbids the MTS from performing encoded information type conversion on a message. |
| *Deferred Delivery* | Requests that a message be delivered no sooner than a specified date and time. |
| *Deferred Delivery Cancellation* | Aborts delivery of a message that had been marked for deferred delivery. The message usually must still be in the first MTA, and the deferred delivery time must not have passed. |
| *Delivery Notification* | Requests that the originating User Agent be notified after a message has been successfully delivered. |
| *Disclosure of Other Recipients* | Instructs the MTS to disclose the list of recipients to whom a message was submitted when the message is delivered. |
| *Grade of Delivery Selection* | Requests urgent, normal, or nonurgent handling of a message by the MTA. The default is normal. |
| *Multidestination Delivery* | Specifies more than one recipient for a message. (Distribution lists are *not* included in this service.) |
| *Probe* | Determines whether a message of a given size and/or encoded information type could be delivered to a User Agent at a specified address. |

The following Message Transfer Layer Services are classified additional:

| | |
|---|---|
| *Alternate Recipient Assignment* | Requests that this User Agent be designated as the dead-letter office. Special privileges may be required in order for the request to be honored. |
| *Explicit Conversion* | Specifies that an encoded information type conversion be performed on a message. |
| *Hold for Delivery* | Requests that messages intended for this User Agent be retained in the Message Transfer System until a later time. Messages may be designated for this treatment based on their encoded information types, sizes, and/or priorities. |
| *Implicit Conversion* | Permits the Message Transfer System to perform whatever encoded information type conversion is necessary on a message. |
| *Prevention of Nondelivery Notification* | Suppresses a potential nondelivery notification. |
| *Return of Contents* | Requests that the message content be returned if the message cannot be delivered. |

## User Agent Layer Services

Several of the Message Transfer Layer services are made directly available to a user of the User Agent Layer. These are delineated by category in the following lists.

The following User Agent Layer Services are classified as basic:

| | |
|---|---|
| *IP-Message Identification* | Permits UAs to assign a reference identifier to each message content sent or received. This reference identifier is distinct from the |

identifier assigned by the Message Transfer Layer. The IP-Message identifier may be used later in conjunction with other User Agent Layer Services. ("IP" stands for "Interpersonal." This prefix is used to distinguish the identifier associated with the message content, which is used by Interpersonal Messaging User Agents, from an identifier associated with a message envelope, which is used by User Agents and Message Transfer Agents to refer to a message submitted to the MTS.)

*Typed Body*

Allows the nature and attributes of the body of a message content to be conveyed along with the body. This information may be used later in conjunction with encoded information type conversion services.

The following User Agent Layer services are classified as essential for both origination by UAs and reception by UAs:

*Conversion Prohibition*

Passed through from the Message Transfer Layer.

*Grade of Delivery Selection*

Passed through from the Message Transfer Layer.

*Originator Indication*

Identifies the user that sent the message.

*Primary and Copy Recipients Indication*

Specifies one or more users who are the primary (*To:*) recipients and zero or more users who are the secondary (*Cc:*) recipients of a message content.

*Replying IP-Message Indication*

Specifies an earlier message to which this one is a reply.

| | |
|---|---|
| *Subject Indication* | Describes what the message is about. |

The following User Agent Layer services are classified as additional for origination by UAs and essential for reception by UAs:

| | |
|---|---|
| *Authorizing Users Indication* | Identifies one or more users who authorized a message to be sent. (Equivalent to the *From:* line of a memo.) |
| *Auto-forwarded Indication* | Marks a message as containing an automatically forwarded message content in its body. |
| *Blind Copy Recipient Indication* | Specifies a list of additional recipients whose identities are not to be disclosed to the primary or secondary recipients. |
| *Body Part Encryption Indication* | Marks a message content as containing encrypted information in its body. |
| *Cross-referencing Indication* | Specifies one or more other message contents to which this one is related. This service makes use of IP-Message identifiers. |
| *Disclosure of Other Recipients* | Passed through from the Message Transfer Layer. |
| *Expiry Date Indication* | Conveys a date and a time after which the originator of the message considers it invalid. |
| *Forwarded IP-Message Indication* | Marks a message as containing a forwarded message content in its body. |
| *Importance Indication* | Specifies the level of importance that the originator assigns to the message. Possible values are low, normal, and high. |
| *Multipart Body* | Allows a message body to be partitioned into several distinct |

| | parts. The nature and attributes of each part are conveyed along with it. |
|---|---|
| *Obsoleting Indication* | Indicates that one or more previously sent messages are now considered obsolete by their originator and are superseded by this one. |
| *Reply Request Indication* | Asks that a response be sent to this message. May also indicate a date and a time by which the response should be sent, and one or more users to whom it should go. |
| *Sensitivity Indication* | Specifies the level of sensitivity that an originator associates with this message. Possible values are personal, private, and company-confidential. |

The following User Agent Layer services are classified as additional for both origination by UAs and reception by UAs:

| | |
|---|---|
| *Alternate Recipient Allowed* | Passed through from the Message Transfer Layer. |
| *Nonreceipt Notification* | Requests that the originator be informed if a message is not received by its intended recipient. |
| *Receipt Notification* | Requests that the originator be informed whether a message is or is not received by its intended recipient. |

The following User Agent Layer services are passed through from the Message Transfer Layer. They apply only to the User Agent that requests them. No recipient UA is involved:

| *Essential Services* | *Additional Services* |
|---|---|
| Deferred Delivery | Alternate Recipient Assignment |
| Delivery Notification | Deferred Delivery Cancellation |

Multidestination Delivery                    Explicit Conversion

                                             Hold for Delivery

                                             Implicit Conversion

                                             Prevention of Nondelivery
                                             Notification

                                             Probe

                                             Return of Contents

## 5.5.2  Protocols

The X.400 Recommendations specify $P_c$, $P_1$, and $P_3$ protocols. The particular $P_c$ protocol is called the "P2" protocol. The $P_1$ and $P_3$ protocols are called "P1" and "P3," respectively. All three protocols use the same encoding scheme.

### Encoding

CCITT Recommendation X.409 specifies a binary encoding scheme for representing pieces of information called *data elements*. Each data element has three parts: an *Identifier*, a *Length*, and a *Contents*.

Each Identifier contains an *ID Code* that distinguishes one data type from another. The meaning of the ID Code depends on the *class* of the data type. There are four classes of data types: universal, application-wide, context-specific, and private-use. *Universal types* are intended to be represented the same way in all applications, including electronic mail systems. *Application-wide* types are associated with a specific application. The exact meaning and the interpretation of *context-specific* types vary inside an application and depend on the context in which the data element is found. *Private-use* types are reserved for private (nonstandardized) use.

X.409 defines a number of universal types, including Boolean, Integer, Bit String, Octet String, Null, Sequence, Set, Numeric String, Printable String, S.61 String, S.100 String, IA5 String, UTC Time, and Generalized Time.

There are two forms of data elements: *primitives* and *constructors*. The Contents of a primitive data element is a series of bytes. The Contents of a constructor data element consists of one or more data elements.

The representation of the Identifier is 1 or more bytes long, as follows:

| High-Order Byte and Bit | | Low-Order Byte and Bit |
|---|---|---|
| Single-byte Identifier | c c f i i i i i | |
| Multi-byte Identifier | c c f 1 1 1 1 1 | 1 i i i i i i i . . . 0 i i i i i i i |

| | | |
|---|---|---|
| *cc = Identifier Class* | *f = Data Element Form* | *ii...ii = ID Code* |
| 00 = Universal | 0 = Primitive | The value of an ID Code is the binary |
| 01 = Application-wide | 1 = Constructor | number formed by concatenating |
| 10 = Context-specific | | the i's. ID Codes in the range 0–30 fit |
| 11 = Private-use | | in 1-byte-long identifiers, ID Codes in |
| | | the range 31–127 require 2-byte-long |
| | | identifiers, ID Codes in the range |
| | | 128–16383 require 3-byte-long iden- |
| | | tifiers, and so on. |

The Length tells how many bytes are in the Contents. Three forms are used to express the Length: short, long, and indefinite. The *short* form takes up a single byte, and is used for values in the range 0–127. The *long* form takes up from 2 to 128 bytes, and is used for values in the range $128–(2^{1016} - 1)$. The *indefinite* form takes up 1 byte, and is used in constructor data elements whose exact length is not given. A special End-Of-Contents data element is used to mark the end of a constructor data element whose length is specified as indefinite.

The Length of a data element is encoded as follows:

| High-Order Byte and Bit | | Low-Order Byte and Bit |
|---|---|---|
| Short form | 0 x x x x x x x | |
| Long form | 1 n n n n n n n | x x x x x x x x . . . x x x x x x x x |
| Indefinite form | 1 0 0 0 0 0 0 0 | |

| | |
|---|---|
| *Key:* xx...xx | These bits are concatenated to form an unsigned binary number that is the value of the Length. |
| nnnnnnn | In the long form, these bits are interpreted as an unsigned binary number that is the total number of bytes making up the value (xx...xx) of the Length. |

The Contents contains the specific information that each data element is intended to convey. Its format and interpretation depend on its type.

A human-readable notation to be used in conjunction with its encoding scheme has also been specified. It is used throughout the X.400 series to define protocol data units and their components. The notation has also been used in other standards. Details of the notation and examples of its use are in Recommendation X.409.

## Message Transfer Layer Protocols

Recommendation X.411 defines P1 and P3 protocols, which provide the services of the Message Transfer Layer. Both protocols follow the model discussed earlier in this chapter, with one difference. A distinction is made between *administration* management domains and *private* management domains. Administration management domains are provided by governmental agencies (e.g., PTTs) or by government-sanctioned organizations (e.g., public carriers). Private management domains are provided by private companies and individuals.

The P1 and P3 protocols look to the *reliable transfer server* to provide bulk data transfer between MTAEs or an MTAE and an SDE. The reliable transfer server manages the use of the session layer. Should a session connection unexpectedly terminate before data transfer has been completed, the reliable transfer server calls on the appropriate session services to reestablish the connection and resume data transfer at the last confirmed checkpoint.

### The P1 Protocol

The P1 protocol uses incremental routing and mail-bagging as messages are relayed across the Message Transfer System. If a message cannot be delivered, a *delivery report* is always returned to the originating MTA. Delivery reports are also used to provide notification of successful delivery, if that has been requested.

The P1 protocol defines an *O/R Name* as consisting of a set of attributes. ("O/R" stands for "Originator/Recipient.") O/R Names have two forms: the first identifies a User Agent; the second identifies a telematic terminal. The second form is designed for interworking with Telex, Teletex, or some other telematic service. It consists of an X.121 address and an optional Telematic Terminal Identifier.

An O/R Name for a User Agent is hierarchical. It is composed of a country name, the name of an administration management domain within that country, and one or more additional attributes that distinguish the specific User Agent from all other User Agents in that management domain. These attributes are drawn from one of three sets:

### Variant 1

One or more of the following:
Private domain name
Personal name
Organization name
Organization unit names
Plus domain-defined attributes (optional).

### Variant 2

UA unique numeric identifier
Domain-defined attributes (optional)

*Variant 3*
    X.121 address
    Domain-defined attributes (optional)

Information in the O/R Name provides a direct basis for routing decisions. The country and administration management domain name are acted on first. The additional attributes are not interpreted until the message bearing the O/R Name reaches the management domain specified by the country and administration management domain name.

There are two kinds of protocol data unit used by the P1 protocol: UserMPDUs and ServiceMPDUs. ("MPDU" means "Message Protocol Data Unit.") *UserMPDUs* are used for message transfer. *ServiceMPDUs* are used to report on the result of a message transfer attempt or to support the Probe service. Although their structures and the information they carry are different, all P1 protocol data units are relayed across the MTS using the same general techniques.

A UserMPDU consists of an envelope and a content. The envelope carries the parameters that control and record the actions of the P1 protocol. The P1 protocol does not concern itself with the nature of the content, except to specify the $P_c$ protocol being used (always the interpersonal messaging P2 protocol) and to deal with encoded information type conversion. For the most part, the parameters in the message envelope directly reflect the transfer services that were requested at submission time. For example, the *Priority* parameter sets the transfer priority, the *Deferred Delivery* parameter gives the earliest possible date and time that the message is to be delivered, and so on. However, a few parameters do merit some special attention.

Two mechanisms are used to support mail-bagging. When the route taken by a message branches, and copies of the message are made, each copy goes to a different set of recipients. One approach is to include the entire recipient list with each copy but to flag each member of the list as "active" or "inactive." When a message entered the MTS, there would be one copy with a list of $n$ recipients, all marked active. As the message's route branched, copies of the message would be made, each with a list of $n$ recipients, some marked active, the rest marked inactive. Eventually there would be $n$ copies of the message. Each would still have a list of $n$ recipients, but only the recipient for which it was intended would still be marked active. Another approach is for each copy of the message to carry a list of only the active recipients. Both methods are supported by the P1 protocol. However, the first method must be used if the Disclosure of Other Recipients service was requested. (Note that, for reasons of compatibility, use of the second method requires that all members included in the list be explicitly marked active.)

The P1 protocol includes a "hook" that is used when pairs of management domains agree to provide services that go beyond those specified in Recommendation X.411. The envelope includes an optional *Per Domain Bilateral Information* parameter. If present, this parameter contains one or more pieces of information, each marked with the management domain for which it is intended. There is no restriction on what the information can be.

Originating MTAs must have a way to coordinate delivery reports with previous message submissions. For every message that it accepts from a UA, an MTA should eventually get back as many delivery reports as the message had recipients. The P1 protocol gives originating MTAs two tools to cope with this. First, the originating MTA assigns an unambiguous identifier, called an *MPDUIdentifier*, to every outgoing message. This identifier is returned in subsequent delivery reports, matching them to the original message. Second, the originating MTA assigns a number from 1 to $n$ to each intended recipient of the message. ($N$ is the total number of intended recipients.) This number, which is called an *extension identifier*, is included in the envelope along with its associated O/R Name. Both are returned in the delivery report for its intended recipient's copy of the message. Together, the MPDUIdentifier and the extension identifier provide an easy and convenient way to match messages and delivery reports.

Trace information is added to each message by each management domain as the message is relayed across the MTS. In addition to providing a means of troubleshooting and auditing, the trace information enables the MTS to avoid routing messages in endless circles. Some of the trace information, such as the date and the time when the message was submitted, or the originator's O/R Name, is made available to each recipient UA when the message is delivered. Other trace information, such as the list of management domains through which the message passed, is not given to User Agents. (Note that no record of the particular MTAs that have handled a message is included in the trace information specified in X.411. If this data is recorded within a particular management domain, its representation is up to that management domain, and there is no standard way to pass it to the next management domain.) Trace information is also recorded on ServiceMPDUs.

ServiceMPDUs are used to support the Probe service and to convey delivery reports. There are two kinds of ServiceMPDU: *ProbeMPDUs* and *DeliveryReportMPDUs*. As their name implies, ProbeMPDUs are used for Probe. A ProbeMPDU is similar to the envelope portion of a UserMPDU.

ProbeMPDUs are processed by the MTL in substantially the same manner as UserMPDUs are.

DeliveryReportMPDUs convey delivery reports. They also return the results of Probes. Like a UserMPDU, a DeliveryReportMPDU has an envelope and a content. The envelope is a subset of a UserMPDU envelope. It has only three components: an MPDUIdentifier, which identifies the DeliveryReportMPDU; the O/R Name of the User Agent that submitted the message or Probe to which the delivery report refers; and trace information for the DeliveryReportMPDU. The content of the DeliveryReportMPDU conveys a delivery report or the results of a Probe. It always contains:

- The MPDUIdentifier of the original UserMPDU or ProbeMPDU.
- Information relating to one or more intended recipients of the original UserMPDU or ProbeMPDU, including:
  O/R Name and extension identifier,
  Results of the delivery attempt or Probe request, as well as an explanation if it was a failure.

The DeliveryReport content may include data such as trace information from the original UserMPDU or ProbeMPDU or billing information. If the original message had been redirected, the O/R Name of the actual recipient would appear. If delivery failed and return of contents had been requested, the content of the original UserMPDU would be present.

### The P3 Protocol

The P3 protocol uses remote operations as a basis for interactions between an SDE and an MTAE. Each operation has two phases. First, the specific operation is invoked by the requesting protocol entity. Then, its peer returns a successful result or an error.

When a UAE requests an MTL service via its SDE, the SDE turns that service request into a remote operation that it invokes from its MTAE. When the operation is completed, the SDE conveys the results to its UAE. Likewise, when an MTAE needs to deliver a message or a notification to a UAE via an SDE, the MTAE passes that information as parameters to a remote operation that it invokes from the SDE.

Remote operations are supported by four different classes of protocol data unit: *Invoke PDUs* request operations; *ReturnResult PDUs* convey the results of successfully completed operations; *ReturnError PDUs* describe operations that failed; and *Reject PDUs* report the recipient and rejection of malformed Invoke PDUs. As in P1, unambiguous identifiers are used to correlate each Invoke PDU with its subsequent ReturnResult, ReturnError, or Reject PDU.

## User Agent Layer Protocol

The P2 protocol is the only protocol in the User Agent Layer. Another name for the P2 protocol is the *Interpersonal Messaging* (IPM) protocol. The purpose of the P2 protocol is to provide the cooperating User Agent services that are defined as UAL services. It does this by providing rules for the interactions between pairs of User Agent Entities.

The general form and the structure of the P2 protocol follow those of the $P_c$ protocol in the model. The P2 model uses the same O/R Names as the P1 protocol does. It also permits the use of free-form names that are meaningful to people but not necessarily to computers. An *O/R Descriptor* consists of an O/R Name, a free-form name, or both. In some parts of the P2 protocol, an O/R Descriptor containing either type of information, or both, may be used. In other parts, an O/R Name is required but a free-form name may also appear.

Two kinds of protocol data unit are used in the P2 protocol: *IP-Message UAPDUs* (IM-UAPDUs) convey message contents; *Status Report UAPDUs* (SR-UAPDUs) convey receipt notifications and nonreceipt notifications. If receipt notification or nonreceipt notification was requested for an intended recipient of an IM-UAPDU, an SR-UAPDU may eventually be returned. User Agent Entities send IM-UAPDUs and SR-UAPDUs to each other by submitting them to the MTL. The MTL P1 protocol carries either kind of UAPDU as the content portion of a UserMPDU.

Every cooperating User Agent service that is supported by the P2 protocol is reflected in one or more parameters of an IM-UAPDU. Receipt and nonreceipt notification use IM-UAPDUs and SR-UAPDUs. Certain UAL services are not supplied by the cooperation of User Agents. Instead, they are provided by the MTL and are passed through the UAEs acting individually. In these cases, no action of the P2 protocol is needed to supply the particular service.

An IM-UAPDU is composed of a heading and a body. The *heading* consists of a mandatory "IP Message Identifier" parameter and a number of optional parameters. Each of the optional parameters conveys information that is directly and clearly related to one or more of the essential and additional UAL services. For example, the "Primary Recipients" and "Copy Recipients" parameters are used to provide the primary and copy recipients indication service. The IP Message Identifier is the message-identifier discussed earlier in this chapter. It unambiguously identifies an IM-UAPDU. The value of an IPMessageID assigned to one message may appear later as the value of another parameter in a new message. This allows a UA to match an original message with responses or cross-references to it, or to correlate it with another message that obsoletes and replaces it. The value of an IP Message Identifier belonging to a message is also returned as part of a receipt or nonreceipt notification for that message.

The *body* of an IM-UAPDU consists of a series of zero or more parts. (A body with zero parts might be used in a message that obsoletes, but does not replace, an earlier message, for example.) Each part of the body has a type. This type can be one of the encoded information types, or it can mark the body part as being encrypted, in a nationally defined format, or as a forwarded message. The structure and representation of a body part depend on its type. For example, a body part that contains a forwarded message has two sections. The first section, which is optional, gives information about the original delivery of the forwarded message. The second section contains the original IM-UAPDU, which is the forwarded message itself.

An SR-UAPDU has fewer components and options than an IM-UAPDU. It consists of the IP Message Identifier copied from an earlier IM-UAPDU, plus information that tells if that IM-UAPDU was or was not received by a particular intended recipient, and under what circumstances that determination was made. For example, if the message was received, the SR-UAPDU would say when that happened and would indicate whether the return receipt was explicitly approved by the recipient or automatically generated and sent by the recipient's User Agent. If the message was not received, the SR-UAPDU would tell why. In addition, the SR-UAPDU would contain the original IM-UAPDU, if its return had been requested.

## 5.5.3 Interworking with Other Telematic Services

Because the usefulness of a communications system is at least partially determined by the number of people that it can reach, the X.400 series makes a number of provisions for interworking between message handling systems and other telematic services, such as Telex, Teletex, and Group 3 facsimile. These provisions take two forms: an access method for Teletex terminals is defined, and a set of encoded information type conversion rules is specified.

### Teletex Access

The Teletex service can be very generally characterized as combining the communications aspects of Telex with word processing capabilities. Teletex enables its users to exchange sophisticated final-form documents. These documents are prepared, transmitted, and received using special

Teletex terminals. All Teletex terminals can send and receive documents. The specific processing and document storage capabilities of Teletex terminals vary. Teletex is a relatively new service, defined by the CCITT during its 1976–1980 study period. Teletex is more widely used in Europe than in other regions of the world.

Most Teletex terminals in use today cannot be used as User Agents. Recommendation X.430 defines a logical entity called a *Teletex Access Unit* (TTXAU). The purpose of a TTXAU is to provide document storage and computation facilities to supplement those found in a single Teletex terminal. Working together, a Teletex terminal and its TTXAU provide User Agent capabilities to their users.

A TTXAU is associated with a particular MTA. Both must be located in the same processing system. The interactions between a TTXAU and its MTA are the same as those that might occur between a UA and its MTA in the same system. Delivery of a message to a TTXAU is equivalent to delivery of a message to a User Agent.

The *Teletex Access Protocol* (P5) is used by a TTXAU and its Teletex terminal to coordinate their efforts at providing User Agent services. A P5 interaction between a TTXAU and a Teletex terminal is called an *action*. Each action is transmitted as a series of one or more Teletex documents, which are exchanged by using ordinary Teletex control procedures. Actions are encoded in human-readable form. This makes it feasible for actions to be entered into, or read from, a Teletex terminal by its operator. Fig. 5.14 shows the relationships between P5 and other protocols.

**Fig. 5.14. Relationships between P5 and other protocols.**

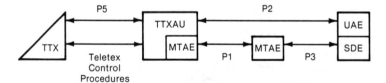

In general, the actions of the P5 protocol correspond to the service primitives of the User Agent and Message Transfer layers. Additional actions provide exception reporting, set the parameters governing the relationship between the Teletex terminal and its TTXAU, and facilitate management and retrieval of messages stored in a TTXAU. An action is composed of one or more *action elements*. Each action element consists of a Teletex document containing control information, followed by zero or more additional documents.

Each action element is represented by units of P5 control information called *elements*. The first element present in the control information for an action element identifies that action element. Subsequent elements convey parameters associated with the action element.

Each element is divided into a number of *fields*. An element is unambiguously identified by an *Element Number* field or by an *Element Name* field, at least one of which must be present in every element. The Element Name field is language dependent. Parameters to each element are encoded as *Element Value* fields. The Element Number and/or Element Name fields always precede the Element Value fields. Element Value fields representing optional parameters or parameters whose values are to be defaulted may be omitted.

Here is an overview of how elements are encoded. Each element is represented as one or more lines of text. If an element takes up more than one line, then the first line must consist of the Element Number and/or Element Name fields only. Elements are separated by one or more empty lines. Fields are separated by a linear white space or a new line sequence. Leading linear white space in a line is ignored. Element Number and Element Name fields are always terminated by a colon character. X.430 treats the specific representation of element value parameters as a national matter.

The format of a Send Action Element, illustrated in the following example, is representative of the general nature and encoding of the P5 protocol. In this example, square brackets ([ ]) enclose element value parameters. An additional set of square brackets encloses a parameter or group of parameters that may appear more than once. The character ⌐ denotes a new line sequence. "SP" denotes the space character.

```
3.1:    SEND:⌐
⌐
62:     QUANTITY OF DOCS:[No. of Documents]⌐
⌐
13:     PRIORITY: [Priority]⌐
⌐
14:     ORIGINATOR:[O/R Name]⌐
        =[Disclosure of Recipients][Alternate Recipient
        Allowed][Date and Time]⌐
⌐
15:     RECIPIENTS:[[O/R Name]⌐
        =[User Report Request][Explicit Conversion]]⌐
⌐
16:     CONVERSION:[Conversion Prohibition]⌐
⌐
17:     CONTENT INFO:[UA Content ID]⌐
        =[Content Return]⌐
⌐
```

18:    CONTENT INDICATORS:[Importance Indicator][Sensitivity
       Indicator][Date and Time]¬

¬

20:    FROM: [Originator/Recipient Descriptor]¬

¬

21:    AUTHORIZED: [[Originator/Recipient Descriptor]]¬

¬

22:    TO: [[Originator/Recipient Descriptor]¬
       =[Receipt Notification Indicator][Nonreceipt Notification
       Indicator][Return Message Indicator][Reply Request
       Indicator]]¬

¬

23:    CC: [[Originator/Recipient Descriptor]¬
       =[Receipt Notification Indicator][Nonreceipt Notification
       Indicator][Return Message Indicator][Reply Request
       Indicator]]¬

¬

24:    BCC: [[Originator/Recipient Descriptor]¬
       =[Receipt Notification Indicator][Nonreceipt Notification
       Indicator][Return Message Indicator][Reply Request
       Indicator]]¬

¬

25:    REPLY:[Date and Time] SP [[Originator/Recipient Descriptor]]¬

¬

26:    SUBJECT: [Subject Identifier]¬

¬

27:    MESSAGE ID: [Message Identifier]¬

¬

28:    CROSS REF: [[Message Identifier]]¬

¬

29:    OBSOLETES: [[Message Identifier]]¬

¬

30:    IN REPLY TO: [[Message Identifier]]¬

¬

The Send Action Element would be followed by a number of Teletex documents that make up the body of the message.

## Encoded Information Type Conversion

The ability to exchange messages between message handling systems and other telematic services requires that the messages generated by one system be understood and rendered by another. Since different telematic services use different methods to encode messages, a set of rules is required for converting information from one encoding scheme to another. An initial set of rules is defined in Recommendation X.408.

Certain kinds of encoded information type conversion are very simple and feasible. For example, mapping one character set into another can be a very straightforward matter. Other potential encoded information type conversions are easy to describe qualitatively but use moderately complex algorithms. Turning text into a facsimile image is an example of this. A third group of potential encoded information type conversions is impractical because the correct mapping would be problematic or because of overwhelming complexity. For example, it is difficult to even describe how a facsimile image should be mapped into a spoken message, while the reverse mapping would be extremely expensive to compute.

Another aspect of encoded information type conversion is the amount of information that is lost going from one encoding to another. For example, character case, font size, and page layout information is lost when transforming Teletex into Telex. Color is lost when converting Videotex into facsimile.

One way to describe a conversion is to differentiate between the format aspect, which describes the two-dimensional appearance of information, and the code aspect, which describes how each piece of information is to be represented. In the case of character and graphic information, the format aspect would govern the lengths of lines, either in absolute terms or as determined by the sizes and numbers of characters they contain, and govern the lengths of pages, either in absolute terms or as determined by the heights and numbers of lines that appear on them.

X.408 represents the start of a serious effort to address these problems. It categorizes conversions among Telex, International Alphabet No. 5 (IA5) text, Teletex, Group 3 facsimile, Text Interchange Format 0 (TIF0), Videotex, Voice, Simple Formattable Document (SFD), and Text Interchange Format 1 (TIF1) as:

- Possible without loss of information.
- Possible with loss of information.
- Impractical.
- For further study (in a few cases where the correct designation is not apparent).

Furthermore, conversions between the following encoded information types are defined:

| From | To | Notes |
|------|-----|-------|
| Telex | IA5 Text | Format conversion is not defined. |
| Telex | Teletex | Format conversion is not defined. |
| IA5 Text | Telex | Format conversion is not defined. |
| IA5 Text | Teletex | |
| SFD | Telex | |
| SFD | IA5 Text | |
| SFD | Teletex | Same code used in both cases. |
| Teletex | Telex | Format conversion is not defined. |
| Teletex | IA5 Text | |
| Teletex | SFD | Same code used in both cases. |
| Teletex | TIF1 | Same code used in both cases. Format conversion is not defined. |

### 5.5.4  Future Standards

The wide initial acceptance of the X.400 series has spawned a variety of efforts within CCITT and other organizations that develop standards. At the time this book was written, the CCITT was engaged in extending and refining the X.400 series. It had also launched a new effort to develop recommendations for a directory service. ECMA and ISO were well along in their work to adopt the X.400 series as their own, with suitable extensions dealing with mechanisms within a single management domain (and thus outside the scope of a CCITT Recommendation) but apparent at the interface between two vendors' products.

# 5.6  References

*CCITT Recommendations*

CCITT. 1984. *CCITT X.400 series recommendations for message handling systems.* (Recommendations X.400, X.401, X.408, X.409, X.410, X.411, X.420, and X.430.) International Telegraph and Telephone Consultative Committee.

CCITT. 1985. *CCITT Special rapporteur on directory systems.* (Question 35/VII.) Version 1 (May). International Telegraph and Telephone Consultative Committee. Geneva. This includes the following draft recommendations for directory systems:

Draft recommendation X.ds0. *Directory systems: Overview and application guidelines.*

Draft recommendation X.ds1. *Directory systems: Model and service elements.*

Draft recommendation X.ds2. *Directory systems: Naming framework.*

Draft recommendation X.ds3. *Directory systems: Protocols.*

Draft recommendation X.ds4. *Directory systems: Standard attribute types.*

Draft recommendation X.ds5. *Directory systems: Standard property types.*

Draft recommendation X.ds6. *Directory systems: Suggested naming practices.*

### Implementation of Electronic Mail Systems

Crocker, D., E. Szwkowski, and D. Farber. 1979. An internetwork memo distribution capability—MMDF. *Proceedings of the Sixth Data Communications Symposium.* IEEE/ACM (November).

### Multimedia Mail

*IEEE Computer Special Issue on Multimedia Mail Systems.* 1985 (October).

Postel, J. 1979. An internetwork message structure. *Proceedings of the Sixth Data Communications Symposium.* IEEE/ACM (November).

### OSI Model

*Basic reference model for open systems interconnection.* ISO 7498.

### Other Standards for Electronic Mail Systems

Crocker, D. 1982. Standard for the format of ARPA Internet text messages. Technical Report RFC 822 (August).

Mockapetris, P. 1983a. Domain names—Concepts and facilities. Technical Report RFC 882 (November). USC/Information Sciences Institute.

____. 1983b. Domain names—Implementation and specification. Technical Report RFC 883 (November). USC/Information Sciences Institute.

Postel, J. 1982. Simple mail transfer protocol. Technical Report RFC 821 (August). USC/Information Sciences Institute.

U.S. National Bureau of Standards. 1983. Specification for message format for computer-based message systems. Federal Information Processing Standards Publication 98 (January). U.S. Department of Commerce.

### Papers about Naming, Addressing, and Message Transfer

Birrel, A., R. Levin, R. Needham, and M. Schroeder. 1982. Grapevine: An exercise in distributed computing. *Communications of the ACM* 25:260–274.

Deutsch, D. P. 1984. Implementing distribution lists in computer-based message systems. From *Computer-Based Message Services.* Edited by H. T. Smith. Amsterdam: North Holland.

Oppen, D., and Y. Dalal. 1981. The Clearinghouse: A decentralized agent for locating named objects in a distributed environment. Technical Report OPD-T8103 (October). Xerox Office Products Division.

Schroeder, M., A. Birrell, and R. Needham. 1984. Experience with Grapevine: The growth of a distributed system. *ACM Transaction on Computer Systems* 2 (February): 1.

Schoch, J. F. 1978. Inter-network naming, addressing, and routing. *Proceedings of COMPCON* (Fall). IEEE Computer Society.

White, J. E. 1984. A user-friendly naming convention for use in communications networks. From *Computer-Based Message Services.* Edited by H. T. Smith. Amsterdam: North Holland.

# 6

# CELLULAR NETWORKS

## Philip T. Porter

Enhanced by the emergence of solid-state electronics, the postwar era has seen the creation of ubiquitous portable broadcast communications. Ever on the move, people today now have at their disposal the means and the choice of "visiting" the recording studio, the concert hall, the stadium, or the newsstand with little inconvenience or expense. Likewise, the ability to communicate selectively in a two-way interchange is now both convenient and practical. The attainment of this level of convenience has not been due solely to advancements in electronics and equipment design, however. A *system plan* for efficient servicing of large areas had to be developed before two-way interchanges truly became practical. The purpose of this chapter is to describe the cellular radio system plan as it has developed throughout the world over the last fifteen years or so, focusing on the North American plan as an example.

In this context, the term *cellular* denotes a method of planning radio frequency assignments. In the past, mobile communication was limited by the number of channels of the radio spectrum made available for this service in a particular city. In the United States, the Federal Communications Commission is responsible for assigning channels. The few channels that were available were assigned a very few at a time. This "large cell" approach assumed that each radio channel was an independent system of its own, and hence it sought to maximize the amount of area covered by one station. This strategy afforded the maximum area of service, which was the community-of-interest of the user, but allowed only one usage of the channel in an entire city. In contrast, the "small cell" or cellular approach views a large block of channels as one system and affords a way to reuse each channel several times in a service area.

In explaining the cellular radio network, this chapter will give a short history of mobile radio, discuss radio propagation, and introduce the cellular concept. Later sections will outline how telephone calls are set up and maintained, look at regulatory aspects of cellular networks, and briefly examine future developments in the cellular concept.

# 6.1   History and Background

Even before the vacuum tube was invented and before the concept of frequency-division multiplexing was understood, early planners in radio appreciated radio's mobile communications potential and applied it to ship-to-shore and ship-to-ship calling. This was consistent with communication needs and technology of the time. Maximum range on an oceanic scale was a major concern, and nature cooperated: the lower frequencies (up to 30 MHz) utilized by the early equipment happened to be ideal for long-distance propagation. By the 1920s, the use of radio for both telegraph and voice in a shipborne mobile context was well established, not only for navigation and safety but also for communications by passengers.

Air-to-ground became the next usage of mobile radio. However, the weight of early equipment and the need to trail an antenna suitable for the long wavelengths used were both concerns. The altitude of the aircraft was a help in achieving the long distances required.

Use of radio to vehicles was slower in coming, since it was possible for the mobile vehicle (e.g., a police patrol car) to stop and call via wireline telephone. Experimental usage began in the 1920s, and some operational systems were set up on a wide-area basis using channels below 30 MHz. But, it was not until after World War II, when the 150-MHz band was opened up, that urban mobile radio usage began in earnest. Since that time, mobile radio frequency bands at 35–42 MHz, 150 MHz, 450 MHz, and now 850 MHz have been progressively opened up by the FCC, even as new technology for public safety, transportation, industrial, and common carrier usage has made use of the new bands possible. In all of these earlier systems, the objective of system designers was to achieve maximum range, consistent with the technology, whether the range was oceanic, state-wide, or city-wide. In contrast, the cellular concept of system design takes the opposite approach. It puts together a large coverage area out of small building blocks, rather than trying to make one base station serve the entire area.

Early systems were predominantly of the *dispatch* variety (coordinate, command, and control, in military terms). System users were considered to be a "fleet"—ships, planes, or vehicles—under control of a common point, for safety, navigation, or business reasons; and the operator was considered to be a professional driver or pilot. The use of radio for common-carrier purposes was slower in emerging. The average person, who was not part of a fleet, did not benefit from mobile communications until later.

Telephone usage began in 1946 and developed along with dispatch usage, but at a much lower level. Private dispatch systems were awarded 50 to 100 times as many channels as the common carriers were, and therefore usage paralleled allocation. In 1980, there were approximately 100,000 common carrier users versus 5 to 7 million dispatch users. Throughout the last thirty-five years or so, common carrier usage of mobile radio in the larger cities has been limited by the allocation available; channels are heavily loaded in most cities in peak hours. The FCC changed this when it approved a cellular allocation. It recognized that the general public would benefit from the availability of a common-carrier system approach.

Over the years the common-carrier system has taken on characteristics of the telephone system to which we are accustomed. Early common-carrier mobile radio operated like a dispatch system. Calling took place via an operator specifically assigned to mobile service. Users were assigned to one specific channel and had to listen first to make sure that the channel was free. Users then keyed the radio transmitter via a "push-to-talk" switch when they wanted to talk. By the mid-sixties, users were able to (1) dial calls themselves, (2) to be called by others, and (3) to use any available assigned channel automatically; thus, no special training was required. Calls were *trunked* over any available channel, just as in wire-line interoffice traffic. This new system was much more efficient, since it eliminated the queuing and call delays experienced by dispatch users.

Planning of channel usage during this evolution reflected the available technology. Frequency modulation became the favored method, since it was easily used on an urban channel where power was changing rapidly (more on this later). Initial channel spacing was 120 kHz, later decreased to 60 kHz, and then to 30 kHz in the heavily used 150-MHz assigned band. Assignment of specific channels to specific cities was generally not thought out in detail, as it had been for uhf television stations, but rather grew on an *ad hoc* basis.

Early in the system design process, even before the development of the transistor, researchers at Bell Laboratories realized that a network of small coverage areas could be put together, in concept, to make a larger service area (Schulte, Jr., and Cornell 1960). The technology and the allocations were not ready, however. A large allocation and a large body of users were needed before the system could be efficient and inexpensive. Radios remotely tunable to many channels had to be feasible, and the system had to use high-speed signalling and logic to control the tuning of the users. The concept lay dormant for twenty years before technical feasibility was demonstrated, and it took another fifteen years of development and regulatory due process before cellular networks were put to use serving the public.

Along the way, in 1968, the former Bell System tested the concept in a linear-cell railroad system by providing telephones for the Metroliner high-speed-train demonstration service between New York and Washington (Paul 1969). This system, a network of ten cells controlled from Philadelphia, reused channels in a limited way and gave designers some confidence that cellular networks were feasible.

# 6.2 Urban Radio Environment (30 MHz to 3 GHz)

The reception of radio signals is affected by the physical world of propagation, the man-made and the natural noise, the echoes encountered, and the statistical correlation of possible detected signals. In order to understand many of the choices made in the design of cellular networks, it is important to understand some of the facts of urban radio propagation (Jakes 1974; Lee 1982; Bullington 1957).

## 6.2.1 Propagation and Path Loss

In free space, the transfer of power between two idealized isotropic antennas is given by:

$$\frac{P_R}{P_T} = \frac{1}{(4\pi)^2(\frac{r}{\lambda})^2}$$

where $P_T$ and $P_R$ are the transmitting and the receiving powers, respectively (copolarized); $r$ is the separation; and $\lambda$ the wavelength.

This is the *inverse-square* law, which follows from the conservation of energy and from elementary geometry. In the real urban world of city streets and buildings, building attenuation, scattering, and ground reflections can be expected to make this prediction statistical in nature so that asymptotically it becomes:

$$\frac{P_R}{P_T} = \frac{k}{r^n} \qquad\qquad (6.1)$$

where the exponent $n$ is experimentally found to be about 4, and $k$

assumes different values depending on the general environment, the frequency, and the antenna heights. When the logarithm of this equation (i.e., the power ratio in decibels as a function of log $r$ is plotted, Fig. 6.1 is the result. $P_R/P_T$ is called the *path loss*. The free-space slope is $-2$; that is, a 20-dB negative change in path loss for a factor-of-ten change in $r$. Beyond one or two miles, the measured slope changes to about $-4$; that is, a 40-dB change per factor of ten. Beyond some value $R_0$, called the *radio horizon*, which depends on the height of the base station antenna above its local terrain, the signal begins to fall off even more rapidly because of the curvature of the earth. For a 100-foot mast, this distance is roughly fifteen miles. Beyond $R_0$, the effects of diffraction and refraction control the process, loss becomes much larger, and signal prediction becomes less reliable.

Fig. 6.1. Local mean path loss along several hypothetical radial paths.

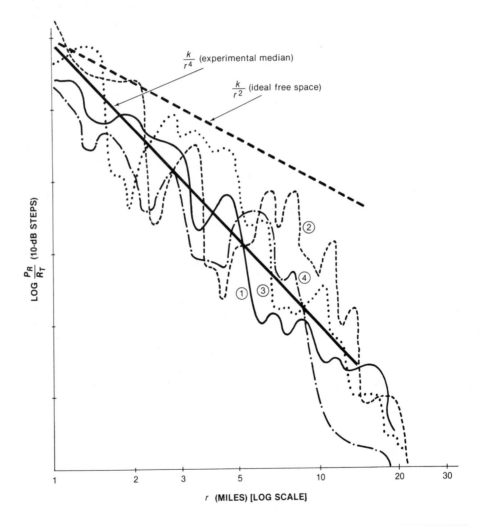

$\frac{k}{r^4}$ (experimental median)

$\frac{k}{r^2}$ (ideal free space)

LOG $\frac{P_R}{R_T}$ (10-dB STEPS)

1    2    3    5    10    20    30

$r$ (MILES) [LOG SCALE]

From Fig. 6.1, another major effect is noted. This figure shows four examples of measurement along a radial. The *local mean value* along any one radial path does not closely follow the prediction[1]. This is because of the hills and clusters of buildings that unpredictably cover the area. If one were to circle many base stations in many cities and circle many areas within those cities at a constant radius, sampling the local mean received power, $P_R$, one would find that $P_R$ measured in decibels would be characterized by a normal distribution. Its mean would be given by the $k/r^n$ law, and its standard deviation $\sigma$ would be found to be about 8 dB (in a vehicular context). Local sample values would be highly auto-correlated, on a scale appropriate to the sizes of the hills and building clusters, and would also be noticeably affected by the street pattern.

The wide standard deviation has profound effects on signal prediction and thus on system design. There is about a 20-dB range (100:1 in power) between the $10^{th}$ and $90^{th}$ percentiles, meaning that the higher 10% of the signals at, say, ten miles exceeds the lower 10% of the signals at three miles.

A final major effect is that caused by multipath interference. For a typical urban situation, where the transmitting antenna is not within clear line of sight, the signal arriving at the receiver is in fact the sum of many signals of random phase and amplitude. Fig. 6.2 shows a situation in which the direct path $r_0$ is blocked, but five indirect paths are present.

This sort of situation can be statistically described as having a Rayleigh amplitude distribution, and a typical plot of the instantaneous signal over a short path of say 40 to 50 wavelengths is shown in Fig. 6.3. Over distances on the order of a half wavelength, the signal may vary over a 10- to 40-dB range. When the signal is instantaneously high, its amplitude and phase will vary slowly with a change of position. When it approaches a null, its rate of change will be quite rapid and its phase will undergo a fast 180° phase change. The latter causes a baseband "click" in an angle-modulated (fm or pm) system, the magnitude depending on the speed of the vehicle through the more or less stationary rf environment. The reader can experience this to some extent by listening to a vehicular fm receiver in a fringe area. (The effect in the 100-MHz fm-broadcast band will be nine times less frequent than at 900 MHz.) At 900 MHz and a car speed of 55 mph, these clicks occur about 150 times per second, and harmonics extend throughout the audio band.

One other aspect of urban propagation worthy of mention is that of Doppler spread. Because of the highly scattered nature of the signal with

---

1. By *local mean*, is meant the value of $P_R$ averaged over several wavelengths. A wavelength is a little over 1 foot at 850 MHz, 6 to 7 feet (2 meters) at 150 MHz.

**Fig. 6.2. Random reflection and scattering—the cause of Rayleigh fading.**

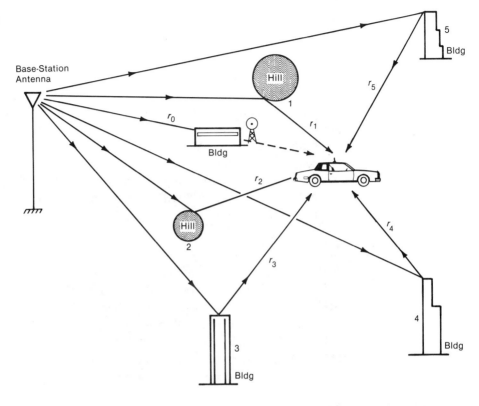

**Fig. 6.3. Typical fading signal measured in a small area.**

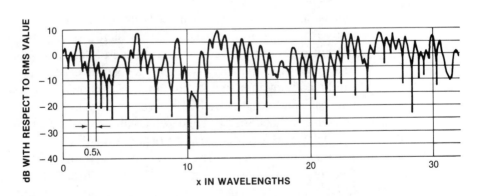

no line-of-sight component, just as many signal components come from behind the moving vehicle as from the front. Thus, the Doppler effect generally causes not a shift, but a broadening of the spectrum. The effect is also sometimes called "random fm." At 850 MHz, this broadening is in the order of 60 to 70 Hz for a 55-mph vehicle and is inseparable from the Rayleigh fading.

To summarize these urban propagation effects: instantaneous path loss $(P_R/P_T)$ can be modeled as the sum (in decibels) of three components:

$$10 \log_{10} (P_R/P_T) = 10 \log_{10} (k/r^n) + X + Y \qquad (6.2)$$

where (1) the distance dependence is contained in the first term, (2) $X$ is a log-normal random variable with a mean of 0 dB and a standard deviation of about 8 dB for vehicles (higher for handheld units carried into buildings) and with high autocorrelation typically spreading over several tens of feet or more, and (3) $Y$ is proportional to the log of a Rayleigh-distributed random variable with a mean of 0 dB and with rapidly changing values over fractions of a wavelength.

## 6.2.2   External Noise

In addition to propagation effects, man-made and natural noise is a significant impairment, even at frequencies as high as 850 MHz. Sources of this noise are: ignition noise, power line harmonics and arcing, unintended harmonics and intermodulation products generated by circuit nonlinearities in other radio services, lightning, galactic noise, etc. The predominantly man-made noise can be a limiting factor in coverage of urban areas, making a low noise figure not a major goal in receiver design.

## 6.2.3   Delay Spread

Echoes (i.e., delay spread) can be a problem in radio (Cox and Leck 1975) just as they are in acoustics. The problem's scaling is, of course, altered, but analogies are useful. Just as echoes in a large auditorium or a stadium can impair intelligibility, whereas acoustical echoes in a small room are rarely bothersome, so too, in radio, reflections in a large coverage area that has been created by high power and a tall mast will cause more problems than reflections in a small cell created with low power and height.

Delay spread is generally not a problem in the usual single-channel-per-carrier systems where maximum deviation and maximum baseband frequency (or bit rate) are low. However, in TDM channels with digital modulation where high rates are attempted while still using omnidirectional antennas, intersymbol interference will cause errors and place an upper limit on the signalling rate. (See Bartee 1985, Chapter 3, for a discussion of this problem in a wire-line context.) The error rate cannot be reduced by increasing the carrier-to-noise ratio ($C/N$) or the carrier-to-interference ratio ($C/I$). Fig. 6.4 is a simple illustrative example (for a direct path and two delayed paths), showing how modulation can be distorted. With an rms value of delay spread in major urban areas around

3 microseconds when high power and a high antenna site are used, an upper limit on symbol rate of about 30 to 50 Kbps results.

**Fig. 6.4. Effect of delay spread on reception of a digital signal.**

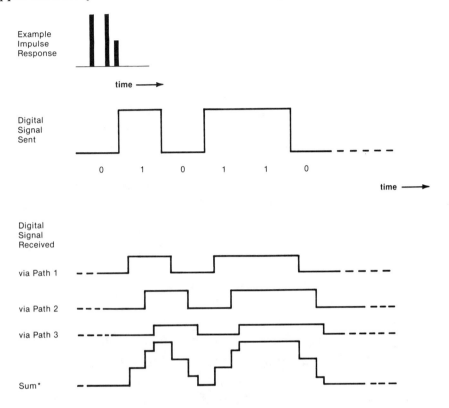

*Bottom line is illustrative only; actually, the information resides in the phase of the sum not the amplitude.*

In wire-line application, adaptive equalization is used to increase the reliable symbol rate. Because of the rapid Rayleigh fading, whereby the channel as seen by a fast-moving vehicle is seldom static for more than a fraction of a millisecond, practical techniques for adaptive equalization have not yet been designed. To achieve high symbol rates requires (1) the use of small cells and (2) the use of masts that do not extend above building tops so that distant reflectors are not illuminated. This is analogous to the use of distributed loud speakers in the acoustic design of auditorium public address systems.

## 6.2.4 Diversity

Any less-than-perfect system (i.e., one whose probability of being "available" is less than 100%) can benefit from replication, so that two uncorrelated failures are required before a system outage occurs. A radio system fits this description in its fringe areas (Jakes 1974; Lee 1982).

Two or more uncorrelated Rayleigh-fading signals (component $Y$ of equation [6.2]) can be created (1) over different frequencies (frequency diversity), (2) with repeats (time diversity), and (3) with separate receiving antennas, either by using different polarizations (polarization diversity) or by spacing them apart (space diversity). Time diversity is ruled out by the real-time demands of a two-way conversation. Frequency diversity is made less attractive, since it uses more bandwidth than is necessary. However, both polarization and space diversity are attractive choices for use in cellular networks.

As far as the log-normal component of propagation ($X$ in equation [6.2]) is concerned, correlations in frequency, time, and polarization are extremely high. Therefore, space diversity is the only effective type of diversity. In one form or another, space diversity has been used in radio systems for decades and involves the use of two or more widely spaced transmitters and receivers in the region to be served. Termed *macroscopic* diversity, it is differentiated from *microscopic* diversity against Rayleigh fading, in which the antennas can be within a wavelength of each other.

Processing of signals received on two or more diversity branches can take one of several forms:

1. It may take place either predetection or postdetection.

2. Either it may select the strongest signal or add the signals using some weighting algorithm; *or* it may stick with one branch until it becomes poor and then switch to another branch.

Predetection processing is not feasible for macroscopic diversity because of the separated receivers, and postdetection addition of signals is not useful because of processing difficulties. The last alternative, switching to another branch, called switch diversity, is generally not very effective either. In general, however, it can be shown that use of postdetection selection diversity is a cost-effective strategy in system planning, for a given engineered quality. This is especially true when the system has a large variability; that is, a large value of standard deviation.

# 6.3   The Cellular Concept

The manner in which base stations are planned is fundamental to the concept of cellular networks. This section will introduce these ideas and contrast them to earlier planning.

### 6.3.1  Fundamentals

The ability of squares and hexagons to pack a plane area efficiently is well known and obvious (tiled floors, honeycombs, etc.). The packing of base stations in a square or hexagonal pattern over a certain area in order to provide uniform service quality is an outgrowth of that observation. Hexagons generally work out slightly better than squares because they more closely approximate circles, the simplest pattern of an antenna (MacDonald 1979).

The base stations, not the cells, are the physical reality. The cell is defined merely as the locus of points closer to that cell's base station(s) than to any other base station. In geometric terms, the cell is called a *Dirichlet region*. See Fig. 6.5. Call the area of the cell *a*. Base stations occupy *lattice points*. The base stations don't have to be in any idealized pattern, but the farther they deviate from the ideal, the wider the distribution of the overall voice quality will be.

**Fig. 6.5. Lattice of base stations.**

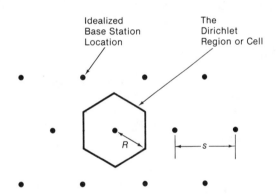

The maximum spacing between base stations, called *s* in Fig. 6.5, determines the probability of getting reliable coverage. It will be a function of the base station's effective radiated power, the sensitivity of the receivers used, the standard deviation that the unpredictability of the environment imposes on path loss, and the reliability/quality that the designer needs to give the users. (A typical criterion used in cellular design is to achieve better than 17-dB local mean carrier-to-noise or carrier-to-interference ratio over 90% of the area. In a moving vehicle, this yields a "fair to good" sounding signal at 850 MHz.)

So far, no mention has been made of radio channels. Obviously, a base station must have an assignment of one or more channels. Of all channels allocated (call this $\Delta$), only a small share or subset can be given to one cell; others must be saved to assign to neighbor cells. In Fig. 6.6, channels are shown to be repeated a distance *D* cells away. The size of this subset is given by $\Delta/N$, where $N$ is the number of subsets needed. Since per-user system cost is decreased as the number of channels at a base station (hence the number of users of a base station) is increased, keeping $N$ as small as possible is important.

**Fig. 6.6. Lattice of co-channel base stations.**

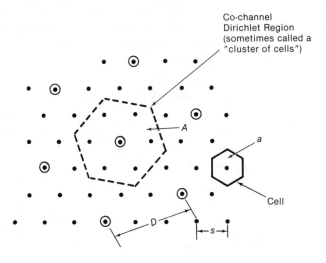

To make this concept clear, let us envision a 20-channel allocation ($\Delta = 20$), split into four subsets ($N = 4$) of 5 channels each.

|  | Channel Set | | | |
|---|---|---|---|---|
|  | A | B | C | D |
|  | 1 | 2 | 3 | 4 |
| Channel | 5 | 6 | 7 | 8 |
| Numbers | 9 | 10 | 11 | 12 |
|  | 13 | 14 | 15 | 16 |
|  | 17 | 18 | 19 | 20 |

In this case, a cell assigned subset B would utilize channels 2, 6, 10, 14, and 18. (A typical North American cellular network uses 312 channels divided into either seven sets of 44 to 45 channels each or four sets of 78 channels each; if a sector plan [see next section] is used, there is a further subdivision into three subsets of 15 channels or six subsets of 13 channels.)

Now let us concentrate on one of the $N$ subsets. If evenness of quality is desired, the subset too will be used in a regular lattice in the same way that base stations in general are laid out; call this the "co-channel lattice." This lattice will also have a well-defined "co-channel Dirichlet region"; call its area $A$. The number of subsets needed to fill space is thus $N = A/a = (D/s)^2$. And it can be shown that only certain values of $N$ are permissible (for these idealized lattices):

For a quadrangular lattice:

$$N_q = (D_q/s)^2 = i^2 + j^2$$

For a hexagonal lattice:

$$N_h = (D_h/s)^2 = i^2 + ij + j^2$$

where $i$ and $j$ are the number of unit steps of length $s$ between co-channel cells, measured along the x and y axes of the lattice, respectively. Thus, $N_h$ can take on the values of 3, 4, 7, 9, 12, 13, etc. All other numbers for $N$ will result in nonisotropic interference conditions.

Considering only one interferer, since from equation (6.1) $C = k/R^4$ and $I = k/D^4$, the ratio of the median desired carrier $C$ to the median co-channel interferer $I$ at the corners of the cells (the point farthest from the base stations) is given by:

$$C/I = (D/R)^4 \propto N^2 \qquad \text{(6.3)}$$

But $R = s/\sqrt{3}$ (for the corner of a hexagonal cell); therefore, the median $C/I$ is given by $9N^2$. Clearly, the more subsets $N$ one creates, the higher $C/I$ will be.

This is, of course, a brief and idealized treatment of a complex topic. In practice, one must take into account some higher percentile than the median. Also, when there is more than one interferer, $I$ is the sum of all the interferer powers.

Clearly, the use of small cells as building blocks permits the designer to tailor the coverage area (a city and its suburbs, for example) more closely to the community of interest. The designer must take into account the fact that users will be anywhere in the area, not near one predictable cell. Also, the coverage area must be extensible, rather than static. The essential features of a cellular network can thus be summarized as:

1. More than one base station per system.

2. Unpredictable, dynamic location of users within the network of base stations.

3. Reuse of rf channel assignments at more than one base station.

4. Growth of both Erlang[2] capacity and area coverage as market demand increases.

## 6.3.2  Sectoring

Besides increasing $N$, another way to increase $C/I$ is to use directional antennas at the base stations, for both transmit and receive (Bell Laboratories 1971). Ideally, the interference should increase in proportion to the beam width of the antennas (i.e., going from 360° to 180° should decrease interference by a factor of two). Layouts of this kind are called *sectoring plans*. Typical sectoring plans used in cellular networks are tied to integer divisions of 360°, for example 120° or 60°. With cells made up of 120° sectors of hexagons, $N$ channel sets are further subdivided into three subsets each (a total of $3N$ subsets), ideally yielding 5 dB more $C/I$; with 60° sectors, $N$ channel sets become $6N$ subsets, ideally yielding about 8-dB improvement in $C/I$. See Fig. 6.7.

Fig. 6.7. Four possible sector plans.

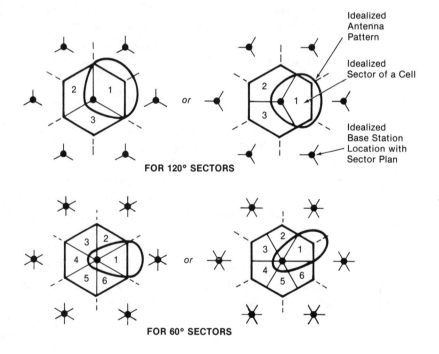

FOR 120° SECTORS

FOR 60° SECTORS

Idealized Antenna Pattern

Idealized Sector of a Cell

Idealized Base Station Location with Sector Plan

---

2. An *Erlang* is the unit of traffic intensity. For example, if an 8-channel server group has an average of 6 channels occupied during the busy hour of the day, it is handling 6 Erlangs of traffic and is 75% efficient.

Sectoring plans thus permit a design trade-off between smaller numbers of radios per base station and smaller sectors, for a given *C/I*. To create a specific *C/I* change, the designer can choose to use a sector plan rather than a larger *N*. Offsetting this savings are the poorer traffic engineering efficiency of the smaller subset and the need to hand off more frequently.

### 6.3.3 Growth

As we said earlier, the *C/I* ratio depends, not on the absolute power in *C* (because one user's *C* is another user's *I*), but on the *D/R* ratio; hence on *N*. As long as this *D/R* ratio is maintained, the cells may be made smaller and thus more traffic per unit area can be handled. This permits a sparse cellular network to be installed initially, with the possibility of *splitting cells* (i.e., making them smaller) as demand increases. Requirements for this growth process are:

1. Minimal elimination of original base stations.
2. Minimal changes in original channel assignments.
3. Gradual, not wholesale, change.
4. In a sector plan, no repointing of antennas.

Fig. 6.8 shows one way to plan for the growth process. If traffic were always uniform over a metropolitan area, the channel assignment task would be easy. Since traffic is not, two or more cell sizes will have to be tolerated within one cell lattice. The lattice shown in Fig. 6.8 is capable of adapting to an increased demand. Maintaining the *C/I* target (hence the *D/R* target) while two different sizes exist means splitting some or all of the *N* channel sets into two pieces. This is done by establishing different locating criteria for the two portions of the channel set, as described later in section 6.6.4. Using different locating criteria further complicates the channel assignment task, however, and makes the maintaining of a reasonable Erlangs-per-channel efficiency goal hard to achieve. Various means for achieving channel efficiency have been proposed (dynamic channel sharing, underlaid cells, power control, back-lobe sharing, etc. [Huff 1985]), but description of these is beyond the scope of this chapter.

# 6.4 Other RF-related Topics

We now turn to other rf-related topics. The next two sections will briefly discuss aspects of multiplexing and modulation and the process of companding.

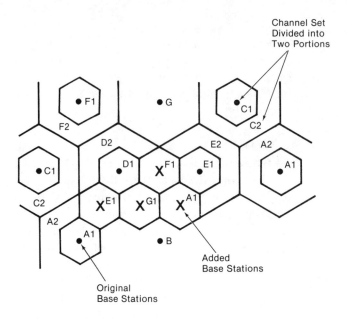

**Fig. 6.8.
Example of a
lattice after
growth has
taken place.**

### 6.4.1  Multiplexing/Modulation

Throughout this chapter, multiplexing has been tacitly assumed to be frequency division. That is because up until now all cellular networks have been designed to use a conventional single-channel-per-carrier frequency-division multiplexing plan. This technology is well understood and efficient. Most importantly, in the mobile-transmit to land-receiver direction, it permits the mobile to be autonomous (i.e., not synchronized with other mobiles). This plan is consistent with well-understood analog modulation methods. However, *future* cellular systems, which could use digital modulation, may adopt time-multiplexing—which has a more complicated synchronization plan but a simplified frequency plan (*Conference Record, Nordic Seminar* 1985). Such a plan would use one complicated receiver per base station rather than many simpler receivers, and could well be cheaper to implement.

All cellular networks currently use a common type of modulation known as frequency modulation. Amplitude modulation is not used because it lacks the "capture effect" of wide-deviation fm. Digital modulation and the means for efficiently encoding the analog human voice into a stream of digits were neither well understood nor economically viable when these networks were being planned during the early 1970s. Again, future state-of-the-art technology and design could move toward digital modulation, as efficient, real-time coding of voice signals into digits becomes possible.

Because cellular networks need only a subset of channels at one base station, they have relaxed adjacent-channel selectivity requirements, as compared with conventional systems. Indeed, the cellular channel assignment plan can be thought of as frequency/space division multiplexing. Space separation augments the frequency separation, since adjacent channels are never used at the same base station. Therefore, the deviation of the fm signal can be set fairly wide; maximum deviation in the North American network is 12 kHz, compared with 5 kHz in conventional usage.

### 6.4.2 Companding

Another new feature, which significantly adds to voice quality, is *syllabic amplitude companding*. This technique, in which speech amplitude is compressed for transmission and expanded after reception by the use of a variable-gain linear amplifier, subjectively improves the perceived signal-to-noise ratio. It is a well-understood technique in wire-line service dating back over thirty years, but only recently could it be applied economically on a per-user basis. The process is diagrammed in Fig. 6.9. Companding serves two main purposes: (1) it suppresses intersyllabic noise, which is subjectively disturbing (the noise is not masked by speech); and (2) it increases the rms deviation of the rf carrier. To be fully effective, however, both the gain characteristics and the time constants of the receiver must be well matched to that of the transmitter; otherwise, on very noisy channels, companding can further decrease, rather than increase, the perceived quality.

## 6.5  Service

Sections 6.3 and 6.4 dealt with the radio aspects of cellular networks. Ideally, the user should be oblivious of the fact that the service is radio-based. The basic service should be as much as possible like ordinary telephone service. Several differences, however, arise because of the use of a radio facility.

One difference is in the statistics of the voice quality. Except when very long wire loops or when noisy analog carrier circuits are encountered, wire-line service is either Excellent (99.9% of the time) or Unusable (0.1% of the time). Service via radio is much more variable. Perhaps 90% of the radio calls are rated Good or Excellent, but roughly 9% may be rated Fair, and 1% Poor or Unusable, depending on the local engineering/planning. In the latter case, a very small fraction of local calls will unavoidably be so poor as to not permit call setup, or, if an active caller

**Fig. 6.9.
Compandor
operation.**

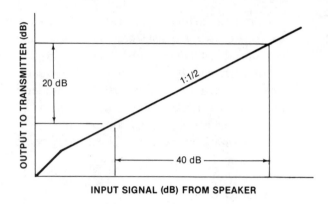

*(A) Compressor gain characteristic.*

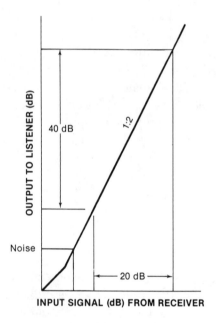

*(B) Expandor gain characteristic.*

moves into such an area, the call may be cut off prematurely. Good engineering can minimize but not completely eliminate these problem areas.

Also, wire-line callers have a dedicated loop to a central office and contend only for trunks between central offices. Radio callers contend for radio channels, which are the equivalent of the local loop. For this reason, and also to minimize the actual holding time, the caller in a cellular network is expected to use *preorigination dialing*; that is, the caller can enter the dialed number into the unit off-line at his or her convenience, prior to seizure, so that when seizure is initiated, the dialed number can be transmitted quickly and controllably to the Central Office (Fluhr and Porter 1979).

Another unique aspect of mobile service is *roaming*. In wire-line service, a telephone is always physically connected at one place in the network. In cellular radio service, a telephone may be anywhere—within the assigned area (the home city), in another area, halfway in-between, or nowhere at all (turned off). The system plan should be able to (and does) accommodate all of these situations gracefully. For example, if the set of the called mobile user is turned on but the user cannot answer, the caller will hear normal ringing; if the unit does not respond (i.e., is not powered up or is out of range), the caller hears a recorded message telling that the user is unavailable. By the way, a cellular user is given a normal dialable telephone number.

Since users don't always know their precise location when dialing, the system must assume that a call dialed with seven digits (i.e., no area code) is destined for the user's home city. For example, a Milwaukee user must always dial Chicago numbers using the 312 area code even when in Chicago, but the user may dial Milwaukee using seven digits (i.e., without the 414 area code) no matter where he or she is.

Also, since users cannot be paged over the whole nation, a choice must be made to narrow the area. The home city is the obvious choice, but allowance must be made to "call forward" (either automatically or manually) if users want to receive calls elsewhere.

Lack of ensured privacy is another service attribute of cellular networks. Any radio-based service is vulnerable to eavesdropping, not only on the conversation but also on the identity of the calling and called parties. In cellular networks, it is not possible to predict what channel of what base station will be used by a specific subscriber. However, using a scanning receiver for random eavesdropping is not illegal and such a law would be virtually unenforceable. Products are available that scramble or otherwise make the speech signal unintelligible. If the modulation were digital, such products could be quite cheap and effective; but, since the modulation is analog, privacy equipment tends to be either expensive or not totally effective.

The characteristics of data service via a cellular radio network will also change. Privacy can be a major concern. If a user follows the normal procedure of using a password to gain access to data files, an eavesdropper can easily learn the password and then access those same files at a later time. For this reason, use of file encryption is recommended.

Strings of symbol errors, caused by handoff or low $C/I$ (due to Rayleigh fading) are another concern. For this reason, error detection coding and protocols for automatic repeating of messages should be designed into the data signalling scheme. The use of standard wire-line data terminals will generally not be satisfactory unless the user is temporarily stationary while using his or her mobile terminal.

# 6.6  How Calling Works

On any common-carrier radio telephone system, three control functions—seizure, signalling, and supervision—must be carefully designed. Discussion of each of these is warranted, since the methods used will differ from usual telephone practice. First, however, let us discuss the infrastructure of a cellular system, the way cellular radio networks allocate some of their resources for control, and the general way that calls are handled.

The basic structure of a cellular network is shown in Fig. 6.10. The vehicle radio sets (typically numbering at least several thousand) are assigned normal ten-digit telephone numbers. These are randomly either turned off, powered on but idle, or active. Users are unevenly spread over the service area. Several base stations (a few or dozens depending on market size and area) house the fixed rf equipment and are deployed over the area in a cellular arrangement, as explained in section 6.3. These in turn connect to the main system control entity, called the Mobile Telephone Switching Office (MTSO).[3] This machine is a digital processor (usually a modified wire-line switch) that provides the ongoing system organization and control and that switches calls between the active users and the network. The network interface can occur at several points, at either local or tandem central offices, to match the local calling areas of the specific city in question. To the rest of the network, the MTSO may appear as either a PBX or a local end office.

Two methods are used in radio systems for controlling the network: (1) use of voice channels for control purposes, at appropriate times in the call, and (2) use of certain channels for voice only, with other channels dedicated to control. Both methods subtract from the total traffic-carrying capacity to a small extent. The North American cellular network uses the second approach, since *a priori* knowledge of where to find control channels helps the mobile unit speed up the seizure process. We will concentrate herein on describing the North American system.

In the system, two operationally identical systems (called A and B)

---

3. Space does not permit an in-depth description of MTSOs; these are implemented differently by different vendors.

Fig. 6.10. The
building blocks
of a cellular
network.

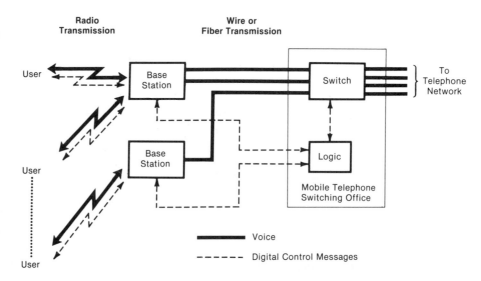

can be used in each city. Each has 333 channels available to it.[4] Of these 333 two-way channels, 21 are permanently dedicated to control purposes: paging called users, initiating seizures from both calling and called users, commanding active units to tune to specific voice channels, and sending "overhead" information (e.g., system identification and characterization). One or more of these control channels are assigned to each base station as required.

Mobile units must be able to tune to any one of 333 or 666 channels, depending on whether one system or both systems A and B need to be used. Tuning is done by means of a programmable channel synthesizer, without having to use a crystal for each channel. Fig. 6.11 shows how this is accomplished conceptually. (In practice, circuits typically perform the synthesis at an intermediate frequency and then mix up to the final output.) When properly tuned, $f_i = nf_{ref}$; $f_{ref}$ is the channel spacing and $n$ is an integer variable in the range of 870 MHz/30 kHz $\simeq$ 29,000. A unit change in $n$ forces an $f_{ref}$ change in $f_i$.

To enter the *idle* state, the control logic, having prior knowledge of $f_i$ for each of the 21 control channels, tunes the receiver to each in turn. The receiver compares the signal strengths on each and then tunes to the strongest of the control signals. It reads the modulation—a digital series of bits formatted into words (paging messages interspersed with tuning and overhead messages)—spaced via synchronization signals. *Paging messages* are the identification numbers of the called units. If a unit finds its own identification number sent out, or if the user initiates a call, the unit

---

4. This may be enlarged to more than 400, depending on future FCC action.

Fig. 6.11. Basic
concept of a
frequency
synthesizer.

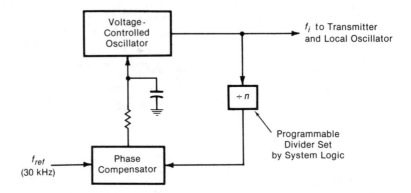

briefly transmits on the mobile-to-land half of the chosen control channel
and sends its identification number. The network responds to the mobile
unit with a voice channel assignment. When the unit has tuned to this
voice channel, it is ready to engage in a two-way call. This process just
described is diagrammed in Fig. 6.12.

## 6.6.1  Seizure

In land-line service, where each user has a dedicated communications
path, the central office can scan these paths in order, or it can arrange
service priorities so that multiple seizures never happen. In a radio system, where users must contend for service on control channels, steps
must be taken to ensure that users do not obliterate each other's signal.
The North American network uses the five following techniques to minimize collisions (Fluhr and Porter 1979):

1.  Every 11th bit of the paging stream is set aside as a "busy/idle" bit.
As long as a base station is processing a seizure from one active user, it
sets this bit to "busy," and other potentially active users wait until the bit
returns to "idle" before attempting their seizures. This is similar to a
technique known as carrier-sense multiple access.

2.  A waiting mobile unit will wait a short but randomized time after
the "busy/idle" bit changes to "idle" before transmitting. This prevents
the bunching-up of call attempts.

3.  The mobile unit, as part of its seizure message, sends a *precursor*, part of which tells the network which specific base station it is
attempting to seize. This is particularly important in the smaller cell
sizes, since the unit's maximum range is more than enough to cover not
only the base station to be seized but also its co-channel base stations
(e.g., for one-mile cells, the co-channel base stations may be only four to
five miles away, easily in range). A base station will ignore a seizure tar-
geted for another base.

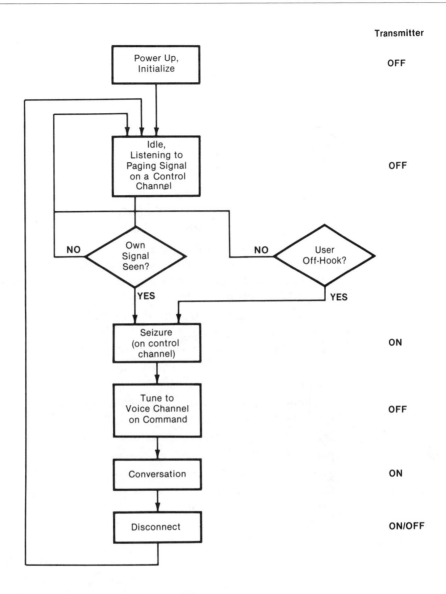

Fig. 6.12. Basic mobile unit sequence.

4. While the mobile is sending its seizure message, it continues to read the condition of the busy/idle bits. If the condition changes to "busy" either too quickly or too late, the mobile unit aborts the call attempt, since the base station must be responding to some other mobile unit.

5. If the initial seizure process is unsuccessful, the mobile unit will automatically try several times. However, to prevent system seizure overload, a limit on reattempts is imposed. The user is prevented from starting a new seizure process until a time-out period has elapsed.

## 6.6.2  Signalling

Reliable signalling in the rapidly fading radio environment is not simple, since errors will sometimes be found in large bursts, even when the overall average error rate is small. Signalling should not waste channel capacity and must meet high standards. Uncompleted calls because of undecodable signalling must be fewer than 1 in 10,000, and mispaged users, wrong numbers, and misidentified users must not occur more than about 1 in $10^7$ messages. The latter stringent requirement comes about because each paging message may be heard by up to 50,000 or so idle units. A $10^{-7}$ falsing rate thus implies that 1 call in 200 will have a false response, which the MTSO will unnecessarily have to use its computing power to deal with.

To combat the bursty nature of the errors while at the same time meeting these requirements, all control messages are encoded via (63, 51) BCH coding and then are repeated several times at the source (see Bartee 1985, Chapter 10). At the receiver, a three-out-of-five majority vote is taken on each bit to determine the code word to be presented to the decoder. The (63, 51) BCH code has the capability, with a syndrome calculation and a simple ROM look-up table, to correct one error and detect at least two more. This provides a good balance between a low miss rate and a low falsing rate.

Digital modulation of the control channel rf carrier uses a method called binary frequency-shift keying (FSK), with discriminator detection; this results in economical detection, consistent with the 30-kHz channel spacing (Arredondo, Feggeler, and Smith 1979). Bits are encoded using a biphase (or Manchester) format. Each logic 1 is encoded as a (0, 1) transition, and each logic 0 is encoded as (1, 0). The peak deviation is 8 kHz, and the basic bit rate is 10 Kbps.

The advantages of this scheme are: (1) the peak of the power spectrum (see Fig. 6.13) is well above the voice band, so that digital signals are easily and quickly distinguished from voice signals (this permits digital signalling for handoffs to take place conveniently on the voice channels); (2) synchronization of the receiver to the incoming bit stream is facilitated by the ever-present biphase transitions, no matter what code word is being sent; and (3) there is no dc component in the power spectrum.

The control information[5] to be sent on the various channels is as follows: for forward control channels—pages, channel designations, mobile power level, and overhead (local parameters); for reverse control channels—mobile identifications and dialed digits; for forward voice channels—orders, channel designations (handoff), and mobile power level;

---

5. *Forward* means land transmitter; *reverse* means mobile transmitter.

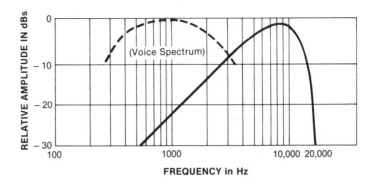

**Fig. 6.13. Power spectrum of signalling modulation.**

and for reverse voice channels—order confirmations and dialed digits (for enhanced services such as three-way calling) (Fluhr and Porter 1979).

Forward signalling uses the (63, 51) BCH word shortened to (40, 28). On the control channel, a continuous stream of words is sent. Two streams, A and B, are word-interleaved. This helps reduce the probability that a burst of errors will obliterate more than two of the five repeats of a word. An idle mobile unit monitors either B or A, but not both, since it is always signalled on only one data stream. With synchronization and busy/idle bits, the stream is shown in Fig. 6.14. Bit sync is a 1, 0, 1, 0, 1, ... sequence; word sync is the 11-bit Barker sequence 1, 1, 1, 0, 0, 0, 1, 0, 0, 1, 0, a sequence with excellent autocorrelation properties.

**Fig. 6.14. Word format (forward control channel).**

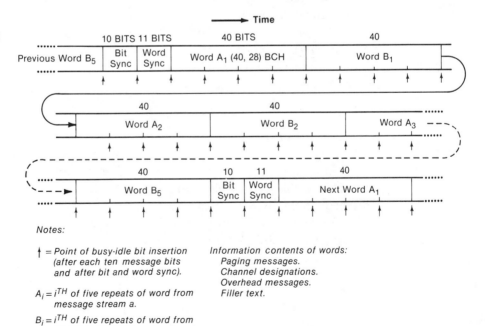

Notes:

$\dagger$ = Point of busy-idle bit insertion (after each ten message bits and after bit and word sync).

$A_i = i^{TH}$ of five repeats of word from message stream a.

$B_i = i^{TH}$ of five repeats of word from message stream b.

Information contents of words:
Paging messages.
Channel designations.
Overhead messages.
Filler text.

The word stream is made continuous, even in times of low traffic, by sending *overhead messages* and *filler text*. The overhead message primarily serves to identify the local system. A mobile unit can compare this message with its own home system identification to determine if it needs to register its presence as a roamer. The overhead message also serves to inform mobile users of certain local parameters, so that users can function properly in any system large or small. Filler text is used to provide a known sequence of bits guaranteed to provide maximum code distance to the sync signal.

On the forward voice channel, where signalling is necessary to permit channel reassignment for handoff purposes, interleaving is not possible; instead, the message is repeated eleven times. If this message is not received properly, a call may be lost. To make matters worse, the message is almost always sent under low $C/I$ or $C/N$ conditions. This message is sent by briefly blanking (interrupting) the voice signal and sending a burst of digital control messages. The mobile user generally does not perceive this interruption.

On the forward (base to mobile) channels, the receiver always sees a carrier signal. On the reverse control channel, this is not the case; the receiver will very often see noise or co-channel interference as the strongest input to the limiter/discriminator. This makes the process of rapid, reliable detection harder. Design of the "precursor" mentioned earlier helps this process. The precursor provides adequate *dotting* or known bit-sync interval, followed by word sync to which the receiver must correlate, and a 7-bit base station identification (sometimes called *color code*). Beyond this point in the message, a base station *not* seeing its own correct 7-bit code must reinitialize and look for a new bit-sync interval, while a proper base station will proceed to determine the incoming code word in the (63, 51) BCH format, shortened to (48, 36).

## 6.6.3  Supervision

In a cellular network, reliable determination of when calls are terminated and when calls are answered (dc loop current in wire-line service provides this function) cannot be done by straightforward rf carrier detection (Fluhr and Porter 1979). Again, a "color code," or characteristic base station mark, must be used. One of three out-of-band tones (5970, 6000, and 6030 Hz) called SAT (Supervisory Audio Tones) is used for this function. In a manner similar to the functioning of dc current in a wire loop, this tone is sent on each active voice channel by the base transmitter and repeated back by the mobile transmitter, as long as the mobile transmitter is on. Another tone (10 kHz), generated by the mobile unit, is used to indicate an on-hook condition. On a land-originated call, when this tone disappears, the mobile user has answered; when it reappears later in the call, the user has hung up.

SAT provides a second function. If the unit enters a deep fade such that interference is greater than carrier, the received SAT will change frequency. When this occurs, the receiver output must quickly be blanked to prevent reception of intelligible crosstalk. If this condition persists for more than five seconds or so, the call is presumed to be terminated.

## 6.6.4  Localization and Handoff

A mobile unit is assigned a channel belonging to one specific cell, yet the unit can physically be located in any cell. The structure of a cellular network requires that an active unit be located with respect to its cell (or base station), both to start calls and to continue calls. This need not be done precisely with respect to all points in the area; merely guaranteeing a minimum allowable *C/I* is the objective. Thus, for example, two units could physically be next to each other but use two different base stations for communication. At any point, there should be more than one allowable base station.

Keeping track of the approximate location of active users can be done differently in different system implementations. Most plans work approximately as follows:

1. "Absolute" signal strength (dBm) on each channel being used for a stable call is periodically measured at the base stations, at intervals of about five to ten seconds, by a specially designed receiver.

2. Each of these measurements is compared to a predetermined threshhold value, which depends primarily on cell size. If this threshold is not exceeded, a handoff may be called for.

3. In this case, the base station, via the MTSO, asks for a measurement of the signal strength on this channel by neighboring base stations. This neighbor list is cataloged in advance for all base stations.

4. When all neighbor base stations have reported the measured absolute level to the base station handling the active call, that base station has the necessary information to tell the MTSO, should a handoff to a specific new base be needed.

5. The MTSO searches its memory to find an idle channel at the new base station. It then tells the mobile unit to retune to the proper channel and tells the new base to expect a new signal on that channel and to assume responsibility for the call.

6. After some bookkeeping and checking (disappearance of a carrier and SAT on the "old" channel, appearance of carrier and SAT on the "new" channel), the call becomes "stable" on the new channel.

Starting a call is somewhat similar. In this case, the mobile unit is solely responsible for the choice of a base station. The mobile unit must tune to each of the complement of twenty-one predetermined channels and choose the strongest. The base station will accept the call if threshold requirements are met, and the call will begin. This process of choosing a base station based on signal strength and of handing off a call to another frequency assignment based on a signal strength measurement is unique to cellular networks. Conventional systems do not need this capability.

# 6.7  Regulatory Aspects

Clearly, the technology of cellular service is heavily influenced by regulatory processes, both the direction of the technology and the speed with which it becomes a reality. This technology has proceeded differently throughout the world.

## 6.7.1  United States

Work toward the design of the cellular network just described began in earnest about 1968 (see Frenkiel 1970; *IEEE Transactions on Communications* 1973; *IEEE Transactions on Vehicular Technology* 1978; and Porter 1971). At that time the FCC initiated Docket 18262. Basically, the docket asked U.S. industry to describe how it would use a large allocation (up to 75 MHz) if one were provided in the bands 806–890 MHz. These bands had previously been allocated to uhf TV but were only lightly used. Several feasibility studies were forthcoming in 1971, notably those of the then-named Bell System (Bell Laboratories 1971) and Motorola. Bell Labs, on behalf of Illinois Bell, subsequently published a tentative specification for a mobile/base radio interface, to which Motorola agreed in principle, and both these entities were awarded developmental licenses by the FCC. The Illinois Bell trial service (dubbed AMPS-Advanced Mobile Phone Service) began operation in late 1978 in Chicago, and the Motorola system started a year or so later in Washington, D.C. (actually, the trial was run by a local radio common carrier). With the technical and market success of these trials, final rule-making took place in 1981 (U.S. Federal Communications Document OST-53).

Using the arguments of (a) system efficiency, (b) technical expertise, and (c) financial resources, the initial proposals called for one system per area, to be run by the licensed operating telephone company in each city. The Radio Common Carriers (RCCs), as expected, objected to this regulatory concept, stressing that they, too, had technical and financial strengths and that competition was the best regulator. In spite of the less efficient channel utilization, this viewpoint won the day, so that the final rule-making set aside two bands, making possible two systems per city, one for the wire-line telephone company and one for any other common carrier establishing technical and financial responsibility. Furthermore, the wire-line telephone "set aside" would expire after a very few years. Competing license applications would be evaluated in comparative hearings to decide the final license. In most areas, the wire-line telephone interests were able to reach settlements, so that only one license application per area was submitted and a speedy resolution took place. The radio common carriers were not always so fortunate, and licensing delays usually meant that the RCC system lagged behind.

Another aspect of the regulatory process was the FCC's procedure for imposing order on the licensing process. It ranked the potential markets in order of size, in groups of thirty, and accepted applications in phases, the largest groups first, arguing that pent-up demand was greatest in the largest cities.

## 6.7.2   International Developments

So far, we have limited our description to U.S. activity. Although Canada was not actively involved in the early work, it followed the events in the United States closely. With the success of the 1979–1980 service trials and with the subsequent FCC rule-making, Canada decided to adopt parallel standards (Canadian Department of Communications Document RSS-118 1983), in order to be fully compatible with the U.S. system. One major difference between the two is that, in Canada, one RCC was awarded a nationwide license, thus avoiding decision-making on a per-city basis.

During 1982, the United Kingdom administration realized that European progress toward a regional standard was slow. It decided to equip what it termed an "interim network," rather than wait for a Pan-European system. Thus, in early 1983, the United Kingdom adopted a plan similar in most of its basic cellular details to the North American AMPS plan. Because the plan had to be based on different rf band and channel spacing, a few changes were made. The basic cellular protocols, signalling, formats, and operational methods remained the same, but the signalling rates, deviation, etc., were scaled down in the ratio 25:30. This system, dubbed TACS, went into operation in early 1985, using the Canadian concept of two nationwide competing systems.

Soon after the former Bell System published its preliminary information on cellular networks in 1971, the Japanese began to plan similar designs. They decided to move forward independently with a system of their own, since the Region 2 (Western Hemisphere) allocations were different from theirs and they did not want to be tied down by the deliberate pace of the FCC. The Japanese cellular system, complete with a handoff capability, was thus expeditiously deployed. The original network has since been supplanted with an enhanced version that improves on earlier techniques.

The state-owned telephone companies of the Nordic countries (Denmark, Finland, Norway, and Sweden) have always been very progressive in the field of communications. This progressiveness has not been limited to technology but has included a spirit of regional cooperation. During the 1970s, they opted to develop, with Ericsson and other support, a cellular network using the 450-MHz band. By prior agreement, they were restricted to using only 5-kHz deviation and fewer channels; but, nevertheless, a high-quality, well-accepted system, which included full regional roaming, was developed for those countries. The Nordic system has been deployed in other parts of the world as well.

Other national networks using cellular concepts are deployed or are being implemented in West Germany, Italy, and France (CCIR Document 8/293-E 1984). However, these various European networks are not interworkable.

During the early 1980s, the European community, as represented by its regional telephone organization CEPT (Conference of European Posts and Telecommunications), began to plan a fully compatible Pan-European mobile telephone system. It would use an allocation between 890 and 960 MHz, would feature international roaming, and would most likely be digital to the maximum extent possible. Current plans are that it will be at least partly operational by the early 1990s. As of this printing, several competing system proposals are being evaluated (*Conference Record, Nordic Seminar* 1985).

# 6.8   Future Possibilities

In the United States, the term *cellular* is often used incorrectly to mean the current specific system plan in use between 825 and 890 MHz. The concept has much broader applicability, however; and further usage of it can be expected if other spectrum allocations are forthcoming. In particular, a marketplace for handheld, personal, portable telephony is emerging, and many market analysts predict that such usage will predominate in future years. Although some usage of portable sets mixed in with the vehicular usage of "AMPS" may take place, for the following reasons the system design is not optimized for a lightweight, pocket-carried radio telephone:

1. Battery capacity (size and weight) and radiation safety preclude power levels comparable with vehicular units.

2. Antenna efficiency for handheld units is generally lower.

3. The rf path loss environment is more extreme for handheld units (by about 20 dB), since they are most often used inside buildings.

4. Offered Erlang load will be typical of wire-line usage, not vehicular.

5. Cellular-to-wire network interface design must reflect the decreased median range of usage by a person, as compared with that of a vehicle.

Certain new techniques could prove useful in future cellular network designs for mobile usage. These are:

1. Digital modulation, which allows easy design for privacy and may permit cheaper equipment and improve the quality of voice when the radio channel is noisy.

2. Time-division multiplexing (TDM), which can eliminate the need to transmit and receive simultaneously (thus saving cost, weight, and volume). TDM would require fewer (but more complicated) base station radios and would simplify transmitter-to-antenna coupling design.

3. System layouts based on reduced base station power and lower antenna height constraints.

4. Possible use of spread spectrum or amplitude-companded single sideband modulation, both of which are even less conventional than TDM.

5. Personal calling response via wide-area directed call pickup, giving the called party a choice of using either the radio system or the wire-line system.

Many of these techniques are being investigated in commercial laboratories and universities in the United States (Cox 1985), Japan (Hirono et al. 1984; Miki and Hata 1984; Suzuki et al. 1984), and Europe (*Conference Record, Nordic Seminar* 1985).

## 6.9  Summary

This chapter has described cellular networks, which have made possible a burgeoning of radiotelephone communications during the 1980s. The early history of radiotelephony was discussed, followed by a description of the major impairments that influence system design. The use of a cellular approach to base station planning was then explained and contrasted with earlier methods, and the related topics of modulation and companding were discussed.

For basic customer service, differences between wire-line service and cellular service were noted, followed by an explanation of the processing of cellular network calls. The regulatory situation in various parts of the world was summarized next. Finally, several future possibilities for system designs that build on the cellular concept were listed. A basic list of references is included.

## 6.10  References

Arredondo, G. A., J. C. Feggeler, and J. I. Smith. 1979. AMPS: Voice and data transmission. *Bell System Technical Journal* 58, no. 1 (January): 97.

Bartee, T. C., ed. 1985. *Data communications, networks, and systems.* Indianapolis: Howard W. Sams.

Bell Laboratories. 1971. High-capacity mobile telephone system technical report. Submitted to FCC in December (Docket 18262). Also see other filings in this Docket.

CCIR Document 8/293-E, Report 742-1. 1984 (August). See also vol. 8, Mobile Services, 1986 issue. Geneva.

*Conference Record, Nordic Seminar on Digital Land Mobile Radiocommunications.* 1985. (February) Espoo, Finland.

Cox, D. C. 1985. Universal portable radio communications. *IEEE Transactions on Vehicular Technology* (August): 2745–50.

Cox, D. C., and R. P. Leck. 1975. Correlation bandwidth and delay spread multipath propagation statistics for 910 MHz urban mobile radio channels. *IEEE Transactions on Communications* COM-23 (November): 1271–80.

Fluhr, Z. C., and P. T. Porter. 1979. AMPS: Control architecture. *Bell System Technical Journal* 58, no. 7 (January): 43.

Frenkiel, R. H. 1970. A high-capacity mobile radiotelephone system model using a coordinated small-zone approach. *IEEE Transactions on Vehicular Technology* VT-19, no. 2 (May): 173–77.

Huff, D. L. 1985. Cellular—Early realities and problems. *Conference Record, National Communications Forum.*

*IEEE Journal on Selected Areas in Communications.* 1984. Special issue on mobile radio communications. See Hirono et al., Miki and Hata, and Suzuki et al. *IEEE Journal on Selected Areas in Communications* SAC-2, no. 4 (July).

*IEEE Transactions on Communications.* 1973. Special joint issue on mobile radio communications. *IEEE Transactions on Communications* COM-21, no. 11 (November).

*IEEE Transactions on Vehicular Technology.* 1978. Special issue on emerging 900-MHz technologies. *IEEE Transactions on Vehicular Technology* VT-27, no. 4 (November).

Jakes, W. C., ed. 1974. *Microwave mobile communications.* New York: Wiley.

Lee, W. C. Y. 1982. *Mobile communications engineering.* New York: McGraw-Hill.

MacDonald, V. H. 1979. AMPS: The cellular concept. *Bell System Technical Journal* 58, no. 7 (January): 15.

Paul, C. E. 1969. Telephones aboard the Metroliner. *Bell Laboratories Record* 25 (March): 65–69.

Pickholtz, R. L. 1985. Modems, multiplexers, and concentrators. Chapter 3 in *Data communications, networks, and systems.* Indianapolis: Howard W. Sams.

Porter, P. T. 1971. Supervision and control features of a small zone radiotelephone system. *IEEE Transactions on Vehicular Technology* 20, no. 3: 75–79.

Schulte, H. J., Jr., and W. A. Cornell. 1960. Multi-area mobile radiotelephone system. *IRE Transactions on Vehicular Communications* 4 (May): 49–53.

United States Federal Communications Commission Document OST-53. 1981. Canadian Department of Communications Document RSS-118, Annex A, October 1983. Also, EIA Interim Standard CIS-3A, available from the Electronics Industries Association, Engineering Department, 2001 I Street, Washington, D.C. 20006.

# 7

# CHALLENGES IN COMMUNICATIONS FOR COMMAND AND CONTROL SYSTEMS

JOHN J. LANE

Efficient communications are a *sine qua non* for an effective command and control system. In fact, the concepts of communications and command and control have become so intertwined that in the United States Department of Defense, to which command and control are so vitally important in protecting the nation's security, both communications and command and control systems are managed by a single entity—the Office of the Assistant Secretary of Defense for Command, Control, Communications $(C^3)$[1] and Intelligence. Military personnel define "command and control" as:

> The exercise of authority and direction by a properly designated commander over assigned forces in the accomplishment of the mission. Command and control functions are performed through an arrangement of personnel, equipment, communications, facilities, and procedures which are employed by a commander in planning, directing, coordinating, and controlling forces and operations in the accomplishment of the mission (Department of Defense 1979, p. 74).

That definition, though couched in military terms, is clearly related to the classic definition of management as consisting of planning, organizing, directing, controlling, etc. In fact, management is defined by military personnel as consisting of "those continuing actions of planning, organizing, directing, coordinating, controlling, and evaluating the use of men, money, materials, and facilities to accomplish missions and tasks" (Department of Defense 1979, p. 205–6). This chapter draws heavily on the author's twenty-three years of experience with military $C^3$ systems,

---

1. "$C^3$" is pronounced "Cee Cubed."

but the communications challenges that must be solved in fielding effective military command and control systems are the same as those that must be addressed in any management environment, though perhaps on a different scale.

A $C^3$ system is difficult to dissect into its three components practically; conceptually, however, a command subsystem consists of those processes and staffs that directly support any decision-maker, military or civilian, in the formulation of a decision. A control subsystem consists of the functions and entities through which both a decision is executed and information is received to facilitate future decision-making and to monitor progress. Communications subsystems interconnect the elements of the command and control subsystems. It is the challenges imposed when attempting to provide those vital communications channels that are the subject of this chapter.

A command and control system is essentially an information-handling entity. It must provide the information necessary for sound decision-making. It thus must be able to interface with any information system that can provide the facts and insight necessary to significantly reduce the risk of making the wrong decision. Such information systems might include marketing, financial, legal, human resources, logistics, intelligence, and other sources of needed information. A fundamental and most critical objective for a command and control $(C^2)^2$ system is thus interoperability with a host of other information systems. For that reason, we will spend a considerable amount of time discussing three interwoven challenges—standards, security, and continuous technological change—all of which must be met in providing effective $C^2$ communications.

# 7.1   A Historical Perspective

Communications in military command and control systems have long been essential to ensure that all forces were acting in concert. Early leaders used semaphore, heliograph, bugles, couriers, signal fires, and a multitude of other techniques to communicate orders and to receive intelligence. As long as organizations were small, members spoke the same language, and rudimentary methods were employed, interoperability was not the most pressing $C^3$ problem. The essentiality of interoperability to success in any

---

2. "$C^2$" is pronounced "Cee Squared."

major organized endeavor, even in the earliest times, was recognized in the biblical recounting of the building of the Tower of Babel. Construction was effectively blocked by eliminating the participants' ability to communicate with one another. To take another example, the War of 1812 began on June 18, 1812 with a declaration of war against Great Britain by the United States. Ironically, two days earlier, the British Foreign Minister had announced the repeal of Orders in Council, a major underlying cause of the war. On Christmas Eve, 1814, in Ghent, Belgium, the United States and Great Britain signed a Treaty of Peace. But fifteen days later, approximately 1500 men died on Chalmette Plantation in the Battle of New Orleans, about fifty years before the first successful transatlantic cable opened in 1866.

The technology underlying command and control communications did not change significantly until the successful demonstration of the telegraph by Samuel F. B. Morse in 1844. Since then, the pace of change has been quickening ever faster. Alexander Graham Bell uttered the first telephone message on March 10, 1876, about three months before George Armstrong Custer's Last Stand. The first transatlantic wireless signal was transmitted on December 12, 1901. By 1907, Lee De Forest's vacuum tube made possible the transmission of the human voice. The early 1920s saw experimentation with television signals. The first communications satellite, Echo I, was launched in 1960. Mark I, the first digital computer, was completed by Harold Aiken in 1944; it employed mechanical and electrical devices. In 1946, ENIAC, the first digital computer using vacuum tubes, was built at the University of Pennsylvania. During the early 1960s, transistors were used to build the second generation of computers. By the mid-1960s, integrated circuits were used to build the third generation of computers.

The distinction between the separately evolving communications and computer fields began to blur both with the introduction of digital processors as integral and controlling parts of telecommunications networks and with the use of those networks to begin to connect together computers and terminals. The pace of this technological change has been phenomenal, and it is this very change that offers both enormous challenge and enormous opportunity to the command and control and communications disciplines. The remainder of this chapter will deal with three aspects of this challenge that are presented by the ever-increasing rate of technological change: adoption of communications standards by command and control organizations; preservation of information security as internetwork communication proliferates; and finally, intertwined with both standards and security, the flexibility of command and control organizations as they adapt to change.

# 7.2   Standards—The Interoperability Glue for C³ Systems

From the advent of the computer revolution, major information systems within the Department of Defense, as in large organizations generally, have been built in "stovepipe" fashion in support of separate functional applications areas. That is, minimal attention has been paid to the need to interoperate with other functional areas. Resource management, logistics, intelligence, transportation, operations planning, personnel, and legal information systems have each been built to satisfy the requirements imposed by each individual system. These systems have, in themselves, been sufficiently complex and large in scope as to present formidable project management challenges in meeting their own integral requirements. Where needs for information interchange were articulated, those needs were frequently met by "air gap" interfaces, in which the information generated by one computer system had to be manually introduced to a second computer system. In the early days of digital computing, the state of the art did not encompass integrated data base management systems and large-scale distributed computing systems. The result was less than satisfactory from the command and control viewpoint. Information that could significantly impact on a decision could not be made readily available to the decision-maker. Instead, requests had to be made for information on an ad hoc basis, and precious time elapsed while that information was obtained, frequently by a series of telephone contacts.

## 7.2.1   The WWMCCS Example

The Worldwide Military Command and Control Network (WWMCCS) is a concrete example of this problem. The system was developed in the late 1960s to facilitate the planning and execution of U.S. military operations. Physically, the system consists of over eighty mainframe computers configured at approximately thirty locations, with many additional minicomputer and terminals located worldwide. The separate locations are linked today by a dedicated data communications network and by transmission capability for formal messages via the AUTODIN network of the Department of Defense (DoD). As a command and control information system, the WWMCCS contains data bases on the readiness status of military forces and contains other data bases and tools necessary to the planning of a military operation, such as the invasion of Grenada. As large and complex as the WWMCCS is, however, it did not originally include all the information necessary to informed planning and decision-making. For example, direct access to intelligence information data bases was not provided, nor was direct access to logistics information systems and data bases. Current intelligence information is, of course, vital to the success of military operations, and logistics information is necessary to provide for the sustainment of forces in the field during prolonged periods of time.

The existence of separate logistics and $C^3$ systems is a reflection of several conditions: the state of the art in automation at the time those systems were initially built, the efficient division of labor in a bureaucracy, and the roles and missions of the major components within the Department of Defense. First, organizationally, the Joint Chiefs of Staff (JCS) and its staff, the Organization of the JCS (OJCS), are the conduit through which flow orders of the National Command Authorities, the President and the Secretary of Defense. The Commanders-in-Chief in the field receive their orders through the JCS, and they are responsible for the conduct of military operations. The planning and execution processes of the WWMCCS support these operational aspects. The Military Departments (Army, Navy, Air Force), on the other hand, are responsible for the provision of military forces, and for their sustainment in the field, while those forces remain responsive to the orders of the National Command Authorities. Thus, national command and control information systems, such as the WWMCCS, are naturally built under the aegis of the JCS, while the logistics sustainment systems are developed and fielded under the watchful eye of each separate military service.

The logistics system of a single service is in itself a very complex information system that must consider and integrate many logistics depots, repair points, and local service supply squadrons or companies around the world. The information systems that support the Air Force's logistics operations, for example, are still not fully integrated even within that single service, despite years of effort. As the Service's logistics information systems evolved from a manual environment to an automated one, hardware and software decisions were made and specific vendor equipment and operating systems were fielded. Applications systems and data bases were designed and implemented using specific programming languages and data base conventions. Each of the four military services developed its own system of hardware and software, with its own set of vendor-specific interfaces and standards. Of course, these evolutions were not totally unconstrained by standards. Each service had its own set of standards; in addition, some Department of Defense–wide and federal government standards were mandated, and some industry-wide commercial standards existed. Nevertheless, four separate noninteroperable logistics systems evolved.

While the logistics systems were evolving, the WWMCCS information system was undergoing its own maturation process and was implemented using its own selected set of hardware, software, and standards. It was thus not directly interoperable with the Service sustainment systems. It should be noted that no criticism of these developments is intended. It is, in the author's estimation, highly unlikely that systems so complex in themselves could have been built to a common set of interoperability standards, even if the necessary software and data base management systems had existed at the time. The point is that the historic evolution of these systems has produced essentially noninteroperable systems that must be made interoperable in some way if the DoD's command and control mission is to be successfully achieved. The need for exchange of information between and among $C^2$ and logistics information systems is but one of many requirements for interoperability. DoD's corporate approach to solving this problem constitutes the essence of this chapter. DoD's miracle interoperability glue is a combination of a common-user data communications network—the Defense Data Network—and a mandated set of standards for information interchange.

## 7.2.2  Defense Data Network

The need for a ubiquitous data communications network with a concomitant standard protocol structure was articulated by the Department of Defense in a study published in 1974. That study itself reflected the network pioneering concepts reduced to practice by the experimental work of the Defense Advanced Research Projects Agency (DARPA) in the now-famous ARPANET, the prototype packet-switching net. In 1982 the Department of Defense selected the technology pioneered by the ARPANET as the basis for its common-user data communications network. To ensure that this new Defense Data Network (DDN) did achieve its potential as a vehicle for Department-wide information system interoperability, the system was mandated for use by every one of the Military Departments and Defense Agencies. This approach has been a painful adjustment for both new and old information systems—but a very necessary one.

## 7.2.3  The Internet

Now we come to the subject of this section—standards. The Defense Data Network employs those concepts demonstrated by DARPA to be important considerations for command and control operations in a military environment. Perhaps the most important of these concepts is that of the *internet*. The ability to communicate between and among different networks is essential both in a military command and control environment and in today's technical environment of proliferating local and global data communications networks. The nature of tactical military operations has traditionally demanded that tactical forces in the field carry their own integral tactical communications and information systems, which are integrated into tactical communications networks. Those tactical information systems and networks must be able to interconnect with the long haul common-user and strategic networks that carry logistics, personnel, and strategic command and control information. The internet concept provides the essential capability for interconnection of tactical and long-haul networks, which support many different functional areas, with global networks. The internet concept is realized in the DDN through use of an internet protocol that allows individual networks to be addressed.

## 7.2.4 Connectionless Protocol Concept

Another useful attribute of a command and control network is the extent to which the network can sustain damage and continue to maintain active communications links. That is, one would like the network to continue to pass information from one point to another even if a node in the currently used path were suddenly lost. Thus, the avoidance of virtual circuits, in which the path is "nailed up," is desirable. A *connectionless protocol*, in which the network nodes possess sufficient intelligence to route a communication around a lost node on a real-time basis, is highly desirable.

## 7.2.5 Layered Protocol Structure

A very useful information transfer concept—the layered protocol structure—was pioneered in the ARPANET, and that work is reflected in the DDN, which has adopted the International Standards Organization (ISO) Open Systems Interconnection (OSI) Seven-Layer Reference Model. Unfortunately, DoD's role as a pioneer in packet switching has caused it continuing stress in the adoption of evolving international standards. DoD has been at least five years ahead of the ISO movement toward adoption of a layered reference model. The DoD protocol standards at each level have been adopted as they matured into operationally proven constructs. The ISO protocols generally differ from the versions at each level developed earlier by the DoD. This has resulted in considerable pressure on the DoD to drop protocol implementations in favor of the new standards. Indeed, DoD has committed itself to the new standards and has in fact adopted the X.25 protocol (which roughly comprises the lower three layers of the ISO model) as the preferred supported protocol. However, DoD has not been able to adopt additional ISO protocols, for the simple reason that the layer four and higher ISO standard protocols are not yet commercially supported by the various hardware vendors. (This statement may no longer be correct by the time this book goes to press, since the adoption of protocol standards is a very rapidly developing area.)

## 7.2.6   Speed of Change

There is a clear command and control lesson here—it is extremely difficult to maintain interoperability in the face of rapidly evolving underlying technology. DoD is necessarily committed to a very conservative approach in adopting a new protocol as a standard. A large bureaucracy moves ponderously. It takes a long time before a newly mandated standard is actually reflected in system procurements, and the government acquisition process in a fully competitive environment is a long one. Where technology is moving rapidly, there is always a better solution on the horizon, and the temptation to wait until that bright new solution arrives is difficult to resist.

## 7.2.7   The TCP Standard

Unfortunately, the "wait till later" approach can result in never achieving an interoperability objective because the means becomes a continually changing target. The transport layer protocol is a clear case in point. DoD's transport protocol standard is the Transmission Control Protocol (TCP). That standard was adopted several years ago as a mandated data communications protocol within DoD. The transport layer provides host-to-host communication. It is built, in the DoD context, upon X.25 (or a costandard—1822) and the DoD Internet Protocol. The transport layer is required to establish a connection between two host computers. It is a necessary, but not sufficient, condition for information system interoperability—the "higher" levels of the ISO reference model are also required for information interchange. Designation of TCP as a DoD standard was not a universally acclaimed action. The mandate was a "top-down" directed action by the Office of the Secretary of Defense and occurred in the face of internal and external countervailing pressures.

Internal pressures were generated by the Military Services and Defense Agencies who were required to include a TCP specification in their Automated Data Processing (ADP) procurements. Most U.S. computer manufacturers, both mainframe and mini, did not support TCP as part of their standard product lines. They would build TCP and provide it to the government as a special feature, but they did not provide standard commercial support. This factor meant that when an operating system was enhanced, the cost of any necessary changes required in support of TCP would have to be borne by the government. Thus, the Service would have to pay more for both the initial acquisition and the life cycle maintenance of the system. Since the program manager's greatest concerns were bringing in a system that met the Service's needs on time and at the lowest cost, and since DoD-wide interoperability concerns typically were not the highest priority in ranking Service requirements, that manager was not anxious to add a protocol, the development of which could both delay the program and increase the cost. The easy alternative for the program manager was use of a winning vendor's native protocol suite.

External pressures came from the hardware vendors, and those pressures aided, abetted, and complemented the natural reluctance of the Services. Each vendor produced a line of processing equipment and provided a total set of native protocol standards that allowed effective information interchange. That vendor's protocol suite had, of course, been optimized for the particular line of supported hardware. From a vendor viewpoint, use of a generalized protocol suite in place of the native one might well decrease the equipment's efficiency. More importantly, use of the native protocol suite tended to lock in the customer to continued procurement of that vendor's systems for system extensions or other interfacing systems.

## 7.2.8  The Dilemma

The proponents for interoperability in a very large government organization are thus caught on the horns of a dilemma. On the one hand, the essence of government contracting is competition—a single vendor is not permitted to become the vendor of choice, for organization-wide reasons, including general interoperability. Components, therefore, are continually acquired from what, in the aggregate, constitutes a range of virtually all available equipment families. On the other hand, specification of a generalized standard protocol suite increases costs and generates internal and external opposition. Vendor opposition can create enormous pressure against interoperability objectives, especially when this opposition is manifested in letters to the highest-level officials within a Department and to congressional representatives. This same opposition is expressed in less formal contacts with the independent management oversight agencies within the government, such as the Government Accounting Office and the Office of Management and Budget. In such an environment, one never wins the interoperability objective; one runs constantly at top speed to keep from losing. The fight for interoperability is nevertheless worth the battle. Achievement of the objectives of a command and control system depends on the ability to interoperate with other information systems. In a DoD context, interoperability is an essential ingredient of our national security.

## 7.2.9   The Solution

It should be obvious that the command and control interests of a large organization are totally aligned with adoption of industry-wide standards. It should come as no surprise to the reader, then, that DoD has adopted the ISO reference model for open systems interchange. The model provides an industry-wide protocol structure that will allow interconnection of any vendor's equipment and the effective exchange of information.

The development of the ISO reference model itself demonstrates that the command and control system advocate cannot sit idly by in the expectation that industry standards will meet command and control requirements. For example, the early inclinations of the ISO community tended to be against inclusion of an internet capability and against inclusion of a connectionless oriented transport capability. For several years it appeared that the ISO reference model, as it was evolving, would not meet the interoperability needs of the DoD command and control systems. That outcome would have been very costly to DoD, both in the additional resources that a DoD unique protocol structure would require and in the technical and manangerial staff resources continually consumed in defending the utility of a DoD-unique protocol suite. Fortunately, and to a large extent through the good offices and excellent work of the Bureau of Standards, the ISO protocol structure now includes most of the functionality needed for command and control purposes. For example, ISO is now specifying an internet protocol (IP), and its transport protocol (TP) provides a connectionless class and contains virtually all the functionality of the DoD TCP.

## 7.2.10   A Continuing Challenge

To agree on standards is not sufficient, however. The vendors must actually adopt and implement those standards as part of their commercially supported product lines. Today, a significant portion of the vendor community appears to have made a commitment to adoption of the ISO reference model. The extent of that commitment is not clear, though—a vendor commitment could take the form either of introduction of the ISO protocol suite as the native supported protocols in the product line, or of provision of various standard protocol conversion schemes. Either approach should provide the interoperability needed by command and control systems, but the overall operating efficiency of the latter could be reduced. In the meantime, eternal vigilance remains the price of continued movement toward interoperability. The siren song of the vendors has become "adopt the ISO specific protocols as costandards on your data communications networks, and that action will accelerate our progress toward a commercially supported ISO protocol suite." Perhaps, but the result of specifying ISO TP and IP could be a vendor provision of TP and IP as special features, rather than as part of the standard commercially supported product line. Such an outcome could act as a brake on the movement to the ISO suite, since scarce expert resources would be directed to building the special feature, the cost of which would need to be recovered over some period of time. The prudent advocate for interoperability in this environment will not depart lightly from an operational internal standard protocol suite but will wait until the ISO protocols are actually available and supported.

This had, in fact, been the DoD strategy. The lower three levels of the ISO model are roughly encompassed by X.25, which itself is roughly equivalent to the earlier DoD standard, the 1822 interface employed on the ARPANET. As X.25 implementations began to be fielded by many manufacturers, and DoD worked toward adoption of X.25 as a costandard, a problem emerged. One vendor's X.25 was not necessarily the same as another's X.25. The extent to which X.25 implementations were actually interoperable depended on whether the vendors had selected the same options and parameters from the menu offered by the standard X.25 specification. In order to ensure interoperability, DoD issued its own X.25 specification, implementations of which would be supported as a costandard on its data communications networks. The DoD specification could be fairly characterized as an intersection of the functionality of the major commercially available implementations. The interoperability lesson learned here is that even vendor adoption of a standard and its implementation in commercially supported form does not necessarily guarantee interoperability with other vendors' equipment. In order to expedite progress toward ISO TP and IP, DoD, with the assistance of the National Bureau of Standards, is working toward issuance of a TP/IP specification in advance of its adoption as a costandard. The specification will be made available to the vendor community in an effort to increase the probability that the various vendor implementations will be interoperable.

### 7.2.11   Dealing with Change

At this point, DoD's approach has resulted in implementation of a common-user data communications network, employing a suite of mandated protocols through the transport layers. New ADP systems are required to use the DDN for their data communications needs. Thus, an excellent foundation for command and control system interoperability is in place. That this situation is a transient state is underscored by several issues that are the product of the rapidly changing technical environment. The first issue is the question of what strategy to take in dealing with the many existing systems that predate the packet-switching evolution. The second is the extent to which packet switching will survive into the future as a dominant and supported data communications technology. A third issue deals with the integration of local area networks into the DDN internet environment.

It is an unfortunate fact of life that when an ADP or telecommunications system is fielded in the Department of Defense, it tends to remain in service for a considerably longer period than it would in the private sector. For example, the AUTODIN system, the DoD common-user formal message system, was originally installed in the early sixties; it is currently scheduled to remain in place until about 1991. The WWMCCS system, mentioned earlier, supports command and control decision-making processes of the National Command Authorities; it was fielded in the very early seventies and is not scheduled for replacement until the end of this decade. The UNIVAC 1050 was the Air Force's standard base-level supply processor for over two decades.

The reasons for the continued retention of older equipment are many, including the length of the government procurement process itself. A major reason is the "second citizen" view of support services carried by many of the Military Department's leaders. ADP and telecommunications officials find it difficult to compete for resources against gun, plane, tank, and ship requirements. If an ADP/telecommunications system continues to chug away doing a job, it may very well not be replaced by more modern equipment until it becomes logistically unsupportable. Thus, a major interoperability initiative such as the DDN must cope with an enormous embedded base of equipment, much of which may be in place for an extended period of time. These older systems must be interfaced if the full promise of interoperability offered by a ubiquitous data communications network is to be realized. Scarce resources must be dedicated to interfacing the various major synchronous protocols, to handling poll select systems, and to providing a standard packet-switching protocol set to the plethora of existing embedded systems. This general problem of dealing with equipment installed for a long duration is exacerbated in today's era of continuous technical change. It increases the set of unique equipment interfaces that must be dealt with. Not only has each vendor handled the same technology in a different way, but the number of different technologies in the field has increased over time.

The DDN is the miracle interoperability glue selected by the DoD. A major challenge is the extent to which that packet-switched network can endure in the face of continuous technological development. Protocols and architectures for the Integrated Services Digital Network are being worked out in the ISO and CCITT (Consultative Committee on International Telephony and Telegraph) communities. The ISDN makes possible a single user end instrument that will provide a range of services—data, voice, and other—to the end user. A system such as the DDN consists of more than just terminals, nodes, and protocol structures. It also includes the physical transmission media and system control facilities, which must be leased from the common carriers or provided by the government. The challenge is to evolve voice and data systems so that they converge on the new international ISDN standards, allowing DoD systems to use the common-carrier-provided facilities.

DoD's systematic approach to this challenge has been to develop a dynamic objective architecture to serve as a road map. That architecture, called the Worldwide Digital Switched Architecture (WWDSA), embodies a series of specific architectural precepts to be employed in new common-user telecommunications systems. The WWDSA is a flexible instrument, designed to be reviewed and changed on a periodic basis. The resultant architecture is a moving target, modified from time to time as new technologies emerge and begin to change our view of the future. This evolution is extremely difficult to achieve practicably. A new standard can be implemented only with great difficulty and expense. It is consequently very difficult to make major adjustments in fielded networks in areas as fundamental as protocol structures. Change sometimes tends, therefore, to be revolutionary rather than evolutionary, if only because the disruption needed to make changes in protocols, for example, seems to incite the fervor of a revolution.

### Electronic Mail

Another element of change is the "culture shock" that it imposes, with the concomitant time lag necessary to assimilate the implications of the change and to adjust business practices so that full benefit may be derived from the new technology. An example is the impact of electronic mail on formal message service.

In their infancy, electronic mail systems have generally been used informally. People use them as a substitute for the telephone, bridging time zone barriers and avoiding the "telephone tag" game, wherein your call finds the recipient unavailable, her return call finds you out of the office, your return call finds her on another line, etc. Formal message service, on the other hand, provides a record copy and usually provides important services, such as guarantee of delivery, multiple levels of precedence and delivery priority, and certain assurances that the message represents a formal position of the organization sending it. Prior to such a formal message being transmitted, it has been passed through several levels of a bureaucracy and been formally "signed off" at each point. The recipient receives a paper copy to read and to file for the record. Under the existing cultural mind set, an electronic mail message may not provide the cachet of authenticity sufficient to allow it to be used to transmit "important" information. In fact, use of electronic mail in a formal message context does present some technical challenges, such as security, authentication of sender, and guarantee of delivery. An even more important consideration would be the guarantee of continued connectivity during any transition period. These challenges are all surmountable, but they cannot be resolved overnight and they add yet another dimension to the problem created by rapidly evolving technologies.

As these challenges are being worked on, numerous side issues that could conceivably impact on the ultimate outcome must be considered. For example, what will be the impact of electronic voice mailboxes on electronic data mailboxes? Voice mailboxes are even more convenient to use—the keyboard is eliminated; one leaves a verbal message, which is stored in digital form. To what extent could voice messages supplant keystroked mail messages? If voice messages become feasible, what would be the impact on any plans to move a formal message service to electronic mailboxes? While considering these questions, keep in mind the length of time required to adopt and implement necessary supporting protocols. Once attained, interoperability is tough to maintain.

## LANs

Another example of the impact of changing technologies is presented by the local area network phenomenon. Six years ago, local area networks (LANs) constituted a new technology with several sets of emerging standards. Today, LANs are installed everywhere, and a variety of technologies and current and proposed standards abound. In many applications, digital PBX technologies give users a choice of installing a broadband coaxial cable plant for a LAN, of using twisted pair for a baseband LAN, of using a digital PBX and the existing twisted pair cableplant, or of using hybrid configurations. Within the DoD, LANs of all kinds are being procured and installed to meet local needs. Once a LAN is installed and operational, interface to a more global communications net becomes an obvious requirement. For example, the WWMCCS modernization program, called the WWMCCS Information System (WIS), will install LANs at each participating command center to accommodate local information exchange requirements. The WIS program will use a subsystem of the DDN to interconnect its community of interest. The same is true of the DoD Intelligence Information System (DODIIS), which will interconnect its LANs via a DDN subsystem. The interface requirements of these major programs can be controlled through their program management offices, but that control may not exist for the preponderance of administrative base-level LANs. The achievement of the full potential of the interoperability promised by a DDN with its internet architecture demands that the various LAN technologies be able to interface with the DDN.

### 7.2.12 An Interoperability Bottom Line

In summary, achievement and maintenance of interoperability in a command and control setting requires the specification of standards and protocols. Adoption and implementation of those standards in an existing multivendor equipment environment is extremely challenging. Maintaining interoperability in the face of continuous technological change adds another dimension of difficulty to interoperability in command and control networks. The next section will consider yet another opportunity to overcome a difficult obstacle in the achievement of interoperability—the security challenge.

## 7.3 The Security Challenge

The way to ensure achievement of interoperability objectives for command and control systems is to put in place the mechanisms that will allow interconnection of the greatest possible number of information systems. That process reduces the risk that significant information is not available to the command and control decision-maker. The way to ensure achievement of security objectives for command and control (and other systems) is to put in place mechanisms that will prohibit interconnection of information systems. That process would reduce the risk of information being obtained by those not authorized for its receipt. The security challenge is to unite these apparently mutually exclusive objectives so that systems can be interconnected in ways that preserve the security of the information within the system.

That this task is formidable is demonstrated by the fact that today the Defense Data Network consists of a number of separate and unconnected subnetworks. Each of those subnetworks has been built using identical hardware, software, and protocol structures, yet they cannot be interconnected. They cannot be interconnected because the security challenge has not been met. Each separate network operates at a different security level. There are separate networks to carry unclassified information, secret, top secret, special compartmented, and nuclear operations information. Today's network security technology will allow any number of users to connect to a network through encryption equipment in which each link on the network is encrypted. In such a network, unencrypted information flows within the network nodes, so that the network nodes must be physically protected to the level of the most sensitive information that can be passed. Such a network is limited to a single community of interest, such as the WWMCCS community or the DODIIS community. That limitation precludes necessary interoperability, such as the ability to pass current intelligence information on-line to operational command and control consumers of that information. This information interchange is accomplished manually at present.

## 7.3.1  Trusted Nodes

Today, the network cannot be trusted to properly segregate levels of information so that only the intended recipient reads the transmission. Network switches interconnecting host processors and terminals are themselves processors that cannot be trusted to internally segregate information. The technology exists to build network processors with trusted computing bases, which can be trusted to so segregate, but to date such processors have not been fielded in an operational network environment. An earlier DoD attempt to field a network with trusted network processors was discarded in 1982 for several reasons, one of which was the unaffordability of that security solution. The cost of the network nodes was so high that, for most users, the cost of joining the common-user network would have exceeded the cost of dedicated secure communications.

## 7.3.2  End-to-End Encryption

An alternative approach to network security, one that does not rely on trusted network nodes, relies on encryption of messages on an end-to-end basis, from sender to receiver. This approach does not require trusted network processors because information flowing through the network nodes remains encrypted. The technology for end-to-end encryption has been proven by its operational use in the ARPANET. A number of Private Line Interface (PLI) encryption devices were built and demonstrated in continuous operational use. They performed very satisfactorily, and that security solution is now employed within the DDN architectural concept. Unfortunately, the PLI devices themselves were prohibitively expensive for use as a general network-wide security solution. An improved PLI device was considered for such use. The Internet PLI (IPLI) was lower in cost, employed more modern design, and added an internet protocol capability. The IPLI, however, had one fundamental limitation—it encrypted communications very well between and among members of a single community, and it allowed those communications to be transmitted securely over a network shared by other communities of interest; but it did not allow communications between the separate communities of interest. Since the IPLI did not meet that fundamental interoperability objective, it also could not serve as the general internet solution.

### 7.3.3 The BLACKER Program

The ultimate solution became known as the BLACKER program. This program will be producing an encryption system to allow dynamic allocation of the encryption key, on a message-by-message basis, so that any BLACKER-equipped network entity can transmit to any other such entity when required. Major elements of the system will include the encryption devices themselves, access control centers, and key distribution centers. In operation, the user requests a connection from the access control center, which verifies that the requested connection is proper and then directs the key distribution center to disseminate the encryption key to both parties. The importance of the security issue is underscored by the fact that the security solution has become the architectural and schedule driver for the DDN. The separate subnetworks cannot be interconnected until the BLACKER equipment has been fielded.

### 7.3.4 Host Security

The BLACKER equipment addresses only part of the security problem, however—it ensures that whatever is delivered to the network will be read only by the intended recipients. It does nothing for security of the host processors that access the network. Traditionally, effective security has been achieved in computer centers by employment of operationally restrictive methods, which generally reduced the efficiency of use of the machines. For example, a center might employ "periods processing," wherein the computer processed "SECRET" level data for one period of time, after which the system was purged for a period of "TOP SECRET" processing, after which the system was purged for a period of unclassified processing. The time required in purging the system was lost, since it could not be used for productive processing. Another method was to obtain several processors and use one for unclassified level, one for SECRET, etc. That method was frequently costly, since more processors were required than the workload itself might have demanded.

To overcome these limitations, various features were specified. The use of these features was felt to sufficiently reduce the risk of information loss so as to allow more than one security level to be processed at a time in a single machine. Machines using this approach were said to be operating in *controlled mode*. Although controlled mode operation was permitted, it was not widely encouraged because the extent to which the risk of information loss was reduced had not been analytically derived and was not well understood. Controlled mode could be used where a user had assessed the value of the information to be processed, had analyzed the risk of loss and the cost of running in a more restrictive mode, and had arrived at the decision that use of controlled mode was warranted. In practice, however, the user in a controlled mode system trusted the system to properly isolate information only to those authorized to receive it. The system supported users at more than one level. Other modes of operation were *dedicated*, in which all authorized users were cleared to the same level and had access to all the contained information, and *system high*, in which all authorized users were cleared to the same level but did not necessarily have access to all contained information. These various modes of operation were defined to apply in an essentially "stand-alone" system environment. That is, the system served a single community of interest—it might be an intelligence system, a command and control system, a logistics system, etc. The communications facilities that served the system were also dedicated and encrypted as necessary.

### 7.3.5   Uniform Accreditation Environment

Assume now that a network, called System A, has been designed to handle and protect SECRET-level information. Assume also that System A has been certified by its owner for SECRET operation as a "system high" system. This designation implies that all connected terminals in the system, even those that might handle only unclassified information, must be protected physically to a SECRET level. Assume now that a System B owner has also certified his system at a SECRET level, but the System B owner chooses to operate in controlled mode, trusting the processor software to properly segregate information. This designation implies that "backside" terminals, processing only unclassified information, do not have to be protected to the SECRET level. Now assume that Systems A and B are both proposed for connection to the SECRET network. The obvious problem is that the System A user may be reluctant to expose information to System B users, having already decided that a more stringent approach was necessary for her system. Thus, the network operator is faced with a situation in which the network is guaranteed to preserve the security of whatever is delivered to it. The System A user has certified her system for secure operation, and the System B user has certified his system for secure operation. Nevertheless, one cannot necessarily securely interconnect these three systems, each of which has been certified for operation at the SECRET level. Part of the solution to this dilemma is a commonly accepted standard by means of which security transitivity can be achieved. Such a guideline, providing a uniform security accreditation environment, will be discussed later.

### 7.3.6 Trusted Computing Base Criteria

The second part of the solution to host processor security is the development of criteria for trusted computing bases, which could serve as a principled basis for the varying degrees of trust to be reposed in a computer operating system. The DoD worked toward such a criteria document for ten years. Finally, in 1986, criteria were adopted as a DoD standard. That document, which specifies the criteria for evaluation of DoD computer security, is well-described in Bartee 1985. In brief, four divisions and seven classes are established, ranging from Division D, minimal protection, to Division A, Class 1, which can be trusted to properly handle information simultaneously ranging from unclassified through top secret. The lower divisions require various features, while the higher levels combine specified features with additional assurances met through a disciplined process employed in building the security-relevant portions of the system. As of this writing, DoD security policy directives are being rewritten to mandate use of the security criteria for every ADP acquisition. Each ADP program manager will be required to assess the level of risk by examining the threat and vulnerability, and by making the conscious decision as to which Division and Class are required for the operating system to be employed.

### 7.3.7 Uniform Application

In deciding on a Division and Class consistent with the risk level, the program manager has an additional problem. The evaluation criteria offer no guidance about which Division and Class should be selected under a given set of circumstances. For example, suppose the system is to allow access to users cleared for unclassified, CONFIDENTIAL, and SECRET information. Suppose further that the system will store information at the unclassified, CONFIDENTIAL, and SECRET levels. Assuming a need to know, the SECRET user may access any information in the system, the CONFIDENTIAL user may access CONFIDENTIAL and unclassified information, and the unclassified user may access only the unclassified information. To what Division and Class must the operating system conform to ensure adequate protection of stored information? It is clearly important that each program manager make the same choice under the same set of conditions, if the "controlled access" problem is to be avoided.

The solution for DoD lay in the generation of a matrix that provided a standard Division-Class decision for each set of user-information security conditions. At this writing, it is DoD's intent to issue that matrix as part of its directive governing security of ADP operations. Enforcement of the matrix as a standard will ultimately result in the establishment of a uniform security environment for DoD networking, an environment in which an accreditation of "SECRET" for one system is equivalent in terms of risk to any other "SECRET" accreditation. As a more precise definition, *certification* refers to a technical judgment that a system will provide a specified level of security, while *accreditation* refers to a decision to actually operate a system at the certified level.

## 7.3.8   A Simple Rule for Interconnection

Let's stop for a moment and take stock of where we are in the generation of an adequate basis for securely interconnecting systems to provide needed interoperability. First, an end-to-end encryption system, such as BLACKER, will ensure that the network itself will not leak information—it will ensure on a message-by-message basis that information is read only by intended recipients. Second, the DoD Computer Security Evaluation Criteria provide a principled basis by means of which the level of trust that can be reposed in a given operating system can be assessed. That is, very precise statements can be made about a processor's degree of trustworthiness. Third, a matrix, such as that under development by DoD, provides a standard for application of the criteria, and this will result in a uniform accreditation environment for DoD networks.

At least one more piece is required. The network is now capable of preserving the security of what has been delivered to it, but it needs a guideline to follow in doing so. A set of simple rules for interconnection, which can be enforced by the network, is required. These rules might be provided as an integral part of an end-to-end encryption system, such as BLACKER; or they might be embodied in the network nodes, where the nodes could be trusted to execute the connection decisions (i.e., the node processors were certified to an appropriate level, using the DoD Computer Security Evaluation Criteria). One formulation of such a rule set has been suggested by LaPadula (1985, p. 2–12):

> For any host processor, $H_i$, to transmit a unit at a given security level, "S-level," it is required that the security level of the intended recipient, $H_j$, dominate (that is, be equal to or higher than) the security level of the transmitted unit and it is required that "S-level" equals one of the values in the accredited security level range of the transmitter.

If the LaPadula Rule were to be applied in a four-host network as indicated in Fig. 7.1, exchange of traffic would be permitted as shown.

**Fig. 7.1.
LaPadula Rule
applied to four-
host network.**

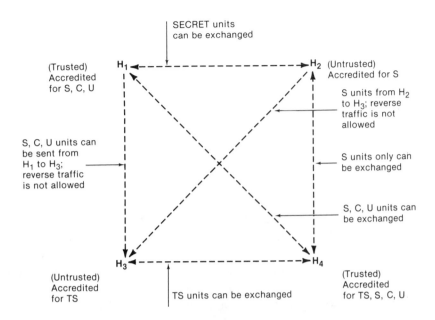

In Fig. 7.1, $H_1$ has been accredited in view of its certification by the Trusted Computer System Evaluation Criteria to process SECRET, CONFIDENTIAL, and unclassified information. Thus, units that it transmits may carry an S-level of S, C, or U. $H_3$, on the other hand, has not been so accredited. Since the system cannot be trusted to properly segregate different classification levels of information, all information handled by the system is treated as if it were TOP SECRET, and all users of the system must possess TOP SECRET security clearances. Thus, it may transmit information only at an S-level of TS. It may receive S, C, and U units from $H_1$, since its security level (TS) does dominate any unit that could be transmitted from $H_1$. However, it may transmit units at the TS level only to a host containing TS in its accredited range.

## 7.3.9  Security Aspects

Until now, the discussion has assumed that compromise of information is the sole security concern. In fact, the DoD Trusted Computer System Evaluation Criteria document is focused primarily on that nondisclosure aspect of information security. The document is grounded in the reference monitor concept that defines all entities as either subject or object and requires the mediation of all requests for access to system objects (resources), by subjects, through a piece of trusted software, the reference monitor. There are, however, two other aspects of information system security—*integrity* and *denial of service*. Integrity deals with assurances that information is not changed in an unauthorized way as it progresses through the system. Denial of service deals with the assurance that the system will be available for use when needed. Denial of service and reliability concerns are closely related. It is not totally surprising that integrity and denial of service are not thoroughly addressed in the Criteria document, which deals with operating system security. In the context of operating system security, integrity and denial of service are not concerns at the same level as avoidance of compromise.

In a network security context, however, those aspects may reach a level, under some circumstances, that surpasses even compromise considerations. In a funds transfer transaction in a financial network, for example, preservation of the decimal point may be far more important than preservation of the privacy of the transaction. In a command and control network, delivery of a notice that the enemy is pouring across the border may be far more important than preserving the secrecy of the message contents. The enemy, after all, already knows it is invading. These examples are somewhat simplistic, but they do serve to make the point that integrity and avoidance of denial of service are important network concerns that must be addressed in any security scheme.

Unfortunately, to date, there does not exist a general theory for network security—similar to that underlying the Trusted Computer System Evaluation Criteria—that encompasses all the necessary elements of network security and permits maximum interoperability. The DDN security approach taken by DoD will provide adequate compromise protections for the range of permitted interconnections, but it is a solution specific to the DDN and does not incorporate denial of service and integrity concerns. Those considerations are dealt with on an ad hoc basis. This comment is not a criticism of the DoD approach, which reflects the state of the art in security and is a practical approach to securely interconnecting networks today. DoD is in the process of addressing the general network security challenge through the development of Network Security Criteria, which will stand as an analog to the processor-oriented existing criteria. The Network Security Criteria will address integrity and denial of service and, in the current draft version, will provide a separate rating for the security posture of a network for each of the three major security aspects.

A point worth noting is that effective security must be provided in computer and network systems for reasons other than protection of national-security-related information. Privacy of information that could reveal personal, financial, legal, and medical facts is important both to comply with legal requirements and to protect against lawsuits. Preventing disclosure of strategic business plans, pricing algorithms, trade secrets, and marketing strategies may be as important to corporate survival as preserving national secrets may be to national survival. Computer security features and assurances also protect against waste, fraud, and other forms of criminal activity.

To summarize, security concerns stand as a major barrier to achievement of the interoperability required in a command and control environment. It is a barrier that is being overcome, but progress is necessarily slow and deliberate.

# 7.4  Evolving Technologies and Command and Control

There are a number of interesting developments that offer promise of enhancing command and control objectives. One of these developments is the introduction of the Ada programming language. As computing evolved, programmers moved from development of applications written in machine language to applications written in assembly language, and from there to the higher-level "programmer friendly" languages such as COBOL and FORTRAN. Real-time applications, however, remained for many years in a state of arrested development. Higher-level languages at the level of a COBOL or FORTRAN were not available for use in communications applications. A higher-level language that would result in efficiently executing object code proved difficult to develop. As costs of internal storage have plummeted and programmer time has become a scarce resource, the major advantages of a high order real-time programming language have become twofold: the more efficient use of the scarce programming resource and the transportability of software from system to system.

## 7.4.1  Ada®

The Ada language was developed to be an efficient high-level language for real-time applications. The language has been adopted by DoD for use in the WWMCCS Information System (WIS), the replacement system for the WWMCCS System mentioned earlier. Estimates indicate that use of Ada for the command and control applications programs will considerably reduce the cost of program development and maintenance. The WIS program is also developing various standards in Ada. For example, the TCP and IP protocols are being coded in Ada. Successful completion of these efforts will produce transportable standards modules that can be readily implemented on many different computer systems. High-level language coding of protocols offers great benefits in reducing the time normally required to react to the changing standards environment. In this instance, a new development that is part of the emerging technology scene can be applied to help solve the problem of coping with rapid change. The security aspects of Ada are also being evaluated through a project sponsored by the National Computer Security Center to build trusted security-relevant software in the Ada language.

## 7.4.2  Supercomputers and AI

Another burgeoning technology offering enormous promise for the future of $C^3$ comprises the related areas of super computers and artificial intelligence (AI). Networked super computers employing expert systems could solve otherwise intractable command and control problems. In a massive assault on the technical challenges, the Defense Advanced Research Projects Agency (DARPA) has embarked on a half-billion-dollar program to achieve order of magnitude breakthroughs in computing technologies. It is hoped that this program will lead to fielding of hitherto unattainable capabilities. The target systems for this development include unmanned vehicles that could operate autonomously in a battlefield or undersea environment, a pilot's associate system in the cockpit to help a pilot cope with the many threats in today's electronic battlefield, and a battle management system that would help a carrier task force cope with the range of threats. Achievement of these objectives requires the development of a number of AI subsystems, such as natural language, vision, navigation, and expert systems, which will be integrated in the configuration needed to achieve a specific capability. These subsystems, however, will place processing demands beyond today's hardware speeds. Accordingly, the DARPA program also includes development work in nontraditional ADP architectures, hoping to achieve the necessary increases in processing speeds through use of parallel processing concepts. Recognizing that the greatest impediment to using parallel traditional architectures has been the lack of the system software needed for efficient parallel operation, the DARPA program includes software development work.

Supercomputers and AI may well be essential for the Strategic Defense Initiative, also termed "Star Wars." The problems involved in detecting, assessing, and taking appropriate action against potentially thousands of missiles and tens of thousands of warheads within an extremely constrained time period demand supercomputer speeds for their resolution. Further, the analysis required to assess whether detected objects constitute a threat will require expert systems for its efficient execution. All of the elements in a Star Wars defense system must be internetted by high-bandwidth circuits to interchange essential data.

The parallel architectures being developed for the new generation of supercomputers offer promise of help in solving network security problems. For example, parallelism concepts could be used to segregate information into streams of differing security sensitivity, easing the so-called trusted-path problem by providing reliable and trusted paths through the processing system.

A major impetus to exploration of parallelism in computing systems has been the fundamental limitations on throughput imposed by the speed of the electron. Components on silicon chips have been packaged more and more densely, which results in decreasing the time required for intercomponent communications. For example, the Very High Speed Integrated Circuits (VHSIC) program is pushing toward half-micron component spacing on a chip. This program has the potential to provide an order of magnitude increase in computational ability, to use only 20% of the power now required, to be one-fourth the size and one-fourth the weight of existing circuits, at one-tenth today's cost. As a spin-off benefit, such densities offer a high payoff in increased reliability of the micro-computer. One of the most frequent causes of computer malfunction is faulty interconnection among the component chips and among the circuit boards. Increased density tends to eliminate boards and chips, cutting down on the connective wiring and reducing vulnerability to fault. As the VHSIC program pushes chip density to its theoretical limits, parallel architecture supercomputers promise increased speed.

Even greater information density may be on the horizon in the field of molecular electronics. The potential for storing information on an individual molecule appears to offer much hope in overcoming the fundamental packing density limits of today's semiconductors. Closely related to that technology is the concept of the biochip—use of genetic engineering techniques to produce living cells that can be used to store information. We can only dimly envision today what such a technology could do for us in terms of simulating the thought processes of the human brain in expert system use—artificial intelligence in the full meaning of the term. The implications in a command and control environment are staggering.

# 7.5 Legal and Regulatory Issues

A watershed event in U.S. telecommunications history occurred with the divestiture of AT&T. The impact of that event, both positive and negative, has been felt as sharply in the command and control areas as in the rest of the nation. On the positive side, the rate of introduction of new telecommunications-related products has certainly increased. Further, the cost of long-haul transmission appears to be decreasing. The negative side has had decidedly more impact, and it remains to be seen to what extent the negative aspects represent a relatively short-term adjustment. First, the length of time required to obtain data communications circuits increased to as long as twenty-four months for 56-Kbps circuits. This lengthening in circuit lead time adversely affected expansion of the DDN, which in the United States relies on use of 56-Kbps circuits. At the time of this writing, circuit lead time had become the pacing factor on DDN network expansion. This factor had reduced flexibility in responding to new requirements.

Another factor impacting negatively on the ability to install and maintain command and control networks has been the loss of a single point of contact for coordination of telecommunications requirements. Prior to divestiture, AT&T acted as an industry point of contact for matters relating to DoD and national requirements. AT&T coordinated requirements on an end-to-end basis, ensuring that modems, access circuits, and trunks were in place and ready when needed, working not only with its own local operating companies but also with the approximately 2500 local telephone entities. That single point of contact relationship has gone with the divestiture winds. The customer must now provide necessary integration and, when difficulties are experienced, must determine which vendor to call—modem, multiplexer, access circuit, or long-line carrier. The wrong choice can result in a longer outage as well as additional cost.

Special mechanisms have been put into place to address this problem. For example, DoD identified a number of priority systems and received approval from the FCC for AT&T to act for those specific systems as a single point of contact. A structure for meeting national security and emergency preparedness requirements on a priority basis has also been put into place. That structure, the National Coordinating Center, is a last resort entity to be used when normal ordering processes cannot meet the requirement. Time will tell whether these structures will do the job. If they do not prove to be responsive in satisfying national requirements, the government may be driven to act as its own end-to-end communications vendor, employing, for example, satellite technology and transportable satellite terminals to ensure rapid connectivity when required.

Another adverse factor springing from divestiture has been the loss of AT&T as the promulgator of Bell standards and practices as de facto industry standards, a practice that helped to ensure system integration. Non–Bell System independents who provided local telecommunications service were, in effect, required to conform to Bell standards in order to obtain access to the long distance network for their connected local subscribers. A necessary by-product of the healthy competition in the long distance carriage market is the potential for divergence of standards and practices. Timing of the introduction of new technology and its concomitant standards is now more difficult to predict. In the competitive arena, such new technology will likely be introduced sooner than later. The task of keeping architectural road maps current, such as the WWDSA, mentioned earlier, becomes more difficult in such a dynamic environment.

Another nondivestiture issue arises out of the oversight of government operations provided by the Congress. The ultimate success of a command and control system is largely dependent on its access to any data base offering information that can bear significantly on the decision to be taken. Since command and control information requirements cannot always be completely predicted in advance, the greater the connectivity of the network the better. The ability to continue to operate in a stressed environment is also extremely desirable. The more communications paths available in the network, the better, from a command and control viewpoint. From a national perspective, the greater the degree of interoperability between and among separate government agencies, the greater the ability of government to carry on the most essential operations in the event of natural or other disasters. Thus, the maximum possible interoperability is operationally desirable. It is also generally, though not always, the most cost effective solution because of economies of scale achievable in circuit usage and equipment acquisition. In fact, the National Communications System is an organization chartered by Executive Order and composed of over twenty government organizations whose primary mission is to achieve interoperability of the separate components' telecommunications systems.

Running counter to that strong impetus for interoperability is the fear, occasionally expressed by congressional and other spokespersons, that extensive government system interoperability could undermine the citizen's right to privacy by allowing information collected by separate government entities to be collated and used against the individual. That concern must be addressed effectively with appropriate privacy safeguards. Many of the computer security techniques can be employed to good effect to relieve fears that "Big Brother" is around the corner.

## 7.6   Summary

Command and control systems must be able to obtain any information that bears significantly on a decision to be taken. Thus, interoperability with other information systems is essential in a command and control environment. Of the many challenges in the achievement of interoperability, probably the most difficult to overcome are the adoption and maintenance of standards and the gaining and maintenance of effective information security. A major complicating factor in this effort is the impact of the accelerating rate of change in the technical, legal, and regulatory environments. At the same time new technologies, such as the Ada language and expert systems, may be useful in overcoming the technical interoperability problems. Finally, interoperability is not universally considered to be a benefit, and objections to interoperability propounded by advocates of individual privacy must be effectively addressed in system solutions.

# 7.7   References

Bartee, T. C., ed. 1985. *Data communications, networks, and systems.* Indianapolis: Howard W. Sams.

LaPadula, L. J. 1985. Some thoughts on network security: A working paper. *Proceedings of the Department of Defense Computer Security Center Invitational Workshop on Network Security.* Washington, D.C.: DoD Computer Security Center.

United States Department of Defense. 1979. *Department of Defense dictionary of military and associated terms.* Joint Chiefs of Staff Publication No. 1. Washington D.C.: Joint Chiefs of Staff, June 1.

# 8

# DIGITAL CODING OF SPEECH

N. S. JAYANT

Digital representations of speech play a key role in telecommunications and information storage. Techniques for digital coding have to be evaluated in terms of digitized speech quality, the information rate in the digital representation, and the complexity of the coding algorithm. This chapter will provide a review of speech coding algorithms in the information rate range of 64 to 2.4 Kbps (kilobits per second). The review will be nonexhaustive; the algorithms chosen for discussion are representative techniques that are generally well understood.

The focus in this chapter will be on *what* has been accomplished in the coding field rather than on *how* the accomplishments have been engineered. Mathematical digressions will be kept to a minimum, but the physical principles behind various coding algorithms will be highlighted. Likewise, instead of describing specific implementations, the chapter will provide broad quantitative comments on the signal processing complexities that coding algorithms demand.

The first section (8.1) of this chapter will discuss the advantages of digitizing speech. The second section (8.2) will discuss the problem of digital coding and define three parameters—information rate, speech quality, and coder complexity. Section 8.3 will describe the properties of the speech waveform that are relevant to efficient coding, and sections 8.4, 8.5, and 8.6 will discuss classes of techniques that are appropriate for high-quality speech coding at 64, 32, and 16 Kbps, respectively. Section 8.7 will examine the 8- to 4-Kbps range of bit rates, an important area for current research and high-communications-quality applications. Section 8.8 will be a brief account of a vocoding technique that currently provides synthetic-quality speech at a bit rate in the order of 2 Kbps. Section 8.9 will provide an overall three-dimensional perspective involving the three parameters of speech quality, bit rate, and complexity.

Most of the material in this chapter has been treated in depth in a previous text on waveform coding (Jayant and Noll 1984) and in earlier review articles on the subject (Jayant 1974, 1976; Flanagan et al. 1979). The topics in sections 8.7 and 8.8 are discussed mainly in scattered literature in the fields of speech processing and speech technology. Readers seeking in-depth treatments of speech and speech processing are referred to Flanagan 1972 and Rabiner and Schafer 1978.

# 8.1   Advantages of Digitizing Speech

The advantages of digitizing an analog signal are well known (Aaron 1979). Historically, digital representation of the speech signal has permitted robust communication over noisy media as well as efficient encryption for message privacy. These advantages carry over to contemporary applications such as Integrated Services Digital Networks and cellular radio telephony. Digital speech is also a natural format for packet switching and for information services such as voice mail.

## 8.1.1   Communication over Noisy Media

Physical media for transmission and storage are noisy analog systems. When an analog signal is transmitted over such a medium, the noise introduced by the medium increases as a function of transmission distance. In the case of analog signal storage, the noise increases with time, as in the example of audio recording on analog tape or a conventional gramophone record.

If the analog signal is represented digitally prior to transmission over the same noisy medium, every information-carrying bit can be regenerated periodically so that the final signal quality is no longer a strong function of transmission distance (or storage time). Long-distance commercial telephony as well as multiple-reuse voice-playback depend on this principle; they use digitized speech over inherently noisy analog media.

Even if errors do get inserted into the bit sequence, they can be corrected, or at least detected, by the use of binary error-protection coding, as long as the bit error rate is not too high (Jayant and Noll 1984; Odenwalder 1985). A good example of robust digital communication over a noise-prone medium is the digital compact disc for high-bandwidth audio. In digital speech communication, prime candidates for binary error protection are power-limited channels such as satellite links and cellular radio systems.

### 8.1.2  Message Encryption—Secure Voice

Systems for speech privacy (or secrecy) are based either on analog speech scrambling or on speech digitization followed by digital encryption. Recent advances in signal processing and analog channel equalization have led to significant enhancements in the voice quality and cryptanalytical strength of analog scramblers (Jayant 1982). It is still conceded, however, that digital encryption provides greater levels of security in voice communication. Efficient procedures have been developed and standardized for encrypting a digital sequence and for generating and distributing encryption keys (Beker and Piper 1982). The problem relevant to this chapter is the part of the system that precedes the digital encryption stage. This is the problem of digitizing the speech with a bit rate that is high enough for good speech quality but low enough that the speech-carrying bits can traverse the analog communication channel with an adequately low probability of error.

If the available analog channel cannot accommodate a bit rate high enough for good voice quality, and if a high level of voice security is an overriding requirement, typical practice in communications has been to compromise voice quality by lowering the digital speech bit rate to a substandard value. A good example is the use of 2.4-Kbps vocoding for tactical secure voice over voice-bandwidth telephone links and radio channels.

There are strong emerging needs for higher-quality secure voice systems over low-power, low-bandwidth channels, such as satellite links and cellular radio systems for commercial telephony. The success of these systems depends on advances in two, somewhat orthogonal, communications fields: advances in modulation/demodulation technology, which will make possible the reliable use of *higher* bit rates over a given analog channel, and advances in source coding technology, which will permit the realization of *lower* bit rates for a given level of digital speech quality.

# 8.2  The Problem of Digital Coding

Fig. 8.1 is a simple description of a digital coding system. The figure shows an input speech waveform X, which is an analog signal in that it is *amplitude-continuous* and *time-continuous*. The waveform Y at the output of the digital system is an *amplitude-discrete* and *time-discrete* approximation of X. An analog version of Y (the smooth waveform in the right part of Fig. 8.1) is obtained by passing it through an appropriate output filter. If X is bandlimited to the telephone frequency range of 200 to 3200 Hz, as is assumed in this chapter, the final output filter is a corresponding bandpass device. The characteristics of these input and output filters are well standardized. Waveform prefiltering and sampling are tacitly assumed front-ends of all digital coding systems. The standard sampling rate for telephone-bandwidth speech is 8000 Hz. Lower rates such as 6400 Hz are employed in specialized low-bit-rate applications that involve either a speech bandwidth of less than 3200 Hz or a subband coding framework with specially designed analysis-synthesis filterbanks (Jayant and Noll 1984).

Fig. 8.1. Digital coding of speech.

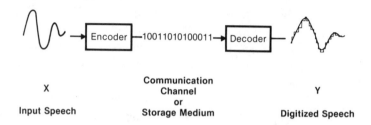

X

Input Speech

Communication
Channel
or
Storage Medium

Y

Digitized Speech

The role of the *encoder* is to map the input X into a sequence of binary digits that, when fed into the *decoder*, result in the output Y. The combination of encoder and decoder operations is a coding system, often referred to simply as the *coder*. In a speech communication system, the binary digits traverse a transmission channel in real time; in a voice response system, the binary digits are stored and used (usually repeatedly) as inputs to a voice-response decoder. In both cases, the *binary digits*, or *bits*, are modulated into unique (analog!) signals that are matched to the physical transmission or storage medium and demodulated back into bits prior to the decoding operation shown in Fig. 8.1. In general, the speech-carrying bits are subject to random inversions or errors, and techniques for redundant error-protection coding (Odenwalder 1985) are sometimes used to provide the speech decoder with an error-free version or a minimal-error version of the bit sequence that leaves the speech coder in Fig. 8.1.

The modulation and error-protection mechanisms just mentioned control the number of bits per second that the physical transmission (or storage) channel can reliably support. This number is upper-bounded, of course, by the so-called channel capacity. In practice, digital speech systems operate with bit rates below this capacity, and with bit error rates that depend on the bit rate as well as the efficiency of the modulation/ error-protection process. The bit error environment can range from independent error rates of, say, $10^{-6}$ to bursty error rates of $10^{-2}$, depending on the application. The role of the error-protection process is to realize lower error rates at the input to the speech decoder. As a very crude rule of thumb, most speech coding algorithms implied in this chapter can work adequately with independent, or declustered, error rates of up to 1 in 1000.

The block diagram in Fig. 8.1 is general enough to describe two important classes: *waveform coders* and *vocoders*. As the name implies, the first category of coders depends on algorithms that seek to realize, at the system output, an explicit approximation to the input speech waveform X, specifically to a time-discrete sample-by-sample description of it. Coders in the second category are based on a compact, highly parameterized description of the speech process; they use this model to distill a compact description of the input X and digitize only the parameters of this compact description. This chapter will discuss both waveform coders and vocoders, but the bulk of the discussion will refer to the waveform coding class of digital speech systems.

The key problem in digital speech coding is to achieve an adequately low *information rate* in the digital representation while realizing an acceptable level of *speech quality.* Later in this chapter a third parameter will be brought into the picture; namely, *coder complexity.* The chapter will conclude by pointing out the trade-offs that exist among the three parameters mentioned previously: bit rate, speech quality, and coder complexity.

How do we define these parameters quantitatively? The answer is simple for the first parameter, bit rate, but important qualifications are needed in discussing measures of quality and complexity.

## 8.2.1 Definition of Information Rate

The bit rate of a speech coding system is measured in bits per second (*bps*), and often described in kilobits per second (*Kbps*). In the simpler examples of waveform coders, the bit rate $I$ can be expressed as the product of the *sampling rate* $f_s$ and the number of *bits per sample* $R$:

$$I = f_s R \tag{8.1}$$

The bit rate $R$ needed to convey the output level information in a simple $L$-level quantizer is

$$R = \log_2 L \tag{8.2}$$

Appropriate generalizations of (8.1) are adequate to define the bit rates of sophisticated waveform coders and vocoders. For example, in subband speech coding, the bit rate is a sum of component bit rates used for the coding of the waveforms in individual subbands of speech. In vocoders, the bit rate is a sum of component rates used for coding the several parameter signals in the vocoder speech model. In coders that employ vector quantization, the number of bits per sample $R$ is only an implied parameter; the quantity that is explicitly specified in these systems is the number of bits per vector, which is simply $R$ times the number of samples in the vector.

Two sets of standard transmission rates are encountered in speech coding practice. One set involves 64 Kbps and submultiples thereof. The other set, derived from data transmission practice, involves 2.4 Kbps and multiples and submultiples thereof. For the purposes of this chapter, it will be sufficient to deal with only the first set of numbers: 64, 32, 16, 8, 4, and 2 Kbps.

Important purposes of this chapter are to specify the maximum level of speech quality that can be expected at a given bit rate and to indicate the complexity of the coding algorithm promising that fundamental ceiling of performance.

## 8.2.2  Definition of Speech Quality

The performance of a digital coding system can be described by two broad classes of measurements: the *objective* and the *subjective* measures of quality.

Objective measurements of quality are usually easier to obtain and to duplicate. These measurements are particularly valuable in the initial simulation and design of a specific coding algorithm. Objective measurements are often valuable in comparing alternate techniques within a broad algorithm class. Good examples of objective measures are the signal-to-(coding-) error ratio, usually referred to as the signal-to-noise ratio, and various refinements thereof (Jayant and Noll 1984). As the terminology implies, the signal-to-noise ratio and related measures describe the closeness of output Y to input X in Fig. 8.1. Such measures of closeness are particularly appropriate in *high-quality* speech coding. Absolute assessments of the perceived quality of Y, without a direct comparison to input X, are more appropriate in systems that offer lower levels of speech quality such as *communications quality* or *synthetic quality*. We will comment again on these three levels of quality, as appropriate.

Other examples of objective measures are *word intelligibility* and *speaker recognizability*. High-quality speech coding implies scores for word intelligibility and speaker recognizability that are very close to the scores for uncoded speech. Communications-quality coding implies noticeable, but not operationally significant, reductions in these scores. Synthetic-quality coding implies significantly lower scores, especially for speaker recognizability.

Subjective measurements of quality are generally more difficult to obtain, but they are also the truly valid criteria of performance from a perceptual viewpoint. The subjective-quality measure that we will use in this chapter (Section 8.9) is an absolute quality measure as measured on a five-point adjectival scale:

**5:** Excellent **4:** Good **3:** Fair **2:** Poor **1:** Bad.

Formal subjective testing implies an adequately large data base of speakers, speech inputs, and listeners; the overall average score is therefore referred to as a *Mean Opinion Score.*

### 8.2.3 Definition of Coding Complexity

Sophisticated algorithms for digital coding tend to be implemented using digital signal processors; specifically, with suitable combinations of general-purpose and special-purpose processors. An appropriate measurement of coder complexity for these implementations is the number of multiply-add operations required for coding 1 second of speech. A closely related number is the number of millions of instructions per second, *mips*. In many classes of digital speech systems, most of the implementation complexity indeed resides in the coding stage. The decoding stage tends to be much simpler, sometimes by an order of magnitude or more. Many coding algorithms, in fact, include a local decoding stage.

A parameter that tends to be monotonically related to coder complexity, though not always so, is *communication delay*. This is the result of delays introduced in the coder and decoder operations of Fig. 8.1. These include delay components that are fundamental to the coding and decoding algorithms, as well as additional delays that may be introduced by specific procedures used to implement these algorithms.

Techniques for speech storage and voice response can afford more sophisticated, and hence lower bit rate, techniques for digital coding. There are two reasons for this. First, most of the complexity of a sophisticated technique usually resides in the encoder, and this is eventually irrelevant in a speech playback system that involves repeated use of only the much simpler decoding operation. Second, the processing delay inherent in a sophisticated coding system is again generally irrelevant in the one-way application of speech playback.

## 8.3  The Speech Waveform

Fig. 8.2 illustrates speech waveform examples that motivate the coding algorithms described in a later section of this chapter.

Fig. 8.2.
Amplitude
waveforms of
speech
(Flanagan et al.
1979).

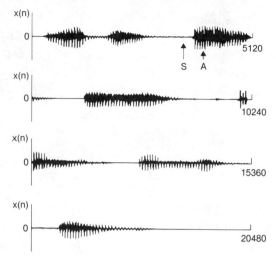

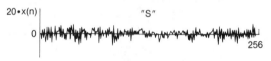

(A) A sentence-length (2.5-second) utterance, "EVERY SALT BREEZE COMES FROM THE SEA." The locations of the sounds "S" and "A" in the word "SALT" are shown by arrows in the first row.

(B) A short-time (32-ms) segment of unvoiced speech sound "A."

(C) A short-time (32-ms) segment of unvoiced speech sound "S."

Part (A) of Fig. 8.2 is the amplitude-versus-time plot of a sentence-length utterance, "EVERY SALT BREEZE COMES FROM THE SEA." The waveform has a total of about 20,000 samples at the sampling rate of 8000 Hz. Therefore, the total duration of the utterance is about 2.5 seconds.

Parts (B) and (C) of Fig. 8.2 are blowups of 32-millisecond waveform segments that belong to the sounds "A" and "S" in "SALT." The waveforms in (B) and (C) are stretched in time, and the waveform for "S" in (C) is also scaled up in amplitude by a factor of 20.

Several properties of the speech waveform emerge from an examination of Fig. 8.2, and these are discussed next.

## 8.3.1  Amplitude Characteristics

The speech waveform contains segments with widely different amplitude levels. *Voiced* speech segments, such as the waveform segment "A," have much greater amplitude levels than *unvoiced* speech segments such as "S." The voiced parts of speech are also sustained longer, and they account for most of the speech duration. The fidelity with which these segments are coded is a major determinant of output speech *quality*. On the other hand, an adequate reproduction of unvoiced segments is critical for speech *intelligibility*.

*High-quality speech coding at 64 Kbps takes explicit account of the amplitude nonstationarity of the speech signal in order to provide perceptually perfect reproduction of both voiced and unvoiced segments* (section 8.4).

## 8.3.2 Power Spectral Density

The voiced and unvoiced segments of speech also differ in terms of their power spectra. The voiced segment in Fig. 8.2(B) is a slowly varying waveform as compared with the rapidly changing unvoiced waveform in Fig. 8.2(C). The preceding time-domain differences have a direct interpretation in the frequency domain. The voiced speech has a preponderance of low-frequency energy, and it can generally be described as a *lowpass* signal. Unvoiced speech is not a lowpass signal in this sense; often, as in the example of the sound in Fig. 8.2(C), unvoiced speech has a preponderance of high-frequency energy, and it can be described therefore as a *highpass* signal.

Fig. 8.3 describes short-time power spectra corresponding to the voiced and unvoiced segments "A" and "S" in Fig. 8.2. Two degrees of frequency resolution are provided. In each case, the two power spectral densities shown are results of analysis with (1) a low resolution in frequency (64-point DFT) and (2) a high resolution in frequency (256-point DFT). Let us examine the low-resolution plots (A1) and (B1) in Fig. 8.3. It is now explicitly clear that the voiced speech spectrum (A1) is lowpass and the unvoiced speech spectrum (B1) is highpass. Another interesting observation emerges from (A1): the envelope of the voiced speech spectrum has distinct peaks, called *formant* frequencies in speech work. These are the resonant frequencies of the vocal tract shape that produces the sound "A." Three of these resonances occur within the telephone band of 3.2 kHz.

The slowly varying nature of the 256-sample waveform in Fig. 8.2(B) shows that this voiced waveform is highly predictable. The lowpass characteristic in (A1) of Fig. 8.3 is an equivalent expression of waveform predictability, or redundancy. The unvoiced signal in Fig. 8.2(C) seems less predictable—and indeed it is—but it too has a nonflat spectrum as shown in Fig. 8.3(B1), and this implies nonzero redundancy. The more nonflat a spectrum, the more predictable or redundant the corresponding wave-

**Fig. 8.3. Short-time power spectral density for speech samples (Jayant and Noll 1984).**

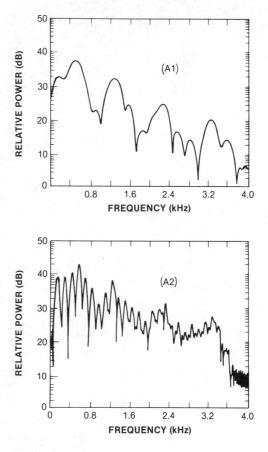

*(A) Voiced speech example in part (B) of Fig. 8.2.*

form, and the lower the information rate needed by a digital coder to reproduce the waveform, provided of course that the coder has an adequate mechanism to exploit the waveform redundancy in the time and/or frequency domains.

*High-quality speech coding at 32 Kbps is based on mechanisms that efficiently utilize speech redundancies by dynamically tracking changes in input power spectra* (section 8.5).

### 8.3.3   Pitch Periodicity in Voiced Speech

The voiced speech waveform in Fig. 8.2(B) has yet another obvious property: it is periodic, or nearly so. There are about four very similar periods in this example, each lasting for about 60 samples, or 7.5 milliseconds. The reciprocal of this fundamental period is in fact the fundamental frequency, sometimes called the *pitch* frequency. The pitch frequency of the (male) speaker who produced the utterance in Fig. 8.2 is about 130 Hz.

Fig. 8.3 cont.

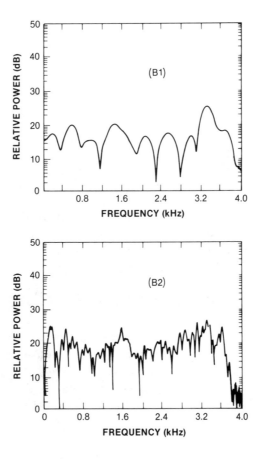

*(B) Unvoiced speech example in part (C) of Fig. 8.2.*

The pitch structure is also evident in the high-resolution voiced spectrum of Fig. 8.3(A2); it is the spacing between adjacent teeth in the spectral fine structure seen in that plot.

The presence of pitch periodicity is yet another source of redundancy in the speech waveform, and, as we have shown, it is manifested explicitly in both time and frequency domains.

*High-quality speech coding at 16 Kbps requires an adequate utilization of the pitch structure in voiced speech* (section 8.6).

### 8.3.4 Coder Complexity and Delay

As implied so far, the attainment of lower bit rates for speech coding requires increasingly greater scrutiny of the speech process. This means greater processing complexity as well as longer amounts of processing delay. For example, incorporation of the pitch structure in a speech coder requires a pitch-measurement window of at least 2 to 3 pitch periods, or a DFT analysis with a 256-sample block as in (A2) and (B2) of Fig. 8.3. Either of these operations implies a coding delay in tens of milliseconds.

### 8.3.5  The Tools of the Trade

*Linear prediction* is a fundamental example of a time-domain mechanism for bit-rate reduction. In the frequency domain, *variable bit allocation* provides the basic mechanism for low-bit-rate coding. An integral part of all digital coding systems is an efficient means for amplitude discretization, or *quantization*. The amplitudes that are discretized are speech samples in simpler examples of speech coders (section 8.4) and various speech-derived signals in more sophisticated coders (sections 8.5 through 8.8). A quantization technique that is extremely efficient in very low-bit-rate work (say 1 bit per sample or less) is *vector quantization* (section 8.7).

## 8.4  High-Quality Speech Coding at 64 Kbps

Perhaps the best-known digital coding system is *pulse code modulation* (PCM), which is merely amplitude quantization followed by a rule that assigns a unique binary codeword to each possible quantizer output. The long-standing international standard for high-quality speech coding is 64 Kbps PCM. This is the result of 8-kHz sampling followed by 8-bit quantization of each sample.

The high quality of the PCM speech output is due to the use of a nonuniform quantizer characteristic in which the fineness of quantization increases as the magnitude of the speech input goes towards zero. This recognizes the fact that there is a preponderance of low amplitudes in speech, both in the voiced and unvoiced regions; and these amplitudes need to be quantized with an imperceptible level of quantization error. Voiced speech also contains many occurrences of very high amplitudes. The nonuniform characteristic of the quantizer provides large quantizing intervals in the high-amplitude regions, and this ensures that the high amplitudes in speech do not overload the quantizing system. As a result, these amplitudes are represented with adequately low fractional error.

Table 8.1 defines an 8-bit or 256-level quantizer used in 64-Kbps PCM (Bellamy 1982). This quantizer is a piecewise linear approximation to the so-called $\mu$-255 logarithmic quantizer. Only the positive half of the characteristic is shown in the table; the negative half has an identical characteristic for nonuniform quantization.

The first column of the table shows input amplitude ranges in eight nonuniformly spaced segments (the major horizontal partitions in the ta-

| TABLE 8.1. | Input Magnitude Range | Step Size | Segment Code | Level Code | Decoder Level Number | Decoded Magnitude |
|---|---|---|---|---|---|---|
| ENCODING/ | | | | | | |
| DECODING | 0 – 0.5 | | | 0000 | 0 | 0 |
| TABLE FOR | 0.5 – 1.5 | | | 0001 | 1 | 1 |
| $\mu$-255 PCM | ⋮ | 1 | 000 | ⋮ | ⋮ | ⋮ |
| | 14.5 – 15.5 | | | 1111 | 15 | 15 |
| | 15.5 – 17.5 | | | 0000 | 16 | 16.5 |
| | ⋮ | 2 | 001 | ⋮ | ⋮ | ⋮ |
| | 45.5 – 47.5 | | | 1111 | 31 | 46.5 |
| | 47.5 – 51.5 | | | 0000 | 32 | 49.5 |
| | ⋮ | 4 | 010 | ⋮ | ⋮ | ⋮ |
| | 107.5 – 111.5 | | | 1111 | 47 | 109.5 |
| | 111.5 – 119.5 | | | 0000 | 48 | 115.5 |
| | ⋮ | 8 | 011 | ⋮ | ⋮ | ⋮ |
| | 231.5 – 239.5 | | | 1111 | 63 | 235.5 |
| | 239.5 – 255.5 | | | 0000 | 64 | 247.5 |
| | ⋮ | 16 | 100 | ⋮ | ⋮ | ⋮ |
| | 479.5 – 495.5 | | | 1111 | 79 | 487.5 |
| | 495.5 – 527.5 | | | 0000 | 80 | 511.5 |
| | ⋮ | 32 | 101 | ⋮ | ⋮ | ⋮ |
| | 975.5 – 1007.5 | | | 1111 | 95 | 991.5 |
| | 1007.5 – 1071.5 | | | 0000 | 96 | 1039.5 |
| | ⋮ | 64 | 110 | ⋮ | ⋮ | ⋮ |
| | 1967.5 – 2031.5 | | | 1111 | 111 | 1999.5 |
| | 2031.5 – 2159.5 | | | 0000 | 112 | 2095.5 |
| | ⋮ | 128 | 111 | ⋮ | ⋮ | ⋮ |
| | 3951.5 – 4079.5 | | | 1111 | 127 | 4015.5 |

*Sources:* Bellamy 1982; Jayant and Noll 1984.

ble). Each of these segments has 16 uniformly spaced quantizing levels, resulting in a total of 128 quantization intervals. Corresponding quantizer outputs are shown in the last column of the table. The third and fourth columns show the seven magnitude bits that code the input magnitude ranges in column 1. These bits, preceded by a sign bit, constitute the 8-bit codewords of 64-Kbps PCM. The decoded magnitudes appear in the last column. Note that the ratio of the largest decoded magnitude to the smallest nonzero magnitude is about $2^{12}$. The ratio of the peak-to-peak decoder range to the smallest decoder output is therefore $2^{13}$. In this sense, the dynamic range of the 8-bit quantizer is equivalent to that of a 13-bit uniform quantizer. This 5-bit gain, which is a result of the nonuniform quantization characteristic, is crucial for the high-quality reproduction of a signal having a very high range of input amplitudes (Fig. 8.2).

The $\mu$-255 quantizer is the basis for 64-Kbps standard adopted in North America and Japan. The European standard for 64-Kbps PCM uses the so-called A-87.56 quantizing system. The subtle differences between the $\mu$-255 and A-87.56 standards are documented in Jayant and Noll 1984.

The 64-Kbps coder attains a Mean Opinion Score (MOS) of nearly 4.5 on a scale of 1 to 5. It preserves an MOS value of at least 4.0 after eight repeated stages of PCM encoding/decoding, and it provides high-quality reproduction of nonspeech signals such as 9.6-Kbps voiceband data signals. For all of these reasons, the 64-Kbps PCM coder is also called a *toll-quality* coder. Lower bit-rate speech coders can be designed to provide high-quality reproductions of speech in a single stage of encoding/decoding, but these coders do not have the robustness or versatility of the 64-Kbps PCM system. This is especially true of speech coders designed for bit rates of 16 Kbps or less.

# 8.5   High-Quality Speech Coding at 32 Kbps

In section 8.3, we mentioned that the speech waveform is in general very predictable. This statement immediately suggests that a system designed to communicate or code *prediction-error* information will be more efficient than one that codes speech amplitudes directly. This is because the power in the prediction-error signal can be much less than that in the speech signal; and efficient quantization of this error signal will therefore require a quantizer with fewer levels or bits, for a given granularity in the representation, or a given quality of speech reproduction. Bit-rate gains relative to 8-bit PCM can be as good as 4 bits per sample, for comparable subjective quality and no increase in processing delay.

A coding system based on the preceding principle is called a *predictive coder*. It is also called *differential PCM* (DPCM) because the quantizer input in this system is the difference between the input speech sample and the prediction of it. The DPCM decoder does an inverse operation similar to integration; it adds the quantized prediction-error signal to the decoder's prediction or estimate of the current speech sample.

In order that the prediction operations at the encoder and decoder be identical, speech amplitudes are predicted on the basis of a recent history of *quantized* signals, rather than on the basis of unquantized past speech samples (which are not available at the decoder). This leads to the *closed-loop* DPCM structure of Fig. 8.4. Note that the DPCM system assumes *linear prediction*, or the estimation of a current speech sample from a weighted linear combination of past quantized samples.

Although the DPCM principle was well known for a long time, standardization of a low-bit-rate DPCM coder did not occur until efficient and

Fig. 8.4.
Differential
pulse code
modulation
(DPCM).

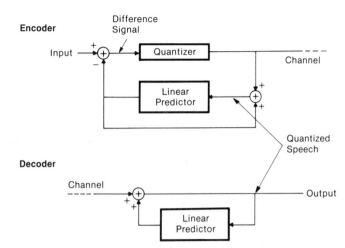

Fig. 8.4.
Differential
pulse code
modulation
(DPCM).

robust algorithms were available for adapting the quantization and pre-diction operations to the nonstationary properties of speech inputs. These algorithms needed to be efficient in the sense that adaptations of quantizer and predictor parameters could be done synchronously at the transmitter (encoder) and the receiver (decoder), without the need for explicit transmission of adaptation information; and the algorithms had to be robust in the sense that they operated in bit-error environments in the order of 1 to 1000 and also in the sense that they provided high-quality coding of 4.8-Kbps (if not 9.6-Kbps) voiceband data waveforms.

At 32 Kbps (8-kHz sampling with 4-bit or 16-level quantization), a well-designed DPCM system, a generalized version of the system in Fig. 8.4, can provide high-quality speech coding with an MOS score exceeding 4.0. It can also provide the robust performance characteristics men-tioned in the previous paragraph. Such a system has been internationally standardized for high-quality telephony (CCITT Report 1983; Jayant and Noll 1984).

# 8.6   High-Quality Speech Coding at 16 Kbps

We have noted that a fairly simple combination of adaptive quantization and adaptive prediction can reduce the bit rate from 8 bits per sample (64 Kbps) to 4 bits per sample (32 Kbps), without providing a significant loss of speech quality. This is a very strong and happy result. The reduc-tion of bit rate from 4 bits per sample to 2 bits per sample (16 Kbps) is a much more difficult task. High-quality coding at 16 Kbps depends on the utilization of the periodicities in voiced speech and on a property of hear-ing whereby a strong speech component can mask low-level components of coding noise in its spectral vicinity.

Fig. 8.5 shows a voiced speech spectrum and three illustrative equal-power spectra of the reconstruction error in digital coding. The flat, or white noise, spectrum in the figure is representative of the error in closed-loop DPCM, as described in section 8.5. The level of this spectrum depends on the DPCM bit rate. As the bit rate is lowered, the level of this error spectrum is raised. Because of the shape of the input speech spectrum, the raising of the noise floor degrades the high-frequency signal-to-noise ratio very rapidly. The coder performance at low frequencies is degraded less, especially because of the following noise-masking property: a strong formant or resonance in the speech spectrum tends to mask the noise in its frequency locality as long as the noise level is below a masking threshold in the order of 15 dB.

**Fig. 8.5.**
**Comparison of**
**reconstruction**
**error spectra in**
**three coding**
**systems (Atal**
**1982).**

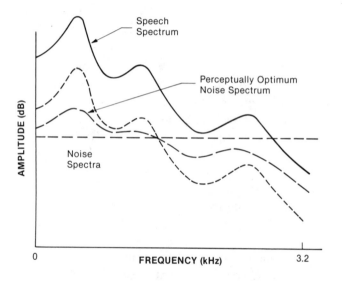

The speech-like noise spectrum in Fig. 8.5 provides a constant signal-to-noise ratio characteristic as a function of frequency. If this ratio is above the masking threshold, it will permit noise masking more uniformly in frequency, rather than only in the low-frequency regions, as in the case of a white noise spectrum.

The perceptually optimum noise spectrum is one that favors the low frequencies in speech but yet does not treat the high frequencies too badly (Atal 1982). It is therefore the intermediate noise spectrum in Fig. 8.5, one that is neither flat nor completely speech-like. High-quality speech coding at 2 bits per sample depends on the use of this property, preferably in the high-resolution sense of Fig. 8.3(A2). This usually implies an appropriate level of pitch processing as well.

One class of such high-quality coders is a DPCM system that includes pitch prediction and quantization error feedback for noise-shaping. Another class is that of frequency-domain techniques such as adaptive transform coding (ATC) and adaptive subband coding (ASBC) (Crochiere 1978; Crochiere et al. 1982). In these systems, noise-shaping is realized by adaptive bit assignment, or the allocation of greater numbers of bits to the perceptually more important frequencies of frequency bands, while preserving an average bit rate of 2 bits per sample.

Fig. 8.6 provides a simplified common description of subband coding and transform coding. A good deal of the complexity of these systems is due to the need for adaptive bit assignment (Honda and Itakura 1984; Soong et al. 1986; Zelinski and Noll 1977). Unlike adaptive quantization and adaptive prediction, adaptive bit assignment algorithms devised so far are based on explicit transmission of the bit allocation information to the receiver, and this takes up a nontrivial fraction of the total bit rate. The use of spectral template matching is a powerful technique for reducing the bit rate needed for bit allocation information. The spectral matching approach is identical to the codeword-matching approach of *vector quantization* (section 8.7.3).

Fig. 8.6.
Subband coding
and transform
coding.

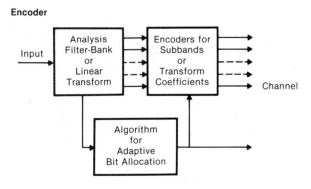

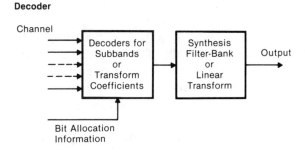

Crucial to the quality of the systems of Fig. 8.6 are the respective signal-processing front-ends, subband filter banks, and linear transforms.

Subband partitions necessarily follow the integer-band sampling constraint that the lowest frequency in any subband is an integral multiple of its bandwidth. This permits a sampling rate equal to the minimum possible value, which is twice the subband bandwidth. Minimization of interband interference (*aliasing*) also requires the use of a special class of filtering techniques, the so-called *quadrature mirror filter* banks (Esteban and Galand 1977; Cox 1986). Typical numbers of subbands range from 4 to 8.

Good examples of linear transforms are the *discrete cosine transform* (Ahmed and Rao 1975) and related transforms. These transforms tend to have fast implementations based on the Fast Fourier Transform (Elliott and Rao 1982). Typical orders of linear transforms range from 64 to 256, implying coding delays of 8 to 32 milliseconds.

High-quality speech coding, with a Mean Opinion Score of 4.0 or higher, has not yet been demonstrated for bit rates *less than* 16 Kbps. But there are no fundamental theorems that rule out high-quality coding at lower bit rates.

In principle, the combination of adaptive noise shaping and pitch processing can lead to high-quality coding at 16 Kbps. But the preceding procedures can lead to significant increases of coder complexity and coding delay, and simpler versions of these coding systems are of great interest, provided that the resulting compromises in quality are small enough to be operationally acceptable. Several examples of such communications-quality systems have been proposed for 16-Kbps applications (Jayant and Ramamoorthy 1986; Yatsuzuka, Iizuka, and Yamazaki 1986).

## 8.7 Communications-Quality Speech Coding at 8 to 4 Kbps

The bit-rate range of 8 to 4 Kbps is an important area of investigation. It is very significant for emerging applications such as narrow-band cellular radio telephony. The realization of robust communications quality at these bit rates is a challenging problem for speech coding research.

A good fundamental basis for these low-bit-rate coders is the speech production model shown in Fig. 8.7(A). The excitation and filter characteristics are assumed to be periodically updated, say once every 10 milliseconds. The idea is that such a description is compact enough to lead to low-bit-rate representations and that, at the same point, it is rich enough to grasp perceptually essential properties of the speech waveform. While the model looks very straightforward in principle, the optimization of the time-varying excitation and filter parameters, as well as the efficient quantization of the optimized parameters, are complex tasks.

**Fig. 8.7. Models for speech production and excitation.**

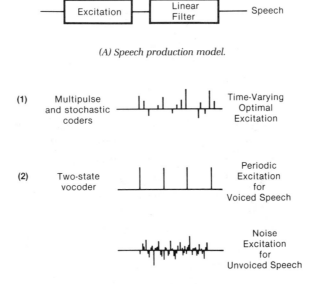

*(A) Speech production model.*

*(B) Speech excitation model.*

Depending on the techniques used to convey the excitation and filter parameters of Fig. 8.7(A), a variety of coding configurations in the bit-rate range of 16 Kbps to 0.5 Kbps can result. Of particular interest is the 8- to 4-Kbps range, which promises interesting levels of communications quality.

## 8.7.1 Coding of the Linear Filter Parameters

The optimal filter is essentially the low-resolution envelope shown in Fig. 8.3(A1, B1). The shape of this filter can be specified either explicitly in the frequency domain or by equivalent descriptions in the time domain. In particular, the optimized parameters of linear prediction, which are matched to the autocorrelation function (acf) of speech by definition, also follow the shape of the Fourier transform of the acf, which is the power spectral density. This observation is the basis of *linear predictive coding* (Atal and Hanauer 1971; Markel and Gray 1976; Makhoul 1975; Rabiner and Schafer 1978).

The linear prediction filter is very similar to the prediction filter in DPCM (Fig. 8.4). An important difference is that for low-bit-rate work it is critical to convey predictor information much more carefully than in medium bit-rate DPCM coding. In the DPCM situation, the instantaneous sample-by-sample corrections provided by the closed-loop coding mechanism tend to compensate for the deficiencies of a poorly specified predictor. The coders we are discussing in this section have, at best, a periodic corrective action, say once every 10 milliseconds.

The bit rate needed for the representation of the linear filter parameters depends on the order of predictor and the efficiency of the quantizing system. A traditional bit rate for this filter is 2 Kbps. Use of vector quantization (section 8.7.3) can lead to significantly lower numbers, such as 1 Kbps or even 500 bps.

## 8.7.2   Coding of the Excitation Parameters

The optimal time-varying excitation depicted in Fig. 8.7(B1) is best derived in a closed-loop search procedure. This procedure provides, once for every parameter-updating period, the best possible excitation for a specified linear filter. The linear filter is also periodically updated to follow the envelope of the short-term power spectrum shown in Fig. 8.3(A1, B1). The search procedure for optimizing the excitation is driven by a perceptually weighted version of the error between the input speech block and a current, synthesized version of it. The perceptual weighting is qualitatively similar to the frequency weighting of errors by the human ear, as implied in the discussion of Fig. 8.5.

The bit rate needed to convey excitation information depends on the time-resolution for the impulses in Fig. 8.7(B1) and on the efficiency of the quantizing system. A comfortable bit rate for the excitation information is in the order of 8 Kbps. Use of vector quantization can result in much lower bit rates, such as 4 Kbps or even 2 Kbps.

## 8.7.3   Vector Quantization

The conventional quantization of a vector of $N$ samples requires $N$ operations of scalar (sample-by-sample) quantization. As in equation (8.2), if each of these operations uses a quantizer with $L$ levels, the information rate is $R = \log_2 L$ bits/sample, and $NR$ bits/vector. In vector quantization, the input vector is quantized all at once rather than sample by sample (Jayant and Noll 1984; Gersho 1977; Gray 1984; Makhoul, Roucos, and Gish 1985). In the simplest case, the quantization is based on a codebook with a number of entries $J$ given by

$$J = 2^{NR}; \quad R = (\log_2 J)/N \tag{8.3}$$

An input vector is compared with each of the $J$ candidate outputs; the closest match to the input is determined on the basis of a suitable fidelity criterion, and this best match is identified for coding by an index in the range of 1 to $J$.

For a given average reconstruction error, the preceding procedure provides lower bit rates than scalar quantization in general. This is true, in particular, for vectors with correlated samples. The price paid for this bit-rate reduction is an increase in computational complexity, storage of candidate vectors, and encoding delay. Vector quantization also provides the flexibility of a nonintegral number for the bits per sample. In contrast to the scalar case of equation (8.2), the division by $N$ in equation (8.3) implies, in general, a noninteger value for $R$.

Crucial to the efficiency of a vector quantizer is the choice of the $J$ candidate vectors. Several deterministic and stochastic procedures are known for populating the *codebook* of a vector quantizer.

Closely related to vector quantization are the sequential, or nonblock, techniques of *tree coding* and *trellis coding* (Jayant and Noll 1984; Fehn and Noll 1982). The deployment of these techniques for communications-quality coding is also an important topic of research.

## 8.7.4 Speech Coding at 8 Kbps

A good candidate for the 8-Kbps bit rate is *multipulse LPC* (MP-LPC) (Atal and Remde 1982). Here, a suitable number of pulses are specified to constitute the excitation sequence for a given speech frame; and the amplitudes and locations of these pulses are optimized, pulse by pulse, in a closed-loop search procedure (Atal and Remde 1982). The bit rate reserved for the excitation information is ideally well in excess of one-half of the total bit rate. If vector quantization is used to convey LPC filter information, more bits will be released to encode the excitation sequence. This will permit more liberal designs for the number of excitation pulses and the accuracy with which these pulses are quantized. The design of multipulse systems is fairly well understood for bit rates of 16, 9.6, and 8 Kbps (Ozawa and Araseki 1986).

## 8.7.5 Speech Coding at 4 Kbps

At a bit rate of 4 Kbps, the information rates allocated for the excitation and filter parameters are likely to be nearly equal. Coding of the excitation sequence with about 2 to 3 Kbps requires a vector quantization approach. This is indeed the approach used in code-excited LPC, or *stochastically excited LPC* (SE-LPC) (Schroeder and Atal 1984). Here, the codebook of candidate excitations is populated by suitable stochastic, or random, sequences, and the best candidate sequence is selected again on the basis of a closed-loop search procedure. If vector quantization is used for the LPC filter as well, as is likely for the total bit rate of 4 Kbps, more bits will be available for conveying excitation information. This will permit a larger number of candidates in the codebook, as well as a larger number of elements in the excitation vector. The design of stochastically excited LPC systems for 4 to 8 Kbps is an important focus of coding research.

# 8.8  Synthetic-Quality Speech Coding at 2 Kbps

An extreme example of the coding philosophy of section 8.7 is a linear predictive *vocoder*. Historically, in fact, this vocoder precedes the types of systems discussed in section 8.7. The vocoder continues to be an important technique for digital voice security over narrow-band channels that can reliably support only about 2 Kbps of digital input.

Fig. 8.7(B2) defines the minimal-complexity excitation model of LPC vocoding. This is a *two-state* model, with a periodic pulse excitation for voiced speech and a noise excitation for unvoiced speech. Specification of this excitation signal requires very little information—a voiced-unvoiced decision needing 1 bit of information per frame, a pitch-period value once per frame, and a single excitation-gain parameter per frame. The low information rate of the excitation signal permits a total bit rate not much higher than 2 Kbps, even without the use of vector quantization.

The traditional LPC-10 vocoder with a tenth-order filter runs at a bit rate of 2.4 Kbps (Tremain 1982). The *word intelligibility* of the system can be quite impressive, but the highly simplified excitation results in a synthetic output quality and a loss of speaker recognizability. The use of vector quantization for conveying filter information can provide even lower bit rates such as 1 Kbps or less with useful levels of intelligibility.

# 8.9  Speech Quality, Bit Rate, and Coder Complexity

We are now ready to summarize trade-offs in speech coding, in terms of parameters that represent speech quality, bit rate, and coder complexity.

## 8.9.1  Quality versus Bit Rate

Fig. 8.8 is a perspective of quality versus bit rate for telephone-bandwidth speech, and it includes all the coder classes discussed in this chapter (Flanagan et al. 1979; Daumer 1982). The vertical axis in the figure has been calibrated with the adjectival description defined in section 8.2.2. The horizontal axis assumes a standard telephone bandwidth of 3.2 kHz and a standard sampling rate of 8 kHz. It also assumes that silences in speech are included as part of the coder input. In systems where silent segments can be eliminated prior to coding, the effective bit rates can be as low as 70% of the values in Fig. 8.8 for the same levels of speech quality.

Fig. 8.8. Speech quality, bit rate, and coder complexity.

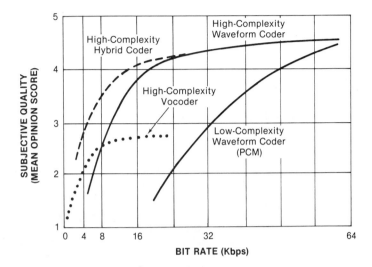

The *low-complexity PCM coder* of section 8.4 attains an MOS value well in excess of 4.0 at 64 Kbps but degrades rapidly with decreasing bit rate. The *medium-complexity coder* of section 8.5 provides high-quality coding at 32 Kbps. The quality-rate curve for this coder would be in-between the two solid curves in Fig. 8.9. The *high-complexity waveform coder* represents the noise-shaping and bit-allocation coders of section 8.6. These algorithms provide an MOS value close to 4.0 at 16 Kbps.

The *high-complexity vocoder* of section 8.8 has a quality-rate perform-ance shown by the dotted line. At a bit rate such as 2 Kbps, where even the high-complexity coder is totally degraded, the vocoder maintains a useful level of performance. The dashed-line characteristic is an expected result from the high-complexity coders of section 8.7. Note that the com-munications quality of these coders is reflected by an MOS value well in excess of 3.0 at 8 Kbps, and a value well in excess of the vocoder MOS even at 4 Kbps. These coders are called *hybrid coders* because they com-bine a vocoder-motivated speech production model—Fig. 8.7(A)—with a closed-loop error minimization philosophy reminiscent of DPCM wave-form coding—Fig. 8.4.

## 8.9.2  Coder Complexity and Delay

The important role of coder complexity is emphasized in Table 8.2. Most numbers in the table are order-of-magnitude estimates, rather than strictly accurate values. The first three rows of the table refer to high-

quality waveform coders, the fourth and fifth rows refer to communications-quality hybrid coders, and the last row refers to a synthetic-quality
vocoder.

TABLE 8.2.
TRADE-OFFS IN
DIGITAL
SPEECH
CODING

| Coder Type | Bit Rate (Kbps) | Computational Complexity[a] (mips) | Communication Delay[a] (ms) |
|---|---|---|---|
| PCM | 64 | $0.0^+$ | 0 |
| ADPCM | 32 | 0.1 | $0^+$ |
| ASBC | 16 | 1.0 | 25 |
| MP-LPC | 8 | 10.0 | 35 |
| SE-LPC | 4 | 100.0 | 35 |
| LPC | 2 | 1.0 | 35 |

[a]The numbers in the third column are order-of-magnitude estimates that characterize specific coder configurations that the author is familiar with. Other, somewhat different,
configurations can have lower or higher values of complexity and delay. For example, an
efficient version of SE-LPC may have a complexity of 50 mips. The numbers in the fourth
column assume efficient quadrature mirror filtering and 4 to 6 subbands in the ASBC system, and a frame length of 10 ms for the three LPC coders. An LPC system with a longer
frame length and with subsystems for parameter interpolation and error protection, such
as the LPC-10 standard in Tremain (1982), can have a total delay in excess of 100 ms.

The computational complexity is measured by millions of instructions
per second (*mips*), anticipating digital signal processor implementations.
It is interesting that in the first five rows of the table, the computational
complexity tends to increase by 10 for every decrease of bit rate by a factor of 2.

In many applications, an algorithm is deemed practical if it can be
implemented on a single signal processor chip. A computational complexity of 10 mips represents the current maximum capability of general-
purpose signal processors. In this sense, the SE-LPC coder is the only
coder in Table 8.2 that cannot be implemented on a single signal processor. A good deal of the storage and computational complexity of this
coder resides in the vector quantization subsystem for determining the
optimum excitation. This subsystem depends, in turn, on the repeated
use of a single primitive function, the computation of the distance
between an input vector and a candidate vector. Because of this repeti-
tive characteristic, the vector quantization subsystem can be realized
effectively on dedicated special-purpose processor chips.

The last column in Table 8.2 shows the one-way communication
delay, the delay due to one stage of encoding and decoding. Coder costs
tend to come down with advances in signal processing technology. But
the delay required by a sophisticated coding algorithm is a fundamental
property of the algorithm, and it can restrict the classes of applications
for which a given coding system is appropriate, especially when the

delay performance of the speech coder is evaluated in the context of other system parameters such as network delays and uncancelled echoes in telephony.

## 8.10 Summary

Starting with a broad statement of the digital coding problem, we described the speech signal and pointed out characteristics of the signal that are relevant to efficient digitization. We then went on to discuss classes of coding algorithms for digitizing speech at successively lower bit rates. We explained how the lower bit-rate algorithms maintained useful levels of speech quality by making use of the properties of the speech signal and the characteristics of speech perception. In section 8.9, we used impressionistic results (Table 8.2 and Fig. 8.8) in an attempt to quantify information that is relevant to speech system designers.

## 8.11 Epilog

The future of human communication
Revolves around an ancient question:
If speech were digits, quality unlessened,
What would that cost in bits per second?

Sixty-four thousand is the fashion still,
But much too often, it is an overkill;
Thirty-two thousand can do the trick.
Sixteen- will soon be ready to pick.

Eight-K is hard for waveform coding,
Despite noise-shaping and water-filling;
Vocoding is frugal but alas! synthetic,
Much too delicate for a real-time critic.

But hybrid coders are full of promise,
Especially those with a vector or trellis;
We are on the right track, we expect;
*Low bit rate* may be *high quality* yet!

## 8.12 References

Aaron, M. R. 1979. The digital (R) evolution. *IEEE Communications Magazine* (January): 21–22.

Ahmed, N., and K. R. Rao. 1975. *Orthogonal transforms for digital signal processing*. New York: Springer-Verlag.

Atal, B. S. 1982. Predictive coding of speech at low bit rates. *IEEE Transactions on Communications* (April): 600–14.

Atal, B. S., and S. L. Hanauer. 1971. Speech analysis and synthesis by linear prediction. *Journal of the Acoustical Society of America* 50 (August): 637–55.

Atal, B. S., and J. R. Remde. 1982. A new model of LPC excitation for producing natural-sounding speech at low bit rates. *Proceedings of International Conference on Acoustics, Speech and Signal Processing*: 614–17.

Beker, H., and F. Piper. 1982. *Cipher systems—The protection of communications*. New York: Wiley.

Bellamy, J. C. 1982. *Digital telephony*. New York: Wiley.

CCITT Report. *See* Study Group XVIII.

Crochiere, R. E. 1978. Sub-band coding. *Bell System Technical Journal* (October): 2927–52.

Crochiere, R. E., R. V. Cox, and J. D. Johnston. 1982. Real time speech coding. *IEEE Transactions on Communications* (April): 621–34.

Cox, R. V. 1986. Quadrature mirror filters for speech and audio processing. *Proceedings of the International Symposium on Circuits and Systems* in San Jose (May): 285–288.

Daumer, W. R. 1982. Subjective evaluation of several efficient speech coders. *IEEE Transactions on Communications* (April): 655–62.

Elliott, D. F., and K. R. Rao. 1982. *Fast transforms—Algorithms, analyses and applications*. New York: Academic Press.

Esteban, D., and C. Galand. 1972. Application of quadrature mirror filters to split band and voice coding schemes. *Proceedings of the International Conference on Acoustics, Speech and Signal Processing* (May): 191–95.

Fehn, H. G., and P. Noll. 1982. Multipath search coding of stationary signals with applications to speech. *IEEE Transactions on Communications* (April): 687–701.

Flanagan, J. L. 1972. *Speech analysis, snythesis and perception*. New York: Springer-Verlag.

Flanagan, J. L., M. R. Schroeder, B. S. Atal, R. E. Crochiere, N. S. Jayant, and J. M. Tribolet. 1979. Speech coding. *IEEE Transactions on Communications* (April): 710–37.

Gersho, A. 1977. Quantization. *IEEE Communications Magazine* (September): 16–29.

Gray, R. M. 1984. Vector quantization. *IEEE-ASSP Magazine* (April): 4–29.

Honda, M. and F. Itakura. 1984. Bit allocation in time and frequency domains for predictive coding of speech. *IEEE Transactions on Acoustics, Speech and Signal Processing* (June): 465–73.

Jayant, N. S. 1974. Digital coding of speech waveforms—PCM, DPCM and DM quantizers. *Proceedings of IEEE* (May): 611–32.

____, ed. 1976. *Waveform quantization and coding*. New York: IEEE Press.

_____. 1982. Analog scramblers for speech privacy. *Computers and Security*: 275–89.

Jayant, N. S., and P. Noll. 1984. *Digital coding of waveforms—Principles and applications to speech and video*. Englewood Cliffs, NJ: Prentice-Hall.

Jayant, N. S., and V. Ramamoorthy. 1986. Adaptive postfiltering of 16 kb/s ADPCM speech. *Proceedings of the International Conference on Acoustics, Speech and Signal Processing*, in Tokyo (April): 16.4.1–4.4.

Makhoul, J. 1975. Linear prediction: A tutorial review. *Proceedings of IEEE* (April): 561–80.

Makhoul, J., S. Roucos, and H. Gish. 1985. Vector quantization in speech coding. *Proceedings of the IEEE* (November): 1551–88.

Markel, J. D., and A. H. Gray. 1976. *Linear prediction of speech*. Berlin: Springer-Verlag.

Odenwalder, J. P. 1985. Error control. Chapter 10 in *Data communications, networks and systems*. Edited by T. C. Bartee. Indianapolis: Howard W. Sams.

Ozawa, K., and T. Araseki. 1986. High quality multipulse speech coder with pitch prediction. *Proceedings of the International Conference on Acoustics, Speech and Signal Processing*, in Tokyo (April): 33.3.1–3.4.

Rabiner, L. R., and R. W. Schafer. 1978. *Digital processing of speech signals*. Englewood Cliffs, NJ: Prentice-Hall.

Schroeder, M. R., and B. S. Atal. 1984. Code-excited linear prediction (CELP): High-quality speech at very low bit rates. *Proceedings of the International Conference on Communications* (June): Amsterdam.

Soong, R. K., R. V. Cox, and N. S. Jayant. 1986. A high quality sub-band speech coder with backward adaptive predictor and optimal time-frequency bit assignment. *Proceedings of the International Conference on Acoustics, Speech and Signal Processing*, in Tokyo (April): 44.8.1–8.4.

Study Group XVIII. 1983. CCITT report. Temporary document, draft recommendation 6.7zz. November 21–25, Geneva.

Tremain, T. E. 1982. The government standard linear predictive algorithm: LPC-10. *Speech Technology* (April): 40–9.

Yatsuzuka, Y., S. Iizuka, and T. Yamazaki. 1986. A variable rate coding by APC with maximum-likelihood quantization from 4.8–16 kb/s. *Proceedings of the International Conference on Acoustics, Speech and Signal Processing*, in Tokyo (April): 56.12.1–12.4.

Zelinski, R., and P. Noll. 1977. Adaptive transform coding of speech signals. *IEEE Transactions on Acoustics, Speech and Signal Processing* (August): 299–309.

# 9

# VIDEO TELECONFERENCING

## Richard C. Harkness

*Teleconferencing* (TC) is the use of telecommunications to link physically separated individuals or groups for purposes of holding a meeting. The technology can range from a simple speakerphone to elaborate custom-designed teleconference rooms equipped with two-way audio, graphics, and video.

Teleconferencing is of interest because it promises to be faster, cheaper, and more convenient than travelling to a meeting location. Besides saving travel time and cost, TC is expected to improve communications and thereby improve both productivity and effectiveness.

Teleconferencing has been around for twenty years, but only since 1980 have costs started to fall greatly. This decrease in costs has spurred a dramatic increase in interest during the last two years. Today many corporations and government agencies are implementing TC systems.

This chapter provides a three-part overview of teleconferencing, but with more emphasis on technology and economics than is usually found. Part I covers nontechnical aspects, including the history of video teleconferencing and research findings. Part II describes premise equipment such as the teleconferencing room and the audio, video, and graphics subsystems. Part III discusses transmission requirements and options.

Good literature on the field is scarce and often obscure, so references are provided. For major books in the field, the reader is referred to: Johansen, Vallee, and Spangler 1979; Kelleher and Cross 1985; Lazer et al. 1983; and Olgren and Parker 1983. For newsletters, see *Telecoms* and *Telespan*.

## PART I

## NONTECHNICAL ASPECTS

The following sections will provide an introduction to video teleconferencing, including definitions of the different forms and the history and benefits of teleconferencing.

# 9.1 Concepts and Definition

Teleconferencing employs electronic communications to facilitate information exchange between physically separated groups. This chapter deals with real time, two-way, audio-video systems that link two (or a few) specially equipped teleconference rooms, often hundreds of miles apart. In today's evolving terminology, this is called interactive or two-way *video teleconferencing,* or *VTC.*

To put VTC in perspective, we note that teleconferencing can be real time or asynchronous, can be one way or two way, and can use one or more of the following technologies: audio, graphics, or video. Let's briefly explain these options.

*Asynchronous conferencing* generally implies a form of computer messaging and is a field unto itself. *Audio conferencing* is perhaps the simplest, lowest-cost form of teleconferencing. The terminal devices can range from speakerphones to custom-designed audio systems in acoustically treated rooms. Networks range from two-point to multipoint conferences established through special conference "bridges" linking up to several hundred locations.

*One-way video* or *"ad hoc" teleconferencing* involves video transmission from a central studio, usually via satellite, to many remote, receive-only locations, often in hotels. This subfield is very different in terms of technology, costs, and applications from two-way video conferencing, so much so that the trade press's inclusion of both under the same label "video teleconferencing" creates much confusion.

*Audio graphics* is the next step up from audio in terms of capability and cost. The objective is to supplement verbal discussions with the exchange of charts, diagrams, documents, or freehand sketches. A variety of devices, including facsimile, electronic blackboards, and still-frame video, can be used.

*Video systems* include the added dimension of real-time motion video. Besides providing cues to help manage verbal exchange, video conveys body language, expression, and feelings of presence and involvement.

The remainder of this chapter deals exclusively with *motion video conferencing.* See Johansen, Vallee, and Spangler 1979; Kelleher and

Cross 1985; and Olgren and Parker 1983 for more details on the other options for teleconferencing.

## 9.2 History

With the exception of isolated demonstrations, video teleconferencing began in the mid-1960s when several large New York City banks implemented analog systems to link their major offices in Manhattan. Quite separately, AT&T began development of a desktop "Picturephone" telephone about 1967. Picturephone was a technical triumph, but a marketing failure. After trials in Pittsburgh, it slowly died about 1970 from lack of demand. AT&T then turned its attention to a service called Picturephone Meeting Service (PMS), which utilized specially equipped teleconference rooms located in central AT&T city offices. These were available to the public on a walk-in basis.

For about $2500 per hour, a group in New York could converse with a group in San Francisco. This service never attracted high usage and was eventually terminated in 1985. Some of the reasons given for its failure were inconvenience, high costs, and inadequate marketing. Meanwhile, a more promising route to success was being pioneered by Dow Chemical, who in the late 1970s established a private system connecting its Michigan headquarters with a large plant in Texas. Only analog techniques were available at the time, and transmission costs were about $2000 per hour. This system lasted several years and pioneered the concept of private, intracompany teleconferencing networks. Private systems gained momentum when organizations like Satellite Business Systems and American Satellite Corporation took over promotional activity in the early 1980s from a temporarily discouraged AT&T. By 1982 Arco, Aetna, American General, Allstate, and several other major firms had successful full-motion systems; at the same time, others, notably IBM, had successful still-frame systems. Transmission had evolved from the expensive analog broadcast TV backup channels used by Dow, to compressed digital transmission at 3 Mbps (Picturephone Meeting Service), and then to satellite transmission at 1.5 Mbps.

By the end of 1985, video teleconferencing was being used by several dozen organizations. NASA, the Department of Defense, and the aerospace community were aggressively implementing systems. End-user articles extolling the benefits of VTC were becoming common in the trade press.

Although itself limited to a common carrier role, SBS stimulated the emergence of firms able to provide complete "turnkey" teleconferencing rooms. Such firms are critical, since room design is sufficiently complex to require such services. (See Brown et al. 1980; McFarlane and Nissen 1983; Midorikawa et al. 1975; and Wilkens and Plenge 1981.) Video codec technology was another key pacer. In 1980 NEC introduced the first commercial codec. This codec compressed video to 6 Mbps, and later improvements compressed the video to 3 Mbps. In 1982 CLI introduced a codec capable of performing well at 1.5 Mbps. (See section 9.11 of this chapter and also Bartee 1985 for a discussion of codecs.)

Most of the research into the effectiveness of VTC versus face-to-face meetings was conducted between 1971 and 1976, principally by Communications Studies and Planning (CS&P) in London (Pye and Williams 1977). CS&P had been funded by the British Government to determine whether teleconferencing might mitigate any "communications damage" resulting from a plan designed to decentralize government agencies away from London (and thus ease traffic congestion). CS&P's pioneering work helped establish on theoretical grounds what has since been found in practice; namely, that teleconferences are an effective substitute for some, but not all, face-to-face meetings. Other human factor aspects have been researched by the Institute for the Future, the University of Wisconsin, Bell Canada, the Annenberg School at USC, and elsewhere. Many of the results are summarized in Johansen, Vallee, and Spangler 1979; Kelleher and Cross 1985; and Olgren and Parker 1983. At such annual "gatherings of the clan" as the University of Wisconsin's spring conference and Telecon, researchers have stressed to systems vendors and end users the importance of human factors and, probably most important, the need for thoughtful user orientation.

Marketing has been, and still is, the key to developing this market. Over the last five years vendors have promoted the concept and explained its benefits through sales calls, conference presentations, and articles. This investment is now paying off with fairly widespread recognition of the concept. Better yet, success stories have begun emanating from end users, bringing hope that this will become a user-driven rather than a supplier-driven market.

Internal marketing is also very important. The success of any particular VTC system depends heavily on the existence of a champion within the end-user organization to promote and educate. The importance of such an individual, who usually gains considerable exposure to top management, cannot be overemphasized.

Altogether, VTC has moved from blue sky concept in the 1960s to emerging industry in the mid-1980s. The chicken and egg nature of multiparty dependencies, where each takes a small step and then waits for the others, accounts for the slow pace at which this field has developed.

As in any other technology-based industry, such as aviation or office automation, the necessary enabling conditions are building toward the critical mass needed for truly rapid growth. The concept and its utility have been verified by pioneering users; the vendors, services, and basic technologies are in place; and the awareness level is adequate. Given the current technology, services, and costs, the field is ready to mature. The main factor that could accelerate growth is a further significant decline in transmission costs. Fortunately that is on the horizon.

## 9.3    Statistics for Business Meetings

Insofar as VTC may serve as a substitute for certain types of business meetings, statistics about meetings provide insight into VTC requirements. Tables 9.1, 9.2, and 9.3 tabulate statistics for duration, distance, number of attendees, and costs for a large manufacturing company. The company's travellers were given comprehensive questionnaires to complete during business trips (Harkness and Burke 1982). Completed in 1979, the survey sampled roughly 760 travellers. Its results correlate fairly closely with similar surveys, using identical questionnaires, that were conducted in another large manufacturing company, an insurance company, and a California bank. Thus, the results are probably representative of Fortune 500 companies.

| TABLE 9.1. MEETING STATISTICS | | |
|---|---|---|
| Duration | 3.9 | hours (average) |
| Duration | 3.0 | hours (median) |
| Total attendees | 5.9 | (average) |
| Attendees who travelled | 3.0 | (average) |
| Percentage at company locations | 40% | |
| Percentage at which all attendees were company employees | 42% | |
| Lost time per meeting hour | 1.6 | hours/travelling attendee (average) |
| Percentage using visuals: | 94% | |
| Typed sheets | 60% | |
| Viewgraphs | 58% | |
| Sketches | 26% | |
| Computer reports | 17% | |
| Reports and books | 14% | |
| Engineering drawings | 9% | |
| Blackboard | 28% | |
| Flip charts | 24% | |
| Photos and slides | 15% | |
| 3-D objects | 7% | |
| Other | 9% | |

<table>
<tr><td>**TABLE 9.2. TRIP STATISTICS**</td><td>Distance</td><td>900 straight-line miles (average)</td></tr>
<tr><td></td><td>Number of business meetings per trip</td><td>1.5 (average)</td></tr>
<tr><td></td><td>Trips to one location only</td><td>85%</td></tr>
<tr><td></td><td>Cost (1985 $) per traveller per trip</td><td>$660 (approximate average)</td></tr>
</table>

<table>
<tr><td>**TABLE 9.3. MEETING COSTS**</td><td>Travel cost per meeting</td><td>$1300 (average)</td></tr>
<tr><td></td><td>Travel cost per meeting hour</td><td>$ 350 (average)</td></tr>
<tr><td></td><td>Cost in lost time per meeting hour</td><td>$ 170[a] (average)</td></tr>
<tr><td></td><td>Total cost per meeting hour</td><td>$ 520/hour (average)</td></tr>
</table>

[a]Assuming a value of $30 per hour for each person's time.

*Source for Tables 9.1, 9.2, and 9.3*: Richard Harkness and P. Burke, "Estimating Teleconferencing Travel Substitution Potential in Large Business Organizations: A Four Company Review," *Proceedings of Conference on Teleconferencing and Electronics Communications*, University of Wisconsin Extension, 1982.

The average trip cost of $660 quoted in Table 9.2 does not come directly from the survey but rather is a separate estimate based on a 900-mile-per-trip distance. It assumes that route miles are 1.3 times straight-line miles and incorporates a Runzheimer International finding that air travel constitutes 44% of total costs for business trips. It is also based on average airline revenue, which has been about 12.5 cents per passenger mile.

## 9.3.1  Implications

The need for a VTC graphics capability is apparent from the data in Table 9.1, since graphics were used in 94% of the meetings. IBM found that a graphics camera was used in 90% of its teleconferences.

The duration of a video teleconference has been reported as being significantly shorter than a comparable face-to-face meeting, typically 30 to 50% shorter (Green and Hansell 1984; Kelleher and Cross 1985, p. 52). Ninety minutes is frequently reported as an average, along with the observation that attendees tend to come better prepared and waste less time on small talk.

Although the average cost of $520 per hour for face-to-face meetings is a rough approximation, it does indicate the cost range that VTC must reach to become an economical substitute for most meetings. It is important, however, to remember that video teleconferences are shorter. If they are 33% shorter, then any VTC costing less than $780 per hour is less expensive than meetings costing $520 per hour. The $520-per-hour cost is, as stated, an average. Actual costs vary widely enough that VTC can find some market at prices well above $780 per hour. A first class coast-to-coast trip to attend a one-hour meeting approaches one cost extreme, whereas travelling from New York to Washington for a week of business meetings represents the other. Obviously the more people who need to travel, the more cost effective VTC becomes.

In justifying VTC, today's VTC users rightly consider productivity benefits even more valuable than travel savings—thereby sometimes justifying VTCs that are even more expensive than travel. However, the author believes that the VTC market will reach its potential only when VTC is inexpensive enough that end users automatically view it as being cheaper than travel. Since the cost of travel is increasing while telecommunications costs are dropping, this circumstance seems inevitable in the near future. Ultimately it comes down to the relative cost of sending digital pulses through glass strands versus the cost of transporting people.

# 9.4 VTC versus Face-to-Face Meetings

This section and the following one (9.5) outline what is known about VTC as a substitute for face-to-face meetings and for travel. The major points are briefly summarized here:

- Experimental research indicates that at least one-half of all face-to-face business meetings could be replaced by teleconferences with no loss in communications effectiveness.
- Attitudes toward VTC, and toward VTC as a substitute for travel, vary widely.
- Users report that specific teleconferences do in fact substitute for specific trips.
- Several large organizations regularly track and report to management the travel dollars displaced by VTC.
- No one has proven that implementing teleconferencing reduces *net* corporate travel costs.
- Even if VTC does not reduce net travel costs, most user organizations are probably unconcerned, since there are offsetting advantages to VTC.

## 9.4.1 Attitude Research

One of the basic questions about teleconferencing has been its suitability as a substitute for face-to-face meetings. Over the years, information on this question has been gathered both with attitudinal research, in which people are asked how they feel about teleconferencing, and with laboratory research, in which experiments measure any differences in a group's ability to conduct certain tasks using audio, video, and face-to-face meetings. In the last two years, this early research, based on subjects with little or no actual VTC experience, has been superseded by studies of actual long-term VTC users.

At least as relevant as substitution is the question of how VTC could be used to supplement, rather than substitute for, travel. It has been suggested, for example, that video teleconferences be held before face-to-face meetings to set the agenda and identify the issues, and afterward to follow up on action items. A careful reading of the VTC-related research into media differences suggests other more subtle possibilities based on exploiting those differences. In a word, VTC is not necessarily an inferior substitute for face-to-face meetings, one to be used only if it is less expensive or more convenient.

Johansen in his book *Electronic Meetings* summarizes the attitudinal research (Johansen, Vallee, and Spangler 1979, Appendix A). The following excerpts from his book typify not only the findings but the manner in which they are stated.

> "Video is perceived as better than audio for interpersonal relations" (Champness 1972a).

> "People are generally more confident in their perceptions of others via video than via audio, but not necessarily more accurate" (Reid 1970).

> "In a survey of Bell Laboratories personnel only 3 percent of travellers would be willing to substitute a system which did not provide moving picture video" (Snyder 1973).

> "Video is perceived as questionable for getting to know someone, bargaining, and persuasion" (Champness, Short 1973).

Results from the study conducted by A. Michael Noll at Bell Laboratories are particularly interesting in showing where the media are most similar and most different. Fig. 9.1 summarizes this study.

Attitudes toward travel itself are another factor. One large study done in 1973 showed that 15% of business travellers want to travel more, 38% want to travel less, and the remainder are content (Kollen and Garwood 1975). Another study showed that those who travelled most frequently were those who most wanted to reduce travel.

## 9.4.2  User Surveys

The preceding findings all deal with forecasts based on early experiments or opinions. Now we have actual end-user experience to rely on.

Fig. 9.1.
Attitudes
toward different
media from
survey of Bell
Labs personnel.

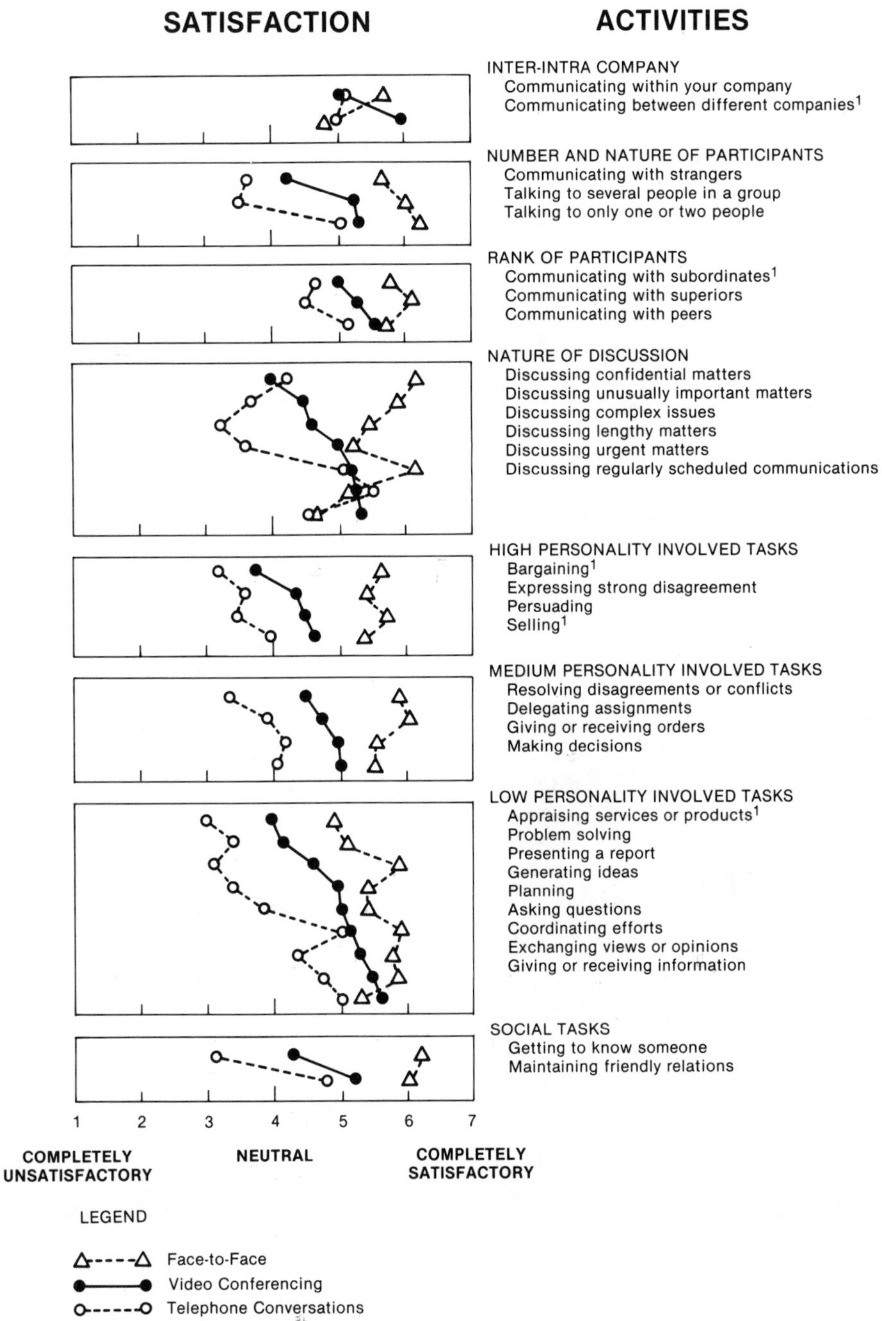

**SATISFACTION**

**ACTIVITIES**

INTER-INTRA COMPANY
  Communicating within your company
  Communicating between different companies[1]

NUMBER AND NATURE OF PARTICIPANTS
  Communicating with strangers
  Talking to several people in a group
  Talking to only one or two people

RANK OF PARTICIPANTS
  Communicating with subordinates[1]
  Communicating with superiors
  Communicating with peers

NATURE OF DISCUSSION
  Discussing confidential matters
  Discussing unusually important matters
  Discussing complex issues
  Discussing lengthy matters
  Discussing urgent matters
  Discussing regularly scheduled communications

HIGH PERSONALITY INVOLVED TASKS
  Bargaining[1]
  Expressing strong disagreement
  Persuading
  Selling[1]

MEDIUM PERSONALITY INVOLVED TASKS
  Resolving disagreements or conflicts
  Delegating assignments
  Giving or receiving orders
  Making decisions

LOW PERSONALITY INVOLVED TASKS
  Appraising services or products[1]
  Problem solving
  Presenting a report
  Generating ideas
  Planning
  Asking questions
  Coordinating efforts
  Exchanging views or opinions
  Giving or receiving information

SOCIAL TASKS
  Getting to know someone
  Maintaining friendly relations

1    2    3    4    5    6    7

COMPLETELY          NEUTRAL          COMPLETELY
UNSATISFACTORY                       SATISFACTORY

LEGEND

△----△   Face-to-Face
●———●   Video Conferencing
○-----○   Telephone Conversations

[1]*Occurred "Never" for 75% or more of the respondents.*
*Source: A. Michael Noll, "Teleconferencing Communications Activities," Bell Laboratories.*

For example, after evaluating their pilot still-frame TC system, IBM published the following user responses (Hagopian 1981):

Right media for conduction of business:
| | |
|---|---|
| Excellent | 16% |
| Good | 52% |
| Fair | 29% |
| Poor | 3% |

Teleconferencing's greatest advantage:
| | |
|---|---|
| Save time | 36% |
| Save dollars | 29% |
| Convenience | 22% |
| Can involve more people | 7% |
| Faster response | 6% |

Accomplishment of meeting objectives as compared with face-to-face meeting:
| | |
|---|---|
| Very satisfied | 42% |
| Moderately satisfied | 46% |
| Moderate to very dissatisfied | 11% |

Would you use TC again for this type of meeting?
| | |
|---|---|
| Yes | 78% |
| No | 22% |

If TC were not available for this meeting, would you or others have travelled?
| | |
|---|---|
| Yes | 66% |
| No | 33% |

The preceding IBM user survey is one of several studies showing that VTC is indeed acceptable as a travel substitute in many, but certainly not all, instances. The answer to the last question in the survey regarding travel is indicative of the percentage of VTC demand that may come from travel substitution.

## 9.4.3  Laboratory Research

The seminal research into the effectiveness of TC versus face-to-face meetings was conducted by Communications Studies and Planning in London in the early 1970s. CS&P conducted a wide variety of experiments to explore the differences between audio conferencing, motion video conferencing, and face-to-face meetings. Typically, individuals or groups were asked to solve problems, negotiate, or complete other meeting objectives when using these three media. Although the results were voluminous and complex, Fig. 9.2 summarizes many of the key findings (Pye and Williams 1977).

Fig. 9.2.
Classification of meetings according to their suitability for various media by Communications Studies Group, London (Pye and Williams 1977).

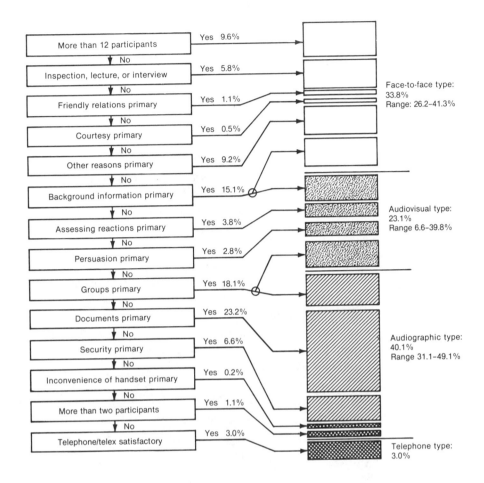

At the time the work was done, meetings having a relatively high emotional or personal content, such as job interviews, were felt to be less suitable for VTC than problem solving and information exchange meetings. Authors of the CS&P report have subsequently stated that they were perhaps too conservative in this respect. AT&T has successfully interviewed prospective employees over their two-way video system, and VTC has been embraced by marketeers as a vehicle for giving clients access to experts at the home office. In Canada lawyers are presenting cases to the Supreme Court via teleconferencing. Only long experience will guide the proper choice of media in delicate situations, just as experience guides the choice of a letter, a phone call, or a visit. In sum, we don't know exactly how many meetings could be replaced by teleconferences, but the best available evidence suggests at least 50%.

# 9.5 Travel Substitution

No one knows if telephones have increased or decreased business travel, or the mail for that matter. To ask if VTC will reduce business travel raises a similar issue and is probably the wrong question. In a budget squeeze or energy crisis, it could be made to do so. In normal times VTC becomes another business tool to be used when appropriate. Because the cost structure and characteristics of VTC are different from travel or telephone, there are instances in which VTC will be preferable to either of these traditional "media." For example, VTC is instantaneous, it has little fixed cost per event, and its cost is insensitive to the number of participants. Also, the gap between travel and telephone is very wide in terms of cost, convenience, and bandwidth. It makes sense to have a new technology available to fill that gap.

If VTC does end up stimulating more travel, there may be offsetting advantages that are far more important. For instance, it may not be cost effective to pursue some remote business opportunities using only existing communication tools. Adding VTC to the mix, as a substitute for what would otherwise have required travel, might change this. We will return to this topic later.

A preoccupation with the viewpoint that VTC is primarily a substitute for travel is probably counter-productive because it diverts attention from the productivity benefits of VTC. Nevertheless, the topic must be addressed, since travel substitution does occur and is used to help justify VTC investments.

## 9.5.1 Travel and Telephone Expenses

Published estimates of business travel expenses are infrequent and often unsupported. The author calculates that in 1982 U.S. scheduled airline revenues were about $36.4 billion. Business travel accounted for about one-half the passenger volume but probably over 50% of the revenues. Assuming a value of 60%, $22 billion was spent on business airfares on scheduled airlines in 1985. The 50,000 private business aircraft in 1978 carried one-third as many business travellers as did commercial airlines. Carrying this ratio forward to 1982 and assuming the same cost per mile adds another $7.3 billion, for a total of $29.3 billion in 1982. Runzheimer International of Chicago estimates that airfares constitute 44% of total business travel costs. Thus, the dollar cost of U.S. business travel in 1982 was about $67 billion (not including the cost of lost time).

A study done by Satellite Business Systems showed that every dollar in direct travel expenses was matched by $.50 to $1.00 in lost time (Harkness and Burke 1982). Assuming $.75, the total cost of business travel by air was roughly $117 billion in 1982. A great deal more was spent on business travel by auto. In addition, U.S. business spent about $13 billion on long distance telephone charges in 1982.

These national travel and telephone expenses measure the importance of communications to the business community. They also indicate future VTC market size, should VTC be used as a substitute for travel or telephone. Replacing certain business trips with VTC could save not only money but also time and effort. However, even the earliest users quickly found that VTC was also useful for holding meetings that simply would not have otherwise occurred. Section 9.6 elaborates on these benefits.

## 9.5.2  Reported Travel Savings

A number of organizations are publicly reporting that VTC saves on travel expenses:

- IBM found that two-thirds of the video conferences in their pilot system saved on travel (Hagopian 1981). Savings of over $1 million per month have been reported unofficially.

- Hercules regularly reports estimated travel savings to management. In the first months of 1984, a savings of $594,000 was reported (Hercules 1985).

- In a survey by Green and Hansell, users listed travel savings as one of the primary benefits of VTC (Green and Hansell 1984).

- Hitachi and Takenaker Komuten (a Japanese company) both report savings on travel costs (*International Management* 1984).

- ARCO estimates travel savings of $21 million per year from its seven-room system (*International Management* 1984).

- Skidmore, Owings, and Merrill reports that teleconferencing has cut down travel (*Telecoms* 1984).

- Aetna estimated a savings of $500,000 in 22 months (*Audio-Visual Communications* 1983).

- Honeywell reported a savings of $150,000 per month for nine VTC rooms in 1982 (Prem and Dray 1983, 74).

It should be noted that organizations claiming travel savings usually estimate them on the basis of user surveys in which respondents are asked how many trips a given teleconference has saved.

### 9.5.3 Net Travel Savings

The evidence just cited is fairly persuasive in showing that VTC does displace certain trips. But it does not necessarily follow that a net reduction occurs. No organization today uses VTC enough to dramatically reduce overall corporate travel budgets and thus incontestably prove the point. Minor variations in travel expenses cannot be traced to VTC, since they may have been caused by economic conditions, major projects, or other factors.

There are two general reasons why some observers claim that displaced trips should not be assumed to represent net travel savings. In both cases, however, there is a positive twist.

The first reason is simply that money saved from displaced trips may have remained in the travel budget and been spent for other trips. This is not necessarily bad unless the new trip was frivolous, in which case the problem is not VTC but rather bad management.

The more significant argument is that increased communications of any sort will lead to establishing more remote business relations. More business will in turn generate more travel. Travel expenses probably will increase, but a simple example shows why increased travel would be beneficial not detrimental. Suppose a $2000 profit could be made in a situation that would require two $1000 sales trips. Clearly that opportunity is not worth pursuing. Now suppose that one trip plus one $500 VTC would suffice. The opportunity is now worth pursuing. Travel and VTC both increase, but so do sales. This example expands to much larger scenarios, such as the ability to manage decentralized organizations or expand overseas.

The theoretical aspects of the travel generation argument were treated in an SRI study, and it was concluded that even if net travel increased due to better communication, the net increase would be more than offset by increased business benefits (Harkness et al. 1977). In short, bringing a new communications tool to bear on the situation improves the communications mix and thus the cost effectiveness of pursuing attractive remote opportunities.

# 9.6 Benefits

During SBS's early promotional days, company executive Tom Rush, when asked what VTC cost, was fond of retorting, "That's the wrong question; tell me first what it's worth." That challenge is as valid today as it was in 1980. Fortunately, end users are beginning to find answers.

As a practical matter, telecommunications managers have usually sought to justify the cost of teleconferencing before recommending it to their superiors. This meant trying to justify VTC on the basis of projected travel substitution and savings. Building on prior work by CS&P, SBS developed and used fairly elaborate models that attempted to forecast travel substitution and associated savings (Harkness and Burke 1982). Other end users developed their own estimates. A number of systems have in fact been sold, at least partly, on anticipated travel savings. Recently, however, rationales justifying VTC have placed more emphasis on productivity and other nonquantifiable benefits and less on travel savings.

One SBS study found that the median delay between recognizing a need to meet and actually travelling and holding the meeting was fourteen days; the average was forty-five days (Harkness and Burke 1982)! Clearly there is considerable "information float" involved here, probably caused by difficulties in scheduling a trip or in accumulating enough agenda items to justify it. Because VTCs require only a short window of mutual convenience, they are easier to schedule. Thus, the ability to address and resolve problems faster is a key benefit to using VTC.

There is also a fundamental difference in the cost structure between travel and VTC. A high fixed cost is associated with travel, but once at the destination the marginal cost of holding a longer meeting is small. With VTC there is little if any fixed cost, but charges are assessed by the minute. This difference in cost suggests that a series of short VTCs will likely replace what would otherwise have been one long face-to-face meeting. Using short VTCs allows a better match between the length of the meeting and the information available for discussion at the time.

The ability to involve more people in meetings, or call them in as needed, is frequently mentioned as a major benefit of VTC. One of the most comprehensive studies of teleconferencing benefits was conducted by Green and Hansell (1984). Their findings are summarized in Fig. 9.3. The user success stories now appearing in the trade press simply reiterate and validate the benefits outlined by Green and Hansell.

Of strategic significance is the probability that VTC benefits will occur in four evolutionary stages:

*Stage 1—Substitution* These direct, immediate benefits arise from the classic substitution of one technology for another. In this case, substitution benefits include faster decisions, better communication, travel savings, and the other benefits just described.

*Stage 2—Individual-Level Change* As VTC becomes a common tool, individuals will reorganize their work patterns and expectations to take advantage of it. In short, they will do what they could not have done before.

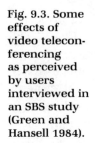

Fig. 9.3. Some
effects of
video telecon-
ferencing
as perceived
by users
interviewed in
an SBS study
(Green and
Hansell 1984).

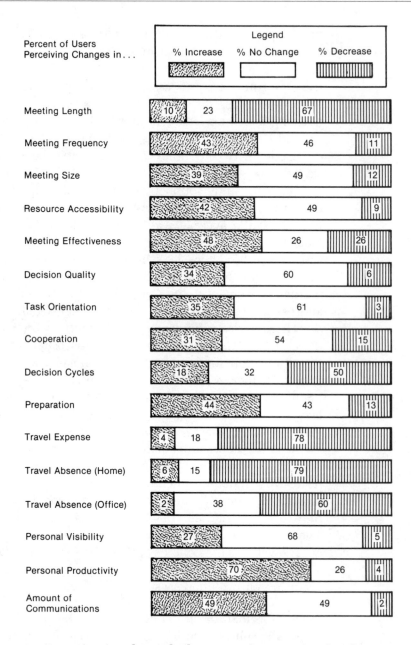

***Stage 3—Organizational-Level Change*** When Stage 2 has become common, the same phenomenon will occur at the group and organizational level. Companies will adjust their degree of centralization, will
relocate employees differently, and will pursue different markets.

***Stage 4—Opportunities for New Business*** Just as computers enabled
the credit card business to flourish and airlines enabled the success

of Federal Express, VTC should ultimately enable the creation of new businesses and of ways of doing business that can hardly be envisioned today. One such related concept, called *telecommuting*, has already been visualized. Telecommuting is frequently associated with working at home via one's personal computer. Actually there are other, probably more appealing, scenarios (see Harkness et al. 1977). All require two types of telecommunication links between a remote worker and coworkers, one for data or "paperwork," another for person-to-person communication. VTC will fill the latter role and in the long run facilitate a radically different organizational structure, one without physical proximity, employing a network of widely separated individuals.

One important implication of the fact that benefits will occur in evolutionary fashion is that companies cannot move from one stage to the next quickly. This means that organizations adopting VTC (or any similar technology like office automation) early will stay ahead of their competitors for a long time in terms of these benefits.

PART II

**PREMISE SYSTEMS**

This part describes the terminal or premise equipment needed for video teleconferencing. Because space precludes any comprehensive treatment, the focus is on a few important topics not well treated elsewhere in the literature.

# 9.7  VTC Rooms

A *VTC room* is a meeting room equipped with the necessary audio, video, and/or graphics equipment for teleconferencing. VTC rooms can range from very informal arrangements in which the electronic devices are placed atop existing furniture, to beautiful boardroom-like settings where all equipment except the displays is concealed. Costs for a full-motion video room, exclusive of codec, range from a few thousand dollars (for a camera, monitor, and speakerphone) to over $500,000 for a custom-designed room with custom cabinetry, furnishings, and a full complement of high-quality equipment. Some of the variation in price results from differences in functionality, such as inclusion of a high-resolution (1000+line) graphics system, PC interfaces, or video tape recorders, but price is mainly driven by quality. Quality can vary greatly in areas such as cameras, large-screen projectors, audio, video codecs, and lighting. Another variable is the human interface. Controls can be limited or powerful; the monitors large or small. Finally, rooms differ in their aesthetics. The cost of site preparation, which includes architectural services, construction, acoustical treatment, air conditioning, power, lighting, and custom furnishings, can easily be as great as the electronics in boardroom-type implementations.

In response to high site-preparation costs, the VTC industry has developed *roll-in* terminals and modular, prefabricated rooms, greatly reducing site preparation costs. Improvements in cameras, audio, power consumption, and other aspects of the technology have allowed environmental specifications to be relaxed somewhat without affecting performance. Nevertheless, one should not expect a roll-in terminal to have the acoustical and picture quality of a well-designed room.

Figs. 9.4 and 9.5 are pictures of video TC rooms. Fig. 9.6 shows a typical floor plan, while Fig. 9.7 is a block diagram of the typical major components. Fig. 9.8 shows a roll-in terminal, and Fig. 9.9 illustrates a desktop terminal. Good technical references on room design are: Brown et al. 1980; Electrical Communications Laboratories 1973; McFarlane and Nissen 1982; Midorikawa et al. 1975; and Wilkens and Plenge 1981.

VTC room equipment can be broken into three major subsystems: audio, video, and graphics; and these are discussed next.

Fig. 9.4. Hambrecht & Quist video teleconferencing room *(Courtesy Pierce Phelps).*

Fig. 9.5. U.S.
Telecom's video
teleconferenc-
ing room,
Atlanta, Georgia
*(Courtesy ESI).*

## 9.8   Audio Subsystem

Audio is probably the most important of the three subsystems. Novices usually assume that good audio is technically easy to accomplish. They find out that the reverse is true. Experience has shown that good-quality audio is as important as it is difficult. Users should not pass judgment after brief demonstrations. Marginal, but understandable, audio has been found to cause fatigue during long teleconferences.

The major audio problem is called *echo* or *feedback*. It occurs when sound from the local loudspeaker is picked up by the local microphone, sent to the remote room where a similar loopback occurs, and finally returned even louder than before. The loud feedback squeal that is heard is similar to the squeal heard in an auditorium when the amplifier is turned up too high. At lower levels, a speaker's voice will return as annoying echo.

These problems are overcome only with difficulty, and as yet not fully. One approach, called *gain switching*, simply reduces local loudspeaker volume when someone is speaking. This, however, clips what-

**Fig. 9.6. VTC room layout** *(Courtesy ESI).*

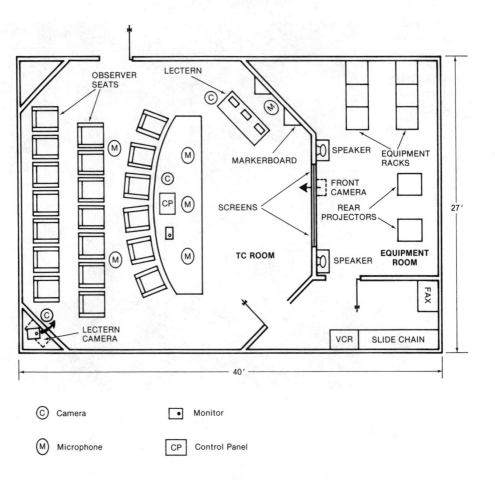

© Camera        ▣ Monitor

Ⓜ Microphone    CP Control Panel

ever speech might be coming from the remote site. Still this brute force technique is inexpensive and works in most untreated rooms. Speaker-phones also use it. Much vendor design effort goes into optimizing the pickup and dropout delays as well as the degree of attenuation.

Better performance is obtained through acoustical treatment of the room, coupled with *echo cancellation*. Prior to modification, the room's acoustical characteristics are measured with instruments. Experts determine the type and amount of acoustical treatment needed. In practice, this often means double layers of wallboard, heavy insulation, absorbent surface treatment, and certain geometrics.

Echo cancellation works by storing the transmitted sound in memory and then subtracting it from returning echo just before that echo would have come through the loudspeaker. The result is phase cancellation. Although simple in concept, echo cancellation has been very difficult to

Fig. 9.7. Block diagram of major equipment for VTC room.

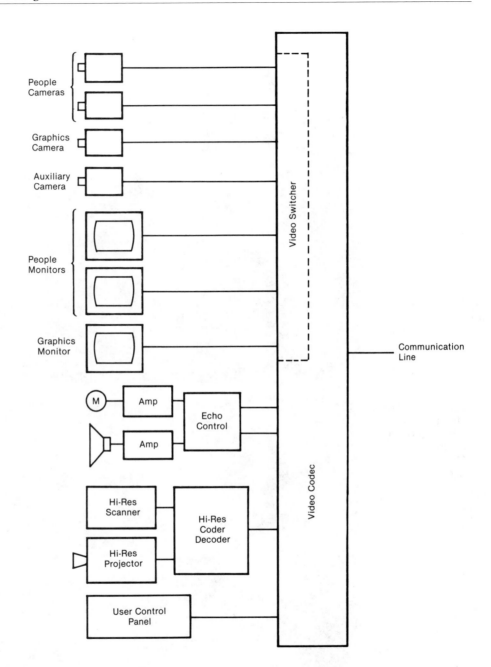

Fig. 9.8. Portable
video
teleconferenc-
ing on "roll-in"
terminal
*(Courtesy ESI).*

Fig. 9.9. Desktop
video
teleconference
terminal
*(Courtesy
Datapoint Corp.).*

achieve. COMSAT® developed some of the first practical echo cancellors for satellite telephone circuits in the mid-1970s. These have been adapted for VTC use.

An effective but not very popular alternative is the use of lavaliere microphones. They greatly reduce the echo problem because the speaker's voice is closer and thus louder than sound emanating from the loudspeaker.

As for bandwidth, telephone- or toll-quality audio is generally considered the minimum requirement. It has a frequency range of about 300 to 3200 Hz. Better than toll quality (meaning 5 to 7 kHz) is much preferred. Incidentally, the video codec described in the next section simply provides a channel of given analog bandwidth for the audio by using ADPCM or PCM coding. The difficult task of echo control is not done within codecs on the market today, largely because audio solutions are situation-specific. Halhed (1986) and Wilkens and Plenge (1981) provide more detail on VTC audio.

# 9.9 Video

Interactive video is by definition the primary focus of video teleconferencing. Its benefits or utility in the communications process span a wide but generally poorly understood range. At one extreme, video provides visual cues that help participants time their verbal remarks. Recognition of expression and body language occupy a midrange. Their importance is fairly obvious, if not quantifiable. Feelings of presence, warmth, anxiety, and so forth are conveyed visually but are even harder to define. The use of a medium also probably implies the importance that the caller places on a communication. Telephone calls are inexpensive and connote little importance. On the other hand, a personal cross-country trip signals much greater importance. VTC and the quality of the facilities used to execute it signal something in-between. However, VTC may imply more urgency than a trip would. Recognition that VTC deals with such complex aspects of communication adds greatly to the challenge of implementing this new medium. However, since few hard conclusions seem evident, the following paragraphs deal only with those very practical matters that novices find relevant but confusing.

## 9.9.1 Video Basics

The United States, Canada, Japan, Korea, and a number of other countries representing about 50% of the free world's gross national product adhere to the NTSC composite video standard, whereas most of Europe and another set of countries use PAL. France uses SECAM. NTSC cam-

eras and monitors work with images composed of 525 scan lines. The
525-line image or *frame* is broken into two fields of $262\frac{1}{2}$ lines each. The
field having even-numbered lines is interlaced with the field of odd-
numbered lines on the display. There are 30 frames per second, or 60
fields per second. The eye blends them together. Of the 525 lines, only
483 actually contain picture information.

NTSC standards were designed so that perceived horizontal and vertical resolution are equal. The resolution of cameras and monitors is usually specified in terms of *lines-per-picture-height* (LPPH), which means the
total number of black *and* white lines that can be resolved by the eye
over a distance on the screen equal to the height of the screen. NTSC has
a vertical and horizontal resolution of 341 LPPH. In order to achieve this
resolution, 483 scan lines and 4.2 MHz of bandwidth is needed. Thus, the
proper way to specify resolution for any black boxes between camera
and monitor, such as codecs or amplifiers, is in terms of bandwidth.
Sometimes pixels are used instead, but inconsistency in how they are
defined by the vendors makes pixels a difficult and perhaps misleading
gauge of quality. Many home TV sets perform below NTSC standards
partly because bandwidth is deliberately limited to save cost. More
expensive sets with comb filters do much better.

PAL signals consist of 625 lines total, of which 575 contain picture
information. PAL operates at 25 frames per second.

In both NTSC and PAL, color information is combined with luminance information into a single *composite* signal for transmission. Quality
is lost in the process. Both 525/30 and 625/25 video can be handled locally
in RGB form, meaning simply that the camera's red, green, and blue sensors are brought out on separate leads rather than being combined into
NTSC or PAL. RGB frequency response is not artificially bandwidth-
limited but rather extends to the full capability of the camera or monitor.
Six MHz is not uncommon. In practice, the end-to-end quality is limited to
the bandwidth of the weakest link, which is usually the codec. Thus, RGB
may be no better than comb-filtered NTSC for teleconferencing graphics.

NTSC monitors have an aspect ratio of 4 to 3. Thus, a 25-inch (diagonal) monitor is 20 inches wide and 15 inches high. If an NTSC video signal
is digitized so that it maintains full broadcast quality, the resulting bit
rate is 80 to 100 Mbps. In contrast, a voice call takes 32 or 64 Kbps. Economical VTC depends on a codec to digitize and compress the amount of
data from 80 Mbps to 1.5 Mbps or below.

## 9.9.2  Geometric Considerations

In order to emulate a face-to-face meeting, various designs are used to
compensate for the limits in today's display technologies. The purpose of

the following discussion is not to disparage these designs, since they are the best that can be done with current technology and users find them acceptable. The discussion will attempt to place today's designs in perspective by considering an ideal display. Both viewing angle and resolution will be considered.

Fig. 9.10 is the author's concept of the "ideal" display. It consists of a screen 16 feet wide and $3\frac{1}{2}$ feet high, placed about 8 feet away. The parameters were selected for the following reasons: The 8-foot distance is typical for people seated across a conference table; a distance of this magnitude is also needed so that multiple attendees can view the screen comfortably. The 90° field of view is less than the 180° often used in a real meeting but is adequate to view about six people seated on $2\frac{1}{2}$-foot centers across the table. The $3\frac{1}{2}$-foot height provides a view from the table to just over the head.

**Fig. 9.10. The "ideal" display.**

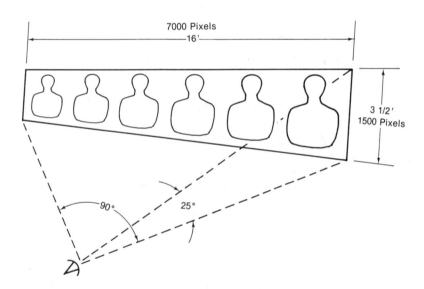

The ideal display needs roughly 1500 by 7000 pixels. It would be nice to have specified this resolution based on some understanding of what is needed to recognize facial expressions, read body language, obtain speaking cues, and perform any other visual missions. Unfortunately, the necessary research is lacking. In its absence, vertical resolution can be specified as the resolution needed to eliminate visible and objectionable line structure. There are studies showing how far home TV watchers prefer to sit from their sets. The answer is six to eight times picture height (Electrical Communications Laboratories 1973). At closer distances the scan lines become visible and objectionable. This viewing distance translates into an 8° viewing angle, which is of course occupied by 480 scan lines. Assume that horizontal and vertical resolution are equal. Thus, if our ideal display were cylindrical, 5500 pixels of horizontal resolution would be needed; but, since it is flat and therefore longer, about 7000 are needed. In short, the ideal display would produce NTSC picture quality but across a much larger field of view.

A final criteria of our ideal display concerns the resolution needed to show a human face adequately. The average human head measures about 10 inches from chin to crown, and from 8 feet away it occupies a viewing angle of 6°. If 480 TV lines are adequate for an 8° angle, then a head needs about 360 lines if shown life-size on a display 8 feet away. As noted previously, the question of how many scan lines are needed over a face to allow recognition of expression, reading of body language, etc., is unknown and is of course different from the number needed to keep the lines themselves from being distracting.

## 9.9.3   Options

Figs. 9.11 and 9.12 illustrate various practical design options used in VTC rooms in relation to the ideal display we have just constructed. Fig. 9.11 shows six different camera shots superimposed over the ideal camera shot, whereas Fig. 9.12 shows how these shots are displayed. Except for Option 6, all shots use the standard NTSC 4-by-3 aspect ratio. The displays, except for the ideal and Option 6, are assumed to be 25-inch-diagonal monitors. The merits of each option are discussed next.

### Option 1: Close-up

In a close-up the camera is manually panned and zoomed to show only one attendee at a time. The 25-inch monitor is large enough to show a life-size facial image at 8 feet. The 10-inch-high head fills two-thirds of the 15-inch screen height and thus covers 320 scan lines, producing very good quality.

### Option 2: Overview

This overview shot is commonly used during teleconferences, though the result is far from ideal. The image must be reduced by a factor of 9.6 so that the 16-foot-wide field of view can fit on a 20-inch-wide monitor. The resulting very small facial sizes are comparable to seating real attendees 77 feet away. Moreover, there are only 33 scan lines per face. Facial expressions are not discernible at this range and resolution combination. Indeed, it is sometimes difficult to recognize people you know.

### Option 3: Switched Cameras

The switched-camera approach was used by AT&T in their PMS rooms. A total of four cameras was required. A crescent-shaped table seating six

Fig. 9.11.
Camera fields of
view relative to
an ideal.

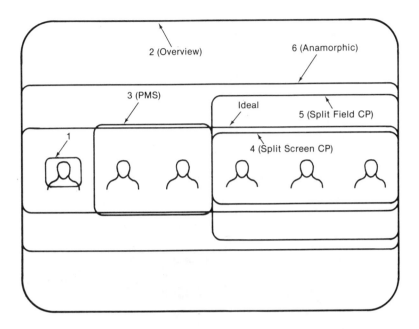

*The ideal is 3½ × 16 feet, with faces on 2½-foot centers.*

was used, with one camera focused on each pair of attendees. The fourth
camera provided an overview shot. Microphones along the table sensed
who was speaking and activated the appropriate camera. Subsequent
research showed that users disliked the frequent camera cuts (Brown et
al. 1980). The people on camera were often not those that the viewers
wished to see. Coughs and other noise could trigger distracting camera
switches. With two people on screen, the 5-foot, life-size field of view is
reduced to 20 inches on the monitor. This reduction by a factor of three
means that people appear 24 feet away. Also, their heads are covered by
only 120 scan lines.

## Option 4: Split Screen

The split-screen option is one of several *continuous presence* (CP) techniques designed to place all attendees continually in view, with the best
resolution and image size that CRT displays allow. Two fixed cameras are
used for split screen: one frames the left set of attendees; the other
frames the right. The 240 center scan lines from each camera are combined into a 480-line stacked image for processing through the codec and
for transmission. At the receive end they can be displayed in stacked
form, but more often they are split apart and placed on two monitors, as
illustrated. The scan lines above and below the image are black. Image

Fig. 9.12. Display
options.

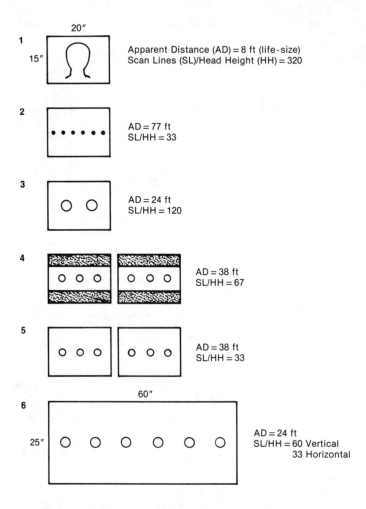

size is reduced by a factor of 4.9 for display (relative to the ideal), so attendees appear about 38 feet away and heads are covered by only 67 scan lines.

### Option 5: Split Field

The split field is a CP technique that provides a full-screen-height picture on both monitors but does so by cutting resolution in half. It is implemented by transmitting the even-line field from one camera and the odd-line field from the other. Scan lines can be interpolated so that each monitor displays 480 lines, but the real information content is only 240 lines. Thus, heads are covered by only 33 information-carrying scan lines. Between Options 4 and 5, it is a matter of trading off the quality of the facial image for views of the table and wall. In addition, some codecs cannot process split-field images, while those that do process them suffer in coding efficiency, since line-to-line correlations are lower.

### Option 6: Anamorphic

For the anamorphic technique, a special anamorphic lens is used to capture a wide aspect (1:2.35) or cinemascope-type image and "distort" it into a standard 3 × 4 television image for camera capture and codec processing. At the receive site the image is restored for display to cinemascope format by adjusting the vertical height control on a monitor or projector. Alternatively, a cinemascope lens can be used if the projector has a single lens output, as does the GE light valve. As seen by the codec, an anamorphic image would have thin, elongated faces. As Fig. 9.11 shows, fewer pixels are wasted on the wall and the table.

For purposes of illustration, assume that the anamorphic image is displayed on a 5-foot-wide projection display. This means that image size is reduced by a factor of about three relative to the ideal, and thus attendees appear about 24 feet away. The heads on the 5-foot display are covered by about 60 scan lines, which is almost twice that achieved by the overview camera in Option 2. The number of horizontal pixels per head is, however, the same as in Option 2, since the pixels are about twice as wide as they are high. Thus, a large-screen anamorphic image is clearly superior to the overview in Option 2, even if the overview is displayed on a similar-size projection screen.

### 9.9.4 Summary

The reader should note that much success has been achieved with displays that fall far short of the ideal constructed here. The effectiveness and attractiveness of VTC as a function of image display size and resolution has received very little study. One can only wonder how much more utility and acceptance VTC would have if displays could approach the ideal. Long-range planning for codec standards should assume that they will, probably inspired by progress in HDTV and projector technology.

On the technical side, the amount of bandwidth needed for compressed video, given constant image quality, depends mainly on the number of pixels *in motion*. Background areas, which do not change from frame to frame, consume very little bandwidth. By this measure it is not surprising that close-ups of a single head are far harder to compress than group overview shots. (A 10-inch-diameter "head" occupies 38% of the screen area, or 73,000 pixels, in Option 1 but only 1.6%, or 3300 pixels, in Option 2 for the six faces combined.) This analysis also shows why poor codec performance can be camouflaged by demonstrations using small or distantly placed monitors. It explains why using large-screen projectors improves the field of view but makes the picture fuzzy. Finally, it suggests that, as teleconferencing popularity grows, there will be increased pressure to abandon the NTSC or PAL format for something with greater resolution and a more appropriate aspect ratio.

# 9.10   Graphics

The capability to transmit images of viewgraphs, typed pages, charts, blackboard sketches, and three-dimensional parts has been found vital for most teleconferencing applications. Table 9.1 showed that 94% of meetings require this capability. There are four general technical solutions to transmitting images, none of which handles all the different requirements. These are: facsimile, TV graphics, high-resolution graphics, and PC graphics.

## 9.10.1   Facsimile

Facsimile has met with almost no user acceptance within the VTC environment because its long transmission time disrupts the meeting flow and because the receiver must convert the paper copy into a viewgraph before the group can view it. Additionally, there is no cursor or pointer capability, something that is essential to any presenter.

Transmission time is a key figure of merit for VTC graphics systems. Long delays disrupt the rhythm of presentations. The time needed to change a viewgraph is about two seconds, and this should be the design goal for teleconferencing.

## 9.10.2   TV Graphics

TV graphics is essentially still-frame video. Typically, the image is captured by a camera pointing down from the ceiling to an area on the conference table or by a camera in a special graphics station where originals can be bottom- or top-lighted. These graphics stations often contain a 35-mm slide projector as well.

The system uses standard TV cameras and monitors. Image capture and transmission are handled by the video codec, such that motion video is interrupted for a fraction of a second when a graphic is sent. The codec contains a still-frame image memory that drives the graphics monitor. The codec usually provides a cursor. Altogether, it costs little to add TV graphics to a motion video codec. Stand-alone still-frame transceivers, such as those manufactured by Colorado Video, Robot, and NEC, are also available and cost between $5000 and $20,000.

A 480-line-by-500-pixel, still, color TV image digitized at 8 bits per pixel consists of 1.9 million bits before compression. Two-dimensional cosine transforms can reduce this by a factor of 1.5 to 16, depending on image complexity. Text will compress by only a small amount, whereas a face compresses by a larger amount. At 16 to 1 compression, it takes 0.1 second to send a 480-by-500-pixel TV graphic at 1.5 Mbps. At 56 Kbps it would take 2.1 seconds. Various schemes have been developed to send a crude but usable image quickly that will then improve in resolution over a period of 30 seconds or so.

Motion video codecs provide higher resolution for graphics than for motion. Graphics almost always use 480 scan lines, and horizontal resolution is typically twice that used for motion. RGB is often provided, though comb-filtered NTSC is almost as good.

The advantages of TV graphics are:

- Color.
- Gray levels.
- Imaging of 3-D objects.
- Relatively low cost.
- Capability for easy zooming and pointing, allowing capture of large drawings, charts, or blackboards.

The disadvantages are:

- Relatively low resolution, precluding the use of full, typed pages.
- High cost of projectors capable of displaying full RGB quality.

The 480-line TV format does not provide adequate resolution to read an entire typed page, or many viewgraphs. On a high-quality system, it is possible to read half a page. In practice, users of TV graphics systems are often given burdensome instructions on how to prepare acceptable presentation graphics.

Resolution is not the only problem. Screen size is a major problem as well. You can easily determine what is really needed by sitting about 8 feet away from a wall and holding up a typed page at a comfortable reading distance before your eyes. This is typically 18 to 24 inches. Without moving the page, refocus on the wall and note the area covered by the page. This is the size the display should be. At 8 feet, this turns out to be about 4 feet square, not coincidentally the image size typically produced by viewgraph projectors.

Thus, even if the TV graphics system has adequate resolution for a typed page, display on a small monitor makes it unreadable. Low-to medium-price TV projectors ($3000 to $15,000) are only a partial answer. In practice, their resolution is not as good as advertised because, when intensity is turned up to the level needed in a VTC room, the beam blooms, thus degrading resolution. The only really adequate projectors for TV graphics are the GE Lightvalves, which are prohibitively expensive ($40,000 to $80,000).

## 9.10.3   High-Resolution Graphics

Unfortunately, vendor hype has resulted in virtually everything being called high-resolution, or "hi-res," particularly in the world of PCs. However, in the VTC arena hi-res has typically been defined as 1000-by-1000-pixel resolution or better and the ability to read a full typed page.

Hi-res systems do not use TV technology; rather, they are based on facsimile technology. The image is usually scanned with a standard facsimile type 1728-element linear CCD array. Today the images are binary black and white only, meaning of course no gray scale and no color. Compression is done with a simple run-length coding technique used for facsimile. For many years the missing link in hi-res was the unavailability of any affordable device that could project the image onto a wall screen. Recently two practical projector technologies have emerged. One, developed by Databeam, uses a cathodo-chromic projector tube of unique design. It is capable of better than 1000-by-1000-line resolution and produces a dark purple trace on white background. The other, developed by Hughes, uses a liquid crystal cell to achieve similar results. Progress has been very rapid in these two technologies within the last year, and truly outstanding hi-res has finally become practical. Color and limited gray scale are under development. The cost of the hardware in some of these hi-res systems is not intrinsically high, and prices should fall as yield improves and volumes increase.

The advantages of hi-res are:

- Legibility of full, typed page or viewgraph.
- Rock steady images without flicker or jitter.

The disadvantages are:

- No color or gray scale (yet).

- Relatively high cost (today).
- Occasional use of facsimile-type scanners, which cannot handle bound reports, oversized objects, or 3-D objects.

### 9.10.4 PC Graphics

In the future many presentation graphics are likely to be composed on personal computers. Users will bring to the conference a floppy disk containing the graphics. At present, this poses a problem: compatibility between PCs. If the users' PCs are configured identically with one in the conference room, the solution is easy, but identical configuration is probably not the usual case. When the configurations are identical, however, the PCs can transmit graphics through a low-speed data circuit or a special scan converter that can be used to convert the PC display format (which is not NTSC) into NTSC for transmission as a TV graphic. More elegant solutions are much needed.

## 9.11  Motion Video Compression

Video codecs are needed to compress video to rates of T1 and below, where transmission costs are reasonable. Most codecs also contain multiplexers that combine audio, graphics, and control signals with the video for transmission through a single long-distance circuit.

Video codecs of the type used for teleconferencing were virtually impossible to implement before VLSI memory chips became available in the 1970s. Codecs are enormously complex devices. The CLI Rembrandt contains about 3000 ICs, about ten times that of the typical PC. When codec sales volume justifies custom VLSI, significant size and price reduction should be possible. Codec prices currently range from $70,000 to $100,000.

Two principal techniques are used for compression of image data: *differential pulse code modulation* (DPCM) and *transform coding*. The first DPCM codecs were developed in the late sixties when fast analog-to-digital converters became available. In the early seventies, when large random-access memories became available and video frame memories could be built, the performance of DPCM encoders was extended.

Transform coding has been developed through the seventies and eighties. A number of transforms have been applied to image coding, and, of these, the cosine has been found to give the best performance. Transform coders for full-motion video were first built commercially in 1980. In this case, the necessary hardware, the high-speed VLSI $16 \times 16$ multiplier, had just become available. DPCM codecs give good performance at higher bit rates, whereas transform coders provide better performance at lower bit rates. Transform coding has the disadvantage of requiring more hardware.

Fig. 9.13 provides a theoretical comparison between the various coding methods. The vertical dimension, bits per pixel, measures the degree of compression. The horizontal scale provides a measure of *mean square error* (MSE) and is a measure of distortion or picture quality. A low MSE indicates a high quality.

**Fig. 9.13. Relative performance of various image compression techniques.**

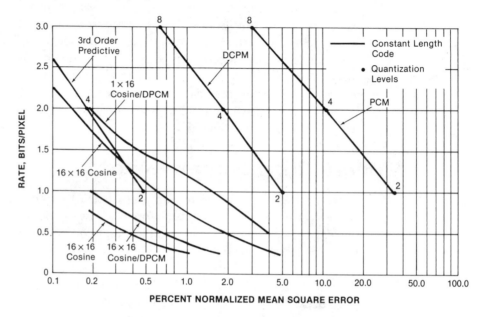

*Reprinted by permission of John Wiley & Sons, Inc., from* Digital Image Processing *by William Pratt, copyright © 1978.*

As can be seen from these comparisons, the highest compression is achieved by using a three-dimensional transform codec (16 × 16 × 16 cosine in Fig. 9.13. However, this codec requires extremely complex hardware and also introduces a large processing delay. It is therefore unsuitable for teleconferencing. A hybrid technique using two-dimensional cosine transform within a single frame and DPCM between successive frames gives performance that approaches the three-dimensional transform, without the large processing delay and complexity. The CLI codec uses this hybrid technique. It is labeled "16 × 16 cosine/DPCM" in Fig. 9.13.

## 9.11.1  DPCM Codecs

DPCM coding converts the analog video signal into digital form and stores the resulting luminescence and chroma values for each pixel in a frame memory. From then on, only frame-to-frame differences are transmitted. The Cost 211 codec built by GEC and others uses this approach.

Both DPCM and hybrid transform coding can be taken a step further by adding *motion compensation.* Motion compensation determines the direction and amount by which each small area of the picture is moving and predicts its probable position in the next frame on this basis. When information from the next frame is obtained, it is compared with the prediction and only the differences need be sent. NEC used motion compensation to reduce bandwidth in their X1 codec from 3 Mbps to 1.5 Mbps.

## 9.11.2 Transform Codecs

Because transform coding is the most powerful technique now available and is likely to become the future standard, it will be the only one explained in detail. The CLI Rembrandt is a convenient example. The CLI Rembrandt hybrid transform/DPCM codec operates essentially as follows.

First, the analog video signal is separated into luminance and chroma components and bandlimited to about 2.75 MHz. Then, the luminance component is digitized at 8 bits per pixel, using a sample rate of 7.2 MHz; the chroma is digitized at 1.7 MHz. Both fields are used, but every other frame is discarded.

Next, the digitized signals are read into a memory containing 480 active lines and 368 pixels per line. The image is then segmented into 690 16 × 16 pixel square blocks, each of which is compressed independently. See Fig. 9.14. The numerical luminance values across one 16-pixel line can be thought of as representing a complex waveform, where, for example, white areas have high values and black areas low values. This "waveform" is transformed using the cosine transform (similar to a Fourier transform), and the result is another sixteen numbers representing the coefficients of a set of frequencies from dc upward.

The first or dc term is the average brightness of the line. The second term can be thought of as meaning the line is brighter on the left than on the right; the third might mean that there are two cycles of brightness and darkness. Each term adds additional detail or resolution. In fact, if all sixteen values are transmitted, the original line can be reconstructed exactly. The more terms that are sent, the closer the original is reproduced. The block of transform coefficients is then transformed a second time, scanning vertically rather than horizontally. The result is a matrix holding 256 terms, many of them being zero in practical situations. The upper left term in that matrix represents the average luminance of the block. If only that term were transmitted, one could produce a very crude image with each of the 690 blocks having a different shade of gray.

**Fig. 9.14. Video compression.**

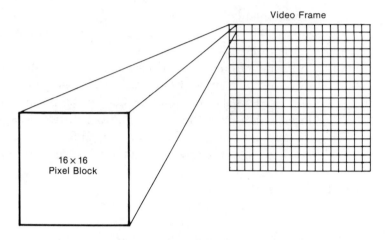

*(A) Frame formatted into 16 × 16 pixel blocks.*

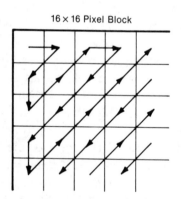

*(B) Zigzag scan of cosine-transformed coefficients.*

*(C) CLI Codec (Courtesy of Compression Labs, Inc.).*

The most important terms in the 256-term matrix are those closest to the upper left corner. In an intraframe codec (without DPCM in the time domain), compression is achieved by simply reading the terms in a diagonal zigzag pattern from the upper left and by sending as many as possible, for all blocks, through the communication line in the fifteenth of a second until the next frame. Compression is achieved because many of the terms are zero and not all the rest need be sent to get a reasonable picture. When DPCM is added to the process, only the differences between coefficients from frame to frame are transmitted. If the image is stationary, meaning there are no luminance changes in the blocks, there is no data to transmit. When the image moves, there comes a point when not all 256 coefficients per block can be sent. In that event, a buffer feedback loop eliminates the least important terms until a manageable level is reached, and the result is some loss in image quality. The more complex the scene and the greater the motion, the fewer the number of terms that can be sent. In practice, a variety of other techniques, such as Huffman coding, are also used to derive the greatest possible compression, but the preceding description is an accurate summation, in lay language, of the essence of hybrid transform coding.

The next step in improving video codecs may well be the addition of motion compensation to hybrid transform coding. This adds considerable hardware and complexity but is expected to yield picture quality at 384 Kbps, about equal to what hybrid transform coding now achieves at 768 Kbps. Since today's 768-Kbps quality is considered acceptable, achieving comparable quality at 384 Kbps will provide acceptable quality with a significant reduction in transmission cost. For this and other reasons, 384 Kbps will become the dominant VTC rate in the next several years.

# 9.12   Standards and Interoperability

Some believe that VTC cannot succeed until costs drop dramatically. Others maintain that standards are needed. Unfortunately, in the next few years these two assertions will continue to conflict, since costs cannot drop dramatically without improving the codec algorithms—thereby making any standards quickly obsolete. If standards are taken seriously, they stifle development; if not, they become meaningless and confusing. The solution to the problem involves balance and timing.

Standards must be viewed not as a unilateral good but rather as a trade-off between ease of interoperation and progress in lowering cost and/or improving quality. As of 1985, approximately 500 video codecs had been shipped; yet market potential is several hundred thousand. On the basis of such miniscule penetration, it is hard to argue that the technology has reached a level where performance and cost are broadly acceptable and should be frozen into standards. In the last five years, video compression technology has advanced from 6 Mbps to 3 to 1.5 to .768 or even lower for what users find acceptable. This trend can continue technically, and will continue commercially, as long as premature standardization does not preclude it.

This does not mean that codec standards will not become practical in the future when circumstances warrant them becoming so. Those circumstances would include: widespread marketplace acceptance of a given picture-quality bandwidth combination, and thus the associated coding technique; a significant demand for VTC between firms as opposed to within firms; a technical plateau that appears to hold for the next several years; a significant number of real-world incompatibility problems; and, finally, more than a very few incompatible brands of codecs each having significant market share. (This last condition would make gateways impractical.)

Today this set of circumstances does not exist. Most applications are concentrating on intracompany VTC within private networks. Where significant interorganizational VTC is envisioned, user groups (the aerospace industry, for example) have informally settled on one codec type. Finally, the creation of *gateways*, in which different codecs are connected back to back, offers a suitable solution, particularly since gateways are often needed anyway to bridge incompatible networks.

Even when it does make sense to set certain standards, users must realize that the standards will have limited lifetimes, just as $3\frac{1}{2}$-inch floppies are replacing $5\frac{1}{4}$-inch, 1-inch professional video tape has replaced 2-inch, 32-Kbps ADPCM voice is replacing 64-Kbps PCM, and so forth.

As a final note on standards, users have found that network incompatibility is a worse problem than codec incompatibility. Lesser but still real incompatibilities can occur between graphics subsystems, room controllers, and audio.

## PART III

## TRANSMISSION

Transmission is the most important aspect of video teleconferencing from a feasibility and cost viewpoint. Fig. 9.15 shows the five-year life cycle costs for a point-to-point VTC system consisting of two VTC rooms connected by a point-to-point transmission circuit. Its purpose is to show the portion of total costs due to room equipment, codecs, and transmission. For this example, the present value of life cycle costs, assuming ACCUNET® 1.5 and four hours per day of use, is about $1.9 million. With

---

The transmission costs for the various VTC services that are quoted in the text and in the illustrations were accurate, to the author's best knowledge, at the time of writing. Nonetheless, the reader should confirm all costs with the carrier before making any decisions.

less expensive transmission via a customer-owned earth station at 384 Kbps, life-cycle costs are about $800,000. Of these totals, two codecs constitute $200,000. This figure shows why it is so important to focus on lowering transmission costs. It can correctly be inferred from the data that it is worth putting considerably more hardware and expense into the codec if the result is the capability to operate at lower bandwidths. For example, assuming ACCUNET transmission prices, the author has calculated that it is *theoretically* worth paying up to *$1,080,000 more* for a codec capable of a suitable picture at 768 Kbps than for one that requires 1544.

Fig. 9.15. Life-cycle costs for room equipment and different transmission options.

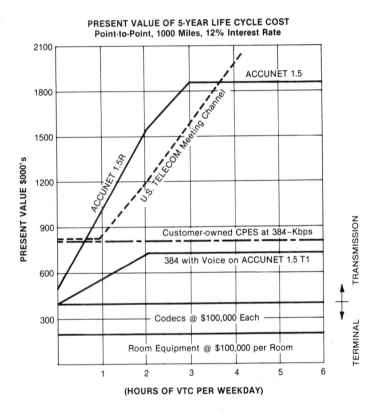

PRESENT VALUE OF 5-YEAR LIFE CYCLE COST
Point-to-Point, 1000 Miles, 12% Interest Rate

ACCUNET 1.5

ACCUNET 1.5R

U.S. TELECOM Meeting Channel

Customer-owned CPES at 384–Kbps

384 with Voice on ACCUNET 1.5 T1

Codecs @ $100,000 Each

Room Equipment @ $100,000 per Room

PRESENT VALUE $000's

TRANSMISSION

TERMINAL

(HOURS OF VTC PER WEEKDAY)

A video teleconference uses far more capacity than a voice telephone call or most computer data links. Thus, it requires special and expensive channels. The high cost and limited availability of such channels have constrained, and continue to constrain, the overall attractiveness and growth of VTC. Fortunately, the transmission situation is improving rapidly. As mentioned earlier, advancing codec technology has reduced the amount of capacity needed for motion video from 90 Mbps to well below 1 Mbps. In parallel, fiber optic and satellite advances have lowered the cost per bit for transmission. Growing competition will accelerate this price reduction.

The transmission needs of video teleconferencing are quite unlike the needs of voice, data, and all other applications of significance. Therefore, VTC cannot simply take advantage of services already in place. And, since VTC is a relatively new application generating relatively small revenues, carriers have made little effort to cater to VTC needs. Instead, they have offered capabilities easily extracted from a voice environment but not necessarily well suited for VTC.

Perceptive users must therefore think of transmission not just in terms of what *is* but also in terms of what *might be*. They must look beyond tariffed offerings to anticipate what could be offered based on existing transmission facilities if carriers developed services to address VTC optimally. Sometimes the carriers need guidance and prodding from their customers in this respect.

# 9.13   Requirements

Before investigating transmission alternatives, users need to clearly define their requirements. The bit rate can range from 56 Kbps to 1.5 Mbps, and this represents a trade-off between picture quality and cost. Codecs normally require a BER of $10^{-6}$ or better. The number of locations to be served and the expected volume of teleconferences are key considerations. It is important to recognize that the least-cost solution for a limited network may not be readily expandable into the least-cost solution for a larger, more heavily used network. Perhaps the initial locations have ready access to terrestrial T1 services, but future plans include connecting remote or small town sites where T1 is not available. For them a satellite solution would be needed. The high-power amplifier in a satellite earth terminal may or may not be sized to accommodate growth. A small antenna may be least costly in start-up or low-use situations, but, as usage increases, a larger dish reduces satellite charges enough to more than pay for itself.

If company-to-company communication is envisioned, then some sort of interface is needed between private networks to ensure compatibility, or both companies must have access to the same public switched network. The fact that bit rates in the public and private networks may differ must be considered.

Research shows that about 30% of meetings involving travel have attendees from three locations. This indicates a need for three-point teleconferencing. Some organizations are so structured that meetings typically involve *n* locations; for example, meetings with regional offices. Multipoint VTC is complex in terms of the network and the room design. Thus, any need for multipoint must be defined early, with particular attention given to reconciling the needs of users with the constraints often imposed by the technology. For example, it may be impractical to have a separate monitor displaying each remote location; so how do users see the people that they wish to see? There are a variety of design approaches that vendors can discuss.

# 9.14 Transmission Options

A user organization has three basic transmission options:

- Lease VTC services from a common carrier.
- Build a private VTC network with owned facilities.
- Integrate VTC along with voice and data on a preexisting corporate private network.

The following sections describe these options. In section 9.15 their costs are plotted for comparison.

## 9.14.1 Leased VTC Services

Perhaps the only common denominator for leased services is that the customer pays by the month or by usage. Otherwise, leased services can be satellite or terrestrial, full-time or occasional use. Satellite services can use earth stations on the customer's premises or can use city earth stations shared by many users and accessed via T1 access lines. Some of the more popular leased services are described next.

*AT&T COMM's ACCUNET® 1.5R*. Formerly known as the PMS network, ACCUNET 1.5 Reserved (R) is a switched 1.5-Mbps service currently available in forty-two cities from AT&T Communications. Customers must lease a full-time T1 access line to reach AT&T COMM's central office, but long-haul circuits are established upon request and billed in half-hour increments. This is the primary public switched network for VTC at this time. Routing can be terrestrial or satellite.

*AT&T COMM's ACCUNET 1.5*. ACCUNET 1.5 is a full-time point-to-point T1 or 1.5-Mbps circuit billed on a monthly basis from AT&T Communications. It is increasingly used for corporate backbone voice/data networks.

*AT&T COMM's SKYNET® Digital Service.* Customers can either lease access lines to the few AT&T-owned earth terminals associated with AT&T COMM's SKYNET Digital service or lease customer-premise earth terminals. These *C* band terminals have 6- to 9-meter antennas. AT&T also provides satellite capacity or "space segment." Channels of 1544 or 768 Kbps are available on a full-time or hourly basis. Channels of 384 Kbps and below are available only full-time at present. With a frequency agility option, the earth terminals can use different carriers for different calls, thus allowing users to construct large networks. Economics favors using the full-time space segment over about two hours per day. When used at 1.5 Mbps, SKYNET terminals can be connected into ACCUNET 1.5R, making this combination the most ubiquitous vehicle for VTC transmission.

*U.S. Telecom's Meeting Channel (ISACOMM).* U.S. Telecom's Meeting Channel service is very similar to SKYNET Digital except that it operates only at 768 Kbps and uses small 3.7-meter Ku band antennas. Typically, the first customer taking service in any given metropolitan area gets the antenna on his or her premises. Subsequent customers often share this antenna via T1 or microwave access lines. Antennas are leased on an hourly basis. Both hourly and full-time space segments are offered, with calls set up on a reservation basis, as with ACCUNET 1.5R. U.S. Telecom plans to integrate this service with its planned 23,000 fiber-optic network.

*American Satellite Company.* AMSAT® offers full-time T1 circuits between its several major earth stations, but it is principally known for constructing and leasing custom private satellite networks using either C or Ku band. AMSAT also leases full-time space segment for these networks. To date, AMSAT has built VTC networks for 1544- and 768-Kbps operation but would provide other rates as well. AMSAT uses special construction tariffs and does not publish prices, so quotes must be obtained.

*MCI (SBS).* SBS, which by the time this is published will be absorbed by MCI, has a national network comprising large earth terminals located in about twenty-one major cities. Designed originally for sophisticated switched data, these terminals can technically support switched or nonswitched data at 56, 64, 128, 224, 448, and 1544 Kbps. To date, SBS has offered only 56 and 1544 Kbps. The user can obtain these as full-time point-to-point circuits or as switched circuits billed by the minute. Although switched 448 Kbps could be very attractive for VTC, SBS has not yet made it available.

Prior to merging with SBS, MCI showed little interest in VTC and its only relevant offerings were leased full-time T1 circuits. When their networks are combined, MCI and SBS will of course be in an excellent position to offer a variety of switched and nonswitched VTC services. MCI is building many miles of intercity fiber optics that would be ideal for offering switched 384, 768, and/or 1544 Kbps.

*Cable & Wireless.* A Washington D.C. subsidiary of the British firm of the same name, Cable & Wireless (C&W) has purchased significant amounts of fiber-optic trunk and is currently offering full-time T1 circuits in several corridors at very attractive rates.

*Intelsat.* Faced with increased threats from undersea cable, Intelsat (International Communications Satellite Organization) has moved aggressively and perceptively with a new offering called IBS. IBS could dramatically accelerate international VTC, since it is priced quite reasonably and theoretically allows communication between small customer-premise earth terminals. Intelsat itself provides only satellite capacity and then only via its national signatories, such as COMSAT® and the European PT&Ts. End users must deal with COM-SAT or a domestic carrier, who in turn deals with Intelsat. COMSAT has made available IBS space segment in increments of 64 Kbps, including 384, 768, and 1544 Mbps. It is available on a full-time, part-time, or occasional use (hourly) basis. Intelsat has also released specifications for a range of earth terminals that various domestic and foreign firms can manufacture. They include a standard "E," which is only 3.6 meters in diameter and will sell for about $100,000. At the time of writing, there still exists much confusion as to whether PT&Ts will allow customers to own earth terminals or place them on their premises. Nevertheless, if politics don't preclude it, a user in, say, Atlanta could link directly from a 3.6-meter dish to a similar customer-premise dish in, say, Germany.

*AT&T COMM's ACCUNET 56.* ACCUNET 56 is a switched 56-Kbps service from AT&T Communications that will be available in sixty-four cities by 1986. Although the long-haul portion is charged on a usage basis (about $48 per hour for distances of 1000 miles or greater), a dedicated access line is needed.

*Vitalink.* Vitalink offers what may be called a hybrid between leased services and owned private networks described in the following section. Vitalink sells 3.7-meter Ku-band owned terminals for about $60,000. Installation costs are around $5000, and site preparation for a simple ground mount costs about the same.

### 9.14.2   Privately Owned Network

Except for short distances, the only practical setup for a privately owned network is satellite transmission using customer-owned, customer-premise earth terminals, along with owned or leased transponder capacity.

Suitable earth terminals range from 3.6 to 9 meters and operate at either C or Ku band. The higher-frequency Ku band allows smaller antennas and avoids the interference problems that make getting FCC clearance more problematic at C band. Earth terminals are available from Harris, Scientific Atlanta, MACOM, Vitalink, Avantek, and others. Fig. 9.16 shows a 3.7-meter product from Avantek that uses an Andrew dish. A nonredundant SCPC terminal capable of supporting a full duplex carrier in the 384- to 1544-Kbps range costs about $50,000 installed, not including site preparation. Site preparation includes a mounting pad, ac power, and cable ducts. It can range from a few thousand dollars for a ground mount to tens of thousands for location atop a high-rise building.

Fig. 9.16.
A 3.7-meter Ku-band antenna used for VTC at Hewlett-Packard *(Courtesy Andrew Corp.).*

Although higher-power satellites will in the next few years allow even smaller antennas, few laypeople understand the experts' admonishment, "You can pay me now or you can pay me later." They mean that using a smaller, less-expensive antenna saves up-front costs but requires more satellite capacity thereafter. Once this becomes apparent, users often opt for a somewhat larger antenna. At Ku band, 3.7- or 4.5-meter antennas are probably most appropriate for VTC.

Transponders sell for several million dollars. Since each can support eight or more full duplex T1's, few organizations purchase a full transponder, preferring instead to lease individual channels from such satellite owners as GTE and AMSAT. By purchasing earth terminals directly from the manufacturer and leasing space segment, organizations can implement VTC networks very economically.

## 9.14.3 Integrated Voice Data

For those organizations large enough to have a private network or leased T1 lines, multiplexing VTC with voice and data offers the lowest-cost way to transmit VTC. There are two ways to accomplish this. The first uses an intelligent T1 multiplexer; the second uses a multiplexer plus a voice compression device.

Fig. 9.17 shows how the T1 multiplexer is used to combine voice, data, and video teleconferencing over a T1 circuit. If VTC service is just beginning, it is probably cheaper to alternate VTC use with voice on part of the T1 rather than to dedicate some portion to VTC full-time. Intelligent muxes can be programmed to begin seizing voice trunks some minutes before the video teleconference begins so that 384 or 512 Kbps will be set aside, ready for the conference. A trunk is seized just after a user hangs up. Experiments show that clearing trunks for a VTC need begin only five to ten minutes ahead of time. During a VTC, the PBX uses a capability called *alternate routing* to route new voice calls via WATS, DDD, MCI, etc. Thus, the phone user experiences no difference in service. The effective cost of VTC using this technique is independent of the amount of usage, so this becomes a very attractive way to start teleconferencing. There are almost no fixed costs. This of course contrasts with private satellite systems, which have a fairly high fixed cost and achieve low, effective hourly costs only when usage is heavy. (See Fig. 9.22 for a cost comparison.)

Beyond one or two hours of VTC per day, it may become less expensive to purchase additional T1 capacity and dedicate 384, 512, or 768 Kbps full-time to VTC rather than continuing to divert voice to WATS, DDD, etc. Exactly when it makes sense to do this is situation-specific, and the decision must take into account whether another T1 would be fully or only partly loaded.

The second technique for obtaining very inexpensive transmission via T1 utilizes voice compression to reduce the bandwidth needed for

Fig. 9.17. Video
teleconfer-
encing
multiplexed on
T1 network
*(Courtesy
Timeplex).*

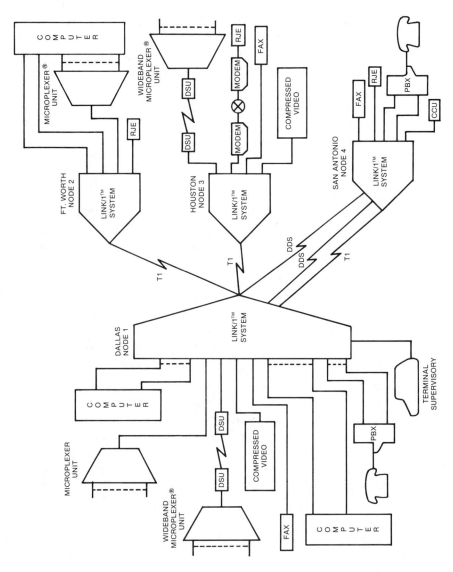

voice trunks on an existing T1 circuit, and thereby free enough capacity for video teleconferencing. Republic Telecom makes a packet-switching device called the RSL-40, which compresses six voice circuits into a single 56-Kbps channel. It can also compress fifteen voice circuits onto two 56-Kbps lines, twenty-five onto three, and forty onto four. If forty-four channels of 32-Kbps voice were normally being carried via a T1 circuit, it would be necessary to displace twelve of them to free 384 Kbps for VTC. The 384 Kbps for VTC, plus two 56-Kbps circuits needed for the twelve packet voice circuits, now occupy 496 Kbps. Sixteen of the original voice circuits are displaced, of which fifteen are restored by the Republic Telecom device. The cost savings are described in the next section.

*A note on 56-Kbps VTC*: This chapter focuses on *full-motion* video teleconferencing. Common industry practice defines full motion as requiring 384 Kbps and above. In recognition of the significantly lower picture quality obtainable at 56 Kbps, teleconferencing at this rate is called *limited motion*. Long distance 56-Kbps switched service from AT&T COMM costs about $48 per hour. Typically a separate telco circuit, costing about $24 per hour, is needed for voice. Thus, 56-Kbps VTC costs about $72 per hour. Access lines average $300 per month. Some codecs transmit at 112 Kbps, using two 56-Kbps lines. The cost of AT&T's switched 56-Kbps transmission is plotted in Figs. 9.19 and 9.22, where it can be compared with the cost of full-motion transmission at 384 Kbps and above.

*A note on the coming of 384 Kbps*: The coming of 384-Kbps transmission services will probably become one of the most significant milestones in VTC history. With 384 Kbps it appears possible to obtain both adequate full motion quality and acceptable costs. For a variety of technical reasons, 384 Kbps is a convenient speed to offer. It has also been selected as an ISDN standard.

# 9.15   Transmission Costs

There are several key variables that most affect transmission costs:

1. *Amount of Usage*. Since there are fixed costs associated with most transmission networks, especially those using dedicated earth terminals, higher usage results in lower hourly costs. Also, heavy usage can justify full-time rather than hourly channels.

2. *Ability to Multiplex VTC with Voice*. If the user has a T1 network capable of multiplexing VTC with voice, and the picture quality of sub T1 rate VTC is acceptable, costs can be much lower. This possibility

is of course limited to intracompany communications between locations large enough to justify T1 interconnection.

  3. *Use of Sub T1 Bandwidths*. Where carriers offer 384-, 512-, and 768-Kbps services, they are of course less expensive than full T1 rate services.

  4. *Competition*. Some carrier offerings are much less expensive than others.

## 9.15.1  Generalizing on Costs

Figs. 9.18 through 9.23 plot the transmission costs for the different options discussed previously. The overall message is that hourly transmission costs for full-motion teleconferencing cannot be generalized and are highly situation-specific. They can run from under $10 per hour to well over $1000 per hour.

  The lowest hourly costs plotted were for teleconferencing at 384 Kbps, for six hours per day, using a 384-channel multiplexed with voice on a T1 circuit. The circuit was leased from Cable & Wireless on their fiber optic from New York to Chicago. In this situation, VTC costs $32 per hour. The $10-per-hour cost would apply from New York to Washington D.C. under otherwise the same scenario.

  Fig. 9.18 compares the three basic AT&T COMM services most relevant to full-motion VTC. Over about two and a half hours per day between two points, a transition from service sold by the hour (like ACCUNET 1.5 Reserved) to a full-time service makes sense. SKYNET Digital is not cost efficient at 1000 miles but would be so for a heavy use coast-to-coast link. At two hours per day, prices are in the $600-per-hour range.

  Fig. 9.19 shows the effect of distance on the three primary AT&T COMM offerings. Each has its area of advantage.

  Fig. 9.20 includes services from SBS, U.S. Telecom, Vitalink, and Cable & Wireless. At two hours per day, T1 prices are in the $400-per-hour range, and half T1 can be obtained for $240 per hour.

  Fig. 9.21 shows the advantage of multiplexing VTC with voice on a private T1 network. Assuming ACCUNET 1.5 pricing for T1's and two hours per day, motion video at 768 Kbps costs about $320 per hour, while 384 Kbps dips to about $160 per hour. A particularly appealing aspect of this scheme is that the cost curve is flat for very light users, since there are virtually no fixed costs. At about two and a half hours per day, it no longer makes sense to divert voice to WATS during teleconferences. Instead, the user would purchase additional full-time T1 capacity. Thus, the downward sloping curves for 384 and 768 simply assume that VTC is charged at one-fourth or one-half, respectively, the

Fig. 9.18.
Comparison of
three basic
AT&T COMM
services.

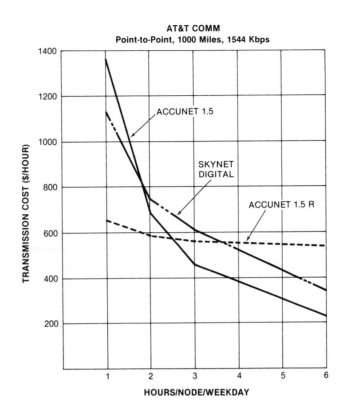

full-time T1 circuit cost. This plot assumes that T1's carry an average of thirty-five voice circuits.

Fig. 9.22 illustrates the three ways to transmit full-motion VTC at lowest cost. For the first way, Fig. 9.22 further elaborates the cost of multiplexing voice with VTC on a T1. The cost of freeing up 384 Kbps for VTC is really the cost of handling via WATS the voice circuits displaced. Where 32-Kbps ADPCM multiplexers are in place, more voice circuits must be displaced than where 64-Kbps PCM multiplexers are in place; thus, the cost to free up 384 Kbps is higher with 32-Kbps multiplexers in place. These two situations plot as two horizontal lines in the figure. The values are of course worst case in that they assume the T1 is fully loaded with voice. In reality, T1's will often operate partly full. If enough capacity is unused, the marginal costs of VTC may be zero. For these reasons and the negligible fixed cost, *multiplexing VTC with voice on T1 networks is extremely attractive*. Fig. 9.22 also differs from Fig. 9.21 in that Cable & Wireless T1 prices are used rather than AT&T prices. If T1 circuits from Cable & Wireless are used, the chart shows that 384-Kbps video teleconferencing costs only $100 per hour at the two-hour-per-day level!

Fig. 9.19. Effect
of distance on
the three
primary AT&T
COMM services.

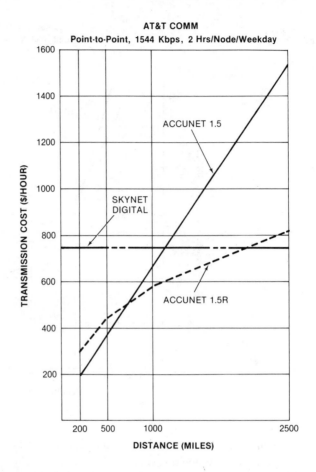

**AT&T COMM**
Point-to-Point, 1544 Kbps, 2 Hrs/Node/Weekday

In comparison, 56-Kbps VTC, with its much lower picture quality, costs about $75 per hour if ACCUNET 56 is used.

For the second way, Fig. 9.22 plots the cost of a privately owned satellite system. The curves assume that the end user purchases a 4.5-meter, SCPC, Ku-band earth terminal directly from a manufacturer (like Avantek, Vitalink, or Scientific Atlanta) for $60,000 and spends another $20,000 for site preparation. On a five-year, 12% loan with no residual value, this $80,000 earth terminal costs only $1780 per month (and is steadily dropping). The 150 WATS EIRP of satellite power needed for a full duplex 384-Kbps circuit could cost about $4650 per month on a five-year contract from GTE. Maintenance should be minor. This alternative provides motion VTC at a very reasonable cost for those without T1 networks.

Finally, Fig. 9.22 shows the cost of obtaining 384 Kbps from a T1 voice circuit using the Republic Telecom packet voice compression device described earlier. The cost plotted is just that of the Republic

Fig. 9.20.
Comparison of
services from
SBS, U.S.
Telecom,
Vitalink, and
Cable &
Wireless.

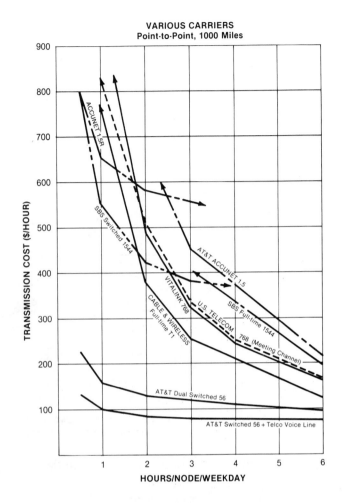

Telecom device, since no additional transmission circuits are needed. The RLX-40 currently sells for $9000 (for both ends) in the fifteen-line size. The monthly cost is $868, or $39 daily, assuming a five-year, 12% loan. This means that for $39 per *day* a user can free up 384 Kbps of capacity for VTC. For two hours per day, the effective cost of VTC transmission is an astonishingly low $20 per hour! The same logic will show that 768 Kbps could be obtained for $78 per day.

From Fig. 9.22 it can be seen that very low cost full-motion VTC is possible at 384 Kbps. The fiber-based Cable & Wireless alternative is, of course, available only in certain major corridors, but comparable offerings should become more ubiquitous as fiber proliferates. The satellite alternative is available anywhere that a 4.5-m antenna can be mounted, and its price is not distance-sensitive.

A heavy VTC user can therefore look forward to teleconferencing

Fig. 9.21.
Advantage of
multiplexing
VTC with voice
on a private T1
network.

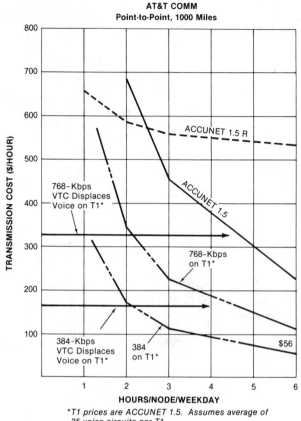

anywhere in the country for well under $100 per hour if these low cost solutions are sought out.

All of the previous figures have dealt with point-to-point situations. Fig. 9.23, however, gives costs for some alternatives appropriate in a multilocation network. For simplicity, solutions based on VTC muxed with voice on T1 are not shown, since their costs depend on how the 384 is tandem-switched through the network. Note that by using privately owned satellite systems, full-motion transmission (384 or 768 Kbps) can be obtained for about the same cost as AT&T's switched 56 Kbps in networks with fairly heavy usage. Clearly, the transmission cost differential between 56-Kbps teleconferencing and full-motion teleconferencing can be much smaller than many realize.

# 9.16   Summary

The preceding paragraphs have outlined the wide variety of transmission systems available for VTC and have clearly shown that costs are highly

**Fig. 9.22. Ways to transmit full-motion VTC at lowest cost.**

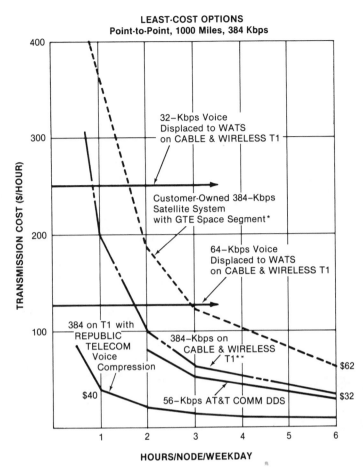

LEAST-COST OPTIONS
Point-to-Point, 1000 Miles, 384 Kbps

*Maintenance not included.
**One-fourth of T1 cost prorated to VTC.

situation-specific. Ways to obtain relatively inexpensive transmission have been identified. One major thrust has been that a combination of new techniques coupled with the ability to compress motion video to lower bandwidths is making VTC far more affordable. A second major thrust is that the cost advantage of 56-Kbps video teleconferencing is being greatly diminished by inexpensive ways to obtain 384 transmission, such that 384 is sometimes less expensive than popular 56-Kbps services. This author believes that the mainstream of VTC is rapidly moving toward 384 Kbps, since acceptable picture quality and acceptable costs will first converge there to trigger widespread growth of this exciting new communications media.

Fig. 9.23.
Relative
costs for
multilocation
network.

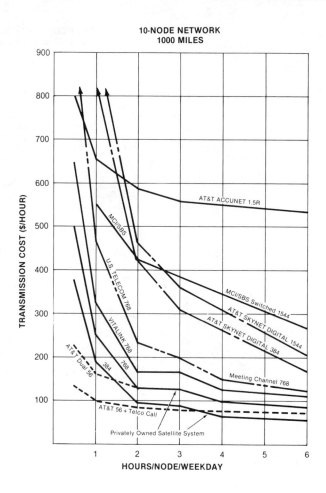

9.17   References

*Audio-Visual Communications.* 1983. Video: Telemeetings and TV dinners.

*Audio-Visual Communications* (August): 14–16.

Bair, James. 1979. Communication in the office of the future: Where the real payoff may be. *Business Communications Review* (January/February).

Bartee, T. C., ed. 1985. *Data communications, networks, and systems.* Indianapolis: Howard W. Sams.

Brown, Earl, V. Geller, J. Goodnow, D. Hoecker, and M. Wish. 1980. Some objective and subjective differences between communication over two video-conferencing systems. *IEEE Transactions on Communications* COM-28, no. 5 (May): 759–64.

Dutton, William, J. Fulk, and C. Steinfield. 1982. Utilization of video conferencing. *Telecommunications Policy* (September): 164–78. Butterworth & Co., Ltd.

Electrical Communications Laboratories. 1973. Video communication performance and transmission standards. *E.C.L. Technical Publication No. 83* (October). Electrical Communications Laboratories, N.T.T., Japan.

Green, David, and Kathleen Hansell. 1984. Videoconferencing. *Business Horizons* (November/December): 57–61.

Hagopian, Harold. 1981. Video conferencing: IBM user survey. Presentation to Institute for Graphic Communications Conference (August 13).

Halhed, Basil R. 1986. Electro-acoustic aspects of video conference room design. *Sound & Video Contractor* (April).

Harkness, Richard, and P. Burke. 1982. Estimating teleconferencing travel substitution potential in large business organizations: A four company review. *Proceedings of Conference on Teleconferencing and Electronics Communications.* University of Wisconsin Extension.

Harkness, Richard, et al. 1977. Technology assessment of telecommunications transportation interaction. Report for National Science Foundation by SRI International, NTIS PB 272-695 (May). Springfield, VA.

Hercules. 1985. World round-up. *Communications International* (April).

*International Management.* 1984. Teleconferencing: A long-heralded technology that is finally finding users. *International Management* (December).

Johansen, Robert, J. Vallee, and K. Spangler. 1979. *Electronic meetings: Technical alternatives and social choices.* Menlo Park, CA: Addison-Wesley.

Kelleher, Kathleen, and T. Cross. 1985. *Teleconferencing: Linking people together electronically.* Englewood Cliffs, NJ: Prentice-Hall.

Kollen, James, and J. Garwood. 1975. Travel/communication trade-offs: The potential for substitution among business travelers. The Business Planning Group, Bell Canada (April).

Lazer, E. A., M. C. J. Elton, J. W. Johnson, et al. 1983. *The teleconferencing handbook.* White Plains, NY: Knowledge Industry Publications.

McFarlane, Robert, and R. Nissen. 1982. Room design and engineering for two-way teleconferencing. *Proceedings of Conference for Teleconferencing and Electronics Communications.* University of Wisconsin Extension. Also published in *Architectural Record*, September, 1983.

Midorikawa, Masahiro, K. Yamagishi, K. Yada, and K. Mima. 1975. TV conference system. *Review of the Electrical Communications Laboratories* 23, nos. 5–6 (May/June). NTT, Tokyo, Japan.

Noll, A. Michael. n.d. Teleconferencing communications activities. Bell Laboratories.

Olgren, Christine, and L. Parker. 1983. *Teleconferencing technology and applications.* Dedham, MA: Artech House.

Parker, Lorne. 1982. *Telecoms Newsletter.* Teleconferencing newsletter (August/September). Center for Interactive Program, University of Wisconsin Extension, Madison, Wisconsin.

Pratt, William. 1978. *Digital image processing.* New York: Wiley.

Prem, David, and S. Dray. 1983. Teleconferencing at Honeywell. *Proceedings of Conference on Teleconferencing and Electronic Communications*. Vol. 2. University of Wisconsin Extension.

Pye, Roger, and Ederyn Williams. 1977. Teleconferencing: Is video valuable or is audio adequate? *Telecommunications Policy* (June): 230–41. IPC Business Press.

*Telecoms Newsletter*. 1984. Case study on Skidmore, Owings, and Merrill. Teleconferencing newsletter (April). Center for Interactive Program, University of Wisconsin Extension, Madison, Wisconsin (608/262-4555).

*Telespan*. Newsletter. Telespan, 50 West Palm Street, Altadena, California (818/797-5482).

Wilkens, H., and G. Plenge. 1981. Teleconferencing design: A technical approach to satisfaction. *Telecommunications Policy*: (September): 216–27. IPC Business Press.

Wilmotte, Raymond M. 1974. Technological boundaries of television. Appendices to vol. 3, PB-241601 (December). Prepared for the FCC, National Technical Information Service, Springfield, VA.

# Acknowledgments

## Chapter 1

All photographs have been provided through the courtesy of AT&T Bell Laboratories. Many of the other illustrations are based on material originally prepared by the author's colleagues at AT&T Bell Laboratories, and the author is indebted to these persons. He also wishes to acknowledge the many colleagues throughout AT&T, interactions with whom have been of inestimable benefit over the years.

*Ira Jacobs*

## Chapter 3

I would like to thank GTE Telenet for its consistent support of my participation in ISDN developments within the United States and on the international level. At times, ISDN has appeared to be more a matter of faith in an eventually useful application for Telenet, rather than an immediate short-term benefit. Time will tell if this faith was well founded.

My most heartfelt appreciation goes to Toni Zimmer. She has put up with the disruptions of international negotiations to our personal lives ever since we first met. Her tolerance has allowed me to both learn and contribute to the advancement of international standards for ISDNs.

Finally, both Toni and I extend our warmest regards to our colleagues and friends in other companies and countries who patiently worked together to bring ISDNs closer to reality.

*Eric L. Scace*

## Chapter 8

I would like to thank Drs. Bishnu Atal, David Goodman, and Juergen Schroeter for their careful reading of my draft manuscript and for their very valuable comments.

*N. S. Jayant*

## Trademark Acknowledgments

All terms mentioned in this book that are known to be trademarks or service marks are listed below. In addition, terms suspected of being trademarks or service marks have been appropriately capitalized. Howard W. Sams & Co. cannot attest to the accuracy of this information. Use of a term in this book should not be regarded as affecting the validity of any trademark or service mark.

ACCUNET and SKYNET are registered trademarks of AT&T.

AMSAT is a registered trademark of Radio Amateur Satellite Corporation.

Apple is a registered trademark of Apple Computer, Inc.

AppleLink is a trademark of Apple Computer, Inc.

CompuServe is a registered trademark of CompuServe Information Services, an H & R Block Company.

COMSAT is a registered trademark of Communications Satellite Corporation.

DEC is a registered trademark of Digital Equipment Corporation.

IBM is a registered trademark of International Business Machines Corporation.

Mailgram is a registered trademark of Western Union Telegraph Company.

Mailway is a registered trademark of Wang Laboratories, Inc.

MCI Mail is a trademark of MCI Corporation.

Telemail is a registered trademark of General Telephone and Electronics Corporation.

# Index

## A

## B